Agriculture Bioinformatics

Agriculture Bioinformatics

Editors

R. Keshavachandran
Bioinformatics Centre
IT-BT Complex
Kerala Agricultural University
Vellanikkara, Thrissur – 680 656

and

S. Raji Radhakrishnan
Bioinformatics Centre
IT-BT Complex
Kerala Agricultural University
Vellanikkara, Thrissur – 680 656

CRC Press is an imprint of the
Taylor & Francis Group, an **informa** business

NEW INDIA PUBLISHING AGENCY
New Delhi – 110 034

First published 2021
by CRC Press
2 Park Square, Milton Park, Abingdon, Oxon, OX14 4RN
and by CRC Press
6000 Broken Sound Parkway NW, Suite 300, Boca Raton, FL 33487-2742

© 2021 selection and editorial matter R. Keshavachandran and S. Raji Radhakrishnan; individual chapters, the contributors

CRC Press is an imprint of Informa UK Limited

The rights of R. Keshavachandran and S. Raji Radhakrishnan to be identified as the authors of the editorial material, and of the authors for their individual chapters, has been asserted in accordance with sections 77 and 78 of the Copyright, Designs and Patents Act 1988.

Reasonable efforts have been made to publish reliable data and information, but the author and publisher cannot assume responsibility for the validity of all materials or the consequences of their use. The authors and publishers have attempted to trace the copyright holders of all material reproduced in this publication and apologize to copyright holders if permission to publish in this form has not been obtained. If any copyright material has not been acknowledged please write and let us know so we may rectify in any future reprint.

All rights reserved. No part of this book may be reprinted or reproduced or utilised in any form or by any electronic, mechanical, or other means, now known or hereafter invented, including photocopying and recording, or in any information storage or retrieval system, without permission in writing from the publishers.

For permission to photocopy or use material electronically from this work, access www.copyright.com or contact the Copyright Clearance Center, Inc. (CCC), 222 Rosewood Drive, Danvers, MA 01923, 978-750-8400. For works that are not available on CCC please contact mpkbookspermissions@tandf.co.uk

Trademark notice: Product or corporate names may be trademarks or registered trademarks, and are used only for identification and explanation without intent to infringe.

Print edition not for sale in South Asia (India, Sri Lanka, Nepal, Bangladesh, Pakistan or Bhutan).

British Library Cataloguing-in-Publication Data
A catalogue record for this book is available from the British Library

Library of Congress Cataloging-in-Publication Data
A catalog record has been requested

ISBN: 978-1-032-02458-5 (hbk)

भारत सरकार
बायोटेक्नोलॉजी विभाग
विज्ञान और प्रौद्योगिकी मंत्रालय
GOVERNMENT OF INDIA
MINISTRY OF SCIENCE & TECHNOLOGY
DEPARTMENT OF BIOTECHNOLOGY
Block-2, 7th Floor C.G.O. Complex
Lodi Road, New Delhi-110003

डा. टी. मदन मोहन
वैज्ञानिक 'जी' / सलाहकार
Dr. T. Madhan Mohan
Scientist 'G' / Advisor

Foreword

Indian Agriculture is poised for quantum jump in production consequent to welfare approach and the scientific support. Polity intervention, Science and Technology, risk taking farmers and improved marketing strategies are the prime reasons for the Green Revolution. Blissful ecology and environment, biodiversity, indigenous farming knowledge, existence of second largest domestic market, globalization and international trade and above all increased purchasing power are the main strengths of India's agriculture. Low productivity, shrinking area under farming, diminishing water availability, scarcity in labour and energy and above all poor infrastructure at farming locations are the weakness. Agriculture provides not only food but also much needed health "Food is thy medicine". Climate change, undependable monsoon and fluctuating market prices are the emerging threats. The science of informatics is providing newer solutions and bringing farmers near to sources of marketing. Space, water and energy saving agriculture – vertical farming, family farming, protected cultivation-is making profound impact on productivity, green crops and clean products. Availability, access and absorption of food are the three pillars of food and nutrition security and assume importance in creating more remunerative jobs and establishing nutri-farms emphasizing on fruits, vegetables, tubers, spices and plantation crops. Availability of quality drinking water is important in absorption of food to digestive human system.

Bioinformatics in Agriculture become relevant to fill up the gaps in productivity, biofortification, bioremediation and above all protecting India's natural heritages including traditional knowledge. The rapid explosion in the science of physics and more particularly computational biology especially use of marker genes and genomics are aiding biotechnological advances in making available true to type plants free from pests and diseases to farmers. Recombinant DNA technology is making transfer of genes among evolutionary different organisms including plants, bacteria and viruses possible. Bt cotton has boosted bole yield to unimaginable level. Business and trade in transgenes are making headlines. GM Technology has potential in many crops whose performances are affected by biotic and abiotic stresses especially drought, salinity and now extreme high acidity. Area like genomics, comparative genomics, proteomics, metabolomics, interactomics and synthetic biology are gaining importance.

Many Universities in India are now offering specialized graduate and post graduate programmes in Bioinformatics. In the revised post graduate syllabi in Agriculture, bioinformatics has received attention. The Department of Biotechnology, Government of India is operating a nationwide major multi-disciplinary program on Bioinformatics as BTISnet. The present book Agriculture Bioinformatics edited by Prof. R. Keshavachandran Former Professor and Ms S. Raji Radhakrishnan, Bioinformatics Centre, Kerala Agricultural University which is as part of the BTISnet, carries 17 chapters authored by 38 eminent scientists working in the area of Bioinformatics. Dr KV Peter, Former Vice-Chancellor, Kerala Agricultural University is the General Editor of the series. Chapters on Bioinformatics databases and Internet Resources; Analysis of Genetic Diversity in Crop Plants using molecular marker data. Plant genomics-A bioinformative perspective and Genome mapping in plants are very comprehensive.

The present updated compilation of 17 chapters has provided reasonable information to Scientists and students in this area. I congratulate the Editors who are basically teachers for compiling the very useful information. I also congratulate New India Publishing Agency, New Delhi for publishing the book.

Dr T. Madhan Mohan
Senior Advisor
Department of Biotechnology
Govt. of India

Preface

The United Nations has declared 2014 as Year of Family Farming considering global significance for food and nutrition security. India has enacted a Food Security Act-2013 making access to food a right rather than a subsidized welfare approach. The production target for 2030 is around 480 million tones of food grains as against 240 million tones in 2013. Doubling the production from dwindling land area under agriculture, depleting water sources and quantum and costly energy and labour, the ways and means to reach the target are quite stupendous. One of the options for increased productivity is use of GM crops proven friendly to environment and ecology especially biodiversity. Use of tissue and cell culture for rapid multiplication of economic crops including ornamentals are now well known and demanded by farmers. Micro nutrients and vitamins deficiency in Indian diet lead to anaemia, goiter, myopia, stunted growth and several physiological disorders in skin and body. About 50% of worlds anaemic due to malnutrition are in India. New Life Style diseases like obeisity, diabetes and cardiovascular diseases are also on rise. Biofortified crops like Golden rice rich in iron and tomatoes rich in lycopene are results of recombinant DNA technology. Weedicide tolerant GM soybean and canola resistant to borers are now in farmers field. With climate change and consequent rise in temperature and oceanic currents, natural disasters like drought and flood are rampant. Hurricanes and typhoons devastating standing crops and making agricultural lands unfit for cultivation are threatening Millenium Development Goals (MDG) especially food and nutrition security. The relevance and need for biotechnological interventions are emphasized in this changing scenario of higher demand and lower supply.

Biotechnological tools supplement various conventional approaches in conservation, characterization and utlilization for increasing production and productivity of agricultural and horticultural crops. The emerging field of bioinformatics is an integrated field arising from merging of biology and informatics. It is a conglomeration of various new frontiers of science like genomics, proteomics, metabolomics etc. The rich warehouse of proteome and genome information nearly doubling every year, has significant implications and applications in various areas of science including agriculture, horticulture, forestry and food science. Cheminformatics is specialized to a range of problems in the field of chemistry. Chemical pesticide reduction is possible by adopting cheminformatics methods to identify naturally occurring chemical compounds in crops which act against pests.

Bioinformatics has transformed the discipline of life science from a purely lab based science to an information science as well. The ICAR has recently launched a National Agricultural Bioinformatics Grid (NABG) to serve as a computational facility in developing national biodatabases and data warehouses.

The present book Agriculture Bioinformatics is a compilation of 17 information packed chapters authored by working scientists in the respective discipline. In addition to the theoretical information, practical and applied aspects to boost productivity and quality of crops are given.

R. Keshavachandran
S. Raji Radhakrishnan

Acknowledgement

We acknowledge all the scientists who contributed to chapters to the book. We acknowledge Dr Madan Mohan, Senior Advisor, Department of Biotechnology Government of India for the Foreword. We acknowledge National Academy of Agricultural Sciences, New Delhi for allowing us to reproduce Policy Paper 72 'Bioinformatics in Agriculture: Way Forward' for the wider coverage. Dr R Keshavachandran thanks his wife Dr Rema Menon for the patience and support. Dr K V Peter acknowledges Prof. P I Peter, Chairman, NoniBiotech, Chennai for the facilities extended and Dr Kirti Singh, Chairperson, World Noni Research Foundation, Chennai for the guidance. Dr K V Peter thanks profusely his wife Vimala Peter and sons Anvar, Ajay and Daughters in law Anu and Cynara, Grand Son Antony Ajay Peter and Grand daughter Anna Vimala Anvar, their smiles and looks are always full of cheer and encouraging.

Contents

Foreword .... *v*

Preface .... *vii*

Acknowledgement .... *ix*

Contributors .... *xiii*

Policy Paper 72 Bioinformatics in Agriculture: Way Forward
Issued by the National Academy of Agricultural Sciences
(NAAS), New Delhi .... *xvii*

1. **Bioinformatics in Agriculture** 1
   H.P. Singh

2. **Bioinformatics Databases and Internet Resources** 9
   R. Amutha, R. Muthukumaran and B. Sangeetha

3. **Analysis of Genetic Diversity in Crop Plants Using Molecular Marker Data** 51
   M.K. Rajesh and S. Jayasekhar

4. **Plant Genomics - A Bioinformatics Perspective** 67
   S. Raji Radhakrishnan and R. Keshavachandran

5. **Integrating Knowledge of Bioinformatics in Medicinal Plant Research** 91
   Gurpreet Kaur, Pritika Singh and Pratap Kumar Pati

6. **Bioinformatics Applications in Plant Biology** 143
   Sharmila Anishetty

7. **Genome Mapping in Plants** 155
   K. Nirmal Babu, S. Asha, V. Jayakumar, D. Minoo and K.V. Peter

8. Inter-Species Conservation of Splice Sites: An Analysis
Using Support Vector Machine Based Pattern Recognition..... 169
Bhumika Arora and Pritish Kumar Varadwaj

9. Applications of Support Vector Machines in
Plant Genomes 187
Shimantika Sharma, Sona Modak and V.K. Jayaraman

10. Protein Structure, Prediction and Visualization 209
Sanjeev Kumar Singh and Sunil Tripathi

11. Mitogenomics: Mitochondrial Gene Rearrangements, its
Implications and Applications 259
Tiratha Raj Singh

12. Spice Bioinformatics 271
Santhosh J. Eapen

13. Application of Bioinformatics in Palm and Cocoa
Research 283
R. Manimekalai, K.P. Manju and S. Naganeeswaran

14. Application of Metagenomics in Agriculture............................. 295
K. Hari Krishnan

15. Biodiversity Informatics 319
P.N. Krishnan, S. Sreekumar, C.K. Biju, and M. Raveendran

16. Omics: What Next? 339
Prashanth Suravajhala and Rajib Bandopadhyay

17. Computational Methods in Plant Genome
Sequence Analysis 359
Archana Pan and Ipsita Chanda

Index 393

Contributors

Amutha, R.
Centre for Bioinformatics
School of Life Sciences
Pondicherry University
RV Nagar, Kalapet, Puducherry – 605 014

Archana Pan
Centre for Bioinformatics
School of Life-Sciences
Pondicherry University, R.V. Nagar, Kalapet
Puducherry – 605 014

Asha, S.
Indian Institute of Spices Research
Calicut – 673 012, Kerala

Bhumika Arora
Indian Institute of Technology-Bombay
Mumbai, Maharashtra

Biju, C.K.
Biotechnology and Bioinformatics Division
Bioinformatics Centre
Tropical Botanic Garden and Research
Institute, Palode
Thiruvananthapuram – 695 562, Kerala

Gurpreet Kaur
Department of Biotechnology
Guru Nanak Dev University
Amritsar – 143 005, Punjab

Hari Krishnan, K.
Environmental Microbiology Lab
Department of Molecular Microbiology
Rajiv Gandhi Centre for Biotechnology
Poojappura, Thycaud P.O.
Thiruvananthapuram – 695 014, Kerala

Ipsita Chanda
Department of Zoology
S.A. Jaipuria College
Kolkata – 700005, West Bengal

Jayasekhar, S.
Division of Social Sciences
Central Plantation Crops Research Institute
(CPCRI)
Kasaragod – 671 124, Kerala

Jayaraman, V.K.
Centre for Development of Advanced
Computing
Pune – 411 007, Maharashtra

Jayakumar, V.
Indian Institute of Spices Research
Calicut – 673 012, Kerala

Keshavachandran, R.
Bioinformatics Centre
IT-BT Complex
Kerala Agricultural University, Vellanikkara
Thrissur – 680 656
Kerala

Krishnan, P.N.
Biotechnology and Bioinformatics Division
Bioinformatics Centre, Tropical Botanic
Garden and Research Institute
Palode, Thiruvanathapuram – 695 562
Kerala

Manimekalai, R.
Sugar Cane Breeding Institute
Coimbatore – 614 007, Tamil Nadu

Manju, K.P.
Sugar Cane Breeding Institute
Coimbatore – 614 007, Tamil Nadu

Muthukumaran, R.
Centre for Bioinformatics
School of Life Sciences
Pondicherry University, RV Nagar
Kalapet, Puducherry – 605 014

Minoo, D.
Providence Women's College
Calicut – 673 008, Kerala

Naganeeswaran, S.
Sugar Cane Breeding Institute
Coimbatore – 614 007, Tamil Nadu

Nirmal Babu, K.
Indian Institute of Spices Research
Calicut – 673 012, Kerala

Pritika Singh
Department of Biotechnology
Guru Nanak Dev University
Amritsar – 143 005, Punjab

Pritish Kumar Varadwaj
Indian Institute of Information Technology
Allahabad – 211 012
Uttar Pradesh

Pratap Kumar Pati
Department of Biotechnology
Guru Nanak Dev University
Amritsar – 143 005, Punjab

Peter, K.V.
World Noni Research Foundations
12, Rajiv Gandhi Road
Perungudi, Chennai – 600 096, Tamil Nadu

Prashanth Suravajhala
Bioclues.org

Rajesh, M.K.
Division of Crop Improvement
Central Plantation Crops Research Institute
(CPCRI)
Kasaragod – 671 124, Kerala

Raveendran, M.
Biotechnology and Bioinformatics Division
Bioinformatics Centre, Tropical Botanic
Garden and Research Institute
Palode, Thiruvanathapuram – 695 562
Kerala

Rajib Bandopadhyay
Department of Biotechnology
Birla Institute of Technology and Science
Ranchi – 835 215
Jharkhand

S. Raji Radhakrishnan
Bioinformatics Centre
IT-BT Complex
Kerala Agricultural University
Vellanikkara, Thrissur – 680 656
Kerala

Sangeetha, B.
Centre for Bioinformatics
School of Life Sciences
Pondicherry University, RV Nagar
Kalapet, Puducherry – 605 014

Sanjeev Kumar Singh
Department of Bioinformatics
Alagappa University
Karaikudi – 630 003
Tamil Nadu

Santhosh J. Eapen
Co-ordinator
Bioinformatics Centre
Indian Institute of Spices Research
Calicut – 673 012
Kerala

Sharmila Anishetty
Centre for Biotechnology
Anna University
Chennai – 600 025
Tamil Nadu

Shimantika Sharma
Biotechnology Department
D.Y. Patil University
Pune – 411018
Maharashtra

Contributors

Singh, H.P.
The Founder and Chairman
Confederation of Horticulture Associations
of India (CHAI)
Sector 18A, Dwaraka
New Delhi – 110 075

Sona Modak
Bioinformatics Department
Pune University, Pune – 4110 073
Maharashtra

Sunil Tripathi
Centre of Excellence in Bioinformatics
School of Biotechnology
Madurai Kamaraj University
Madurai – 625 021, Tamil Nadu

Sreekumar, S.
Biotechnology and Bioinformatics Division
Bioinformatics Centre, Tropical Botanic
Garden and Research Institute
Palode, Thiruvananthapuram – 695 562
Kerala

Tiratha Raj Singh
Department of Biotechnology and
Bioinformatics
JUIT, Waknaghat
Solan – 173 234
Himachal Pradesh

Policy Paper 72
Bioinformatics in Agriculture: Way Forward

Issued by the National Academy of Agricultural Sciences (NAAS), New Delhi

---

Bioinformatics and computational biology are entwined employing techniques and concepts from life sciences, computer sciences and information technology. This interdisciplinary approach maintains close interactions with life sciences to realize its full potential. Bioinformatics applies concepts of information technologies to make the vast, diverse and complex biological data more readable, understandable and usable while, computational biology uses algorithms, mathematical models and computational approaches to address experimental and theoretical queries. In this way, apart from being distinct in functions and approaches, there is significant overlap in their activities to bridge the interface of the science of any biological discipline with informatics. The ultimate task of the bioinformatics in the biological system is to provide a complete computational representation of a cell (Singh *et al.*, 2012).

Because of the complexity among the simplest to the most advanced biological systems, massive multi-dimensional data generation, analysis and interpretation are required for a critical understanding of the cellular processes. With the application of high performance data generating machines (sequencers, mass spectrometers and other instruments) for complex but interdependent biological research, wealth of diverse data are continually accumulating. Massive and fast evolving data often create problems of misleading results and inconclusive interpretations. This has also posed the difficult task of identification, characterization and integration of globally available research information and finally coming out with a conclusive end in addressing typical problems of agriculture and medical sciences, and the environment (Jones *et al.*, 2006).

Priorities for Indian Agriculture

Agriculture is an assemblage of diverse physical, chemical and biological components, which in harmony with each other results in greater productivity. Present day agriculture in India is moving very fast from the *green revolution* to *evergreen revolution*. Basic priorities in the present day agriculture are to keep pace with the increasing population to meet the food requirement of the country, secure its own sustainability in the era of chemicalization and industrialization, coping with the fast emerging technologies and withstand with the unprecedented and abrupt changes in the climate. In the era of knowledge flood, agriculture as a whole is being witnessed as own components of which, plants and soil are the most live interactive concepts followed by other biotic and abiotic interactions to which, crops are regularly exposed (Varshney *et al.*, 2005). The overall impact of the environment can have genotypic and phenotypic short or long-lasting changes in plants and other organisms that usually open the door of the emergence of new genetic, molecular and biosynthetic pathways (Mochida and Shinozaki, 2010). Documentation of the interconnections between the conditions and representative alterations within the system can help to find out the response mechanisms behind the interactions. Equipped with this knowledge, biologists and agricultural scientists can find out the ways to mitigate the challenges of climate change, decline in the crop productivity and the issues like bio-safety and bio-security (Vassilev *et al.*, 2005). By genome analysis, gene function can be predicted and their functional categories be defined with the interventions of bioinformatics tools. Therefore, bioinformatics can facilitate the identification of the genetic basis of agronomically important traits that can accelerate the development of improved varieties.

Why Bioinformatics?

Worldwide, research activities in experimental molecular biology and whole genome sequencing projects have generated large volume of nucleotide and protein sequencing data. Traditional molecular biology research is still carried out making use as advanced machines that has led to increase in generation of large and varied data sets needing the involvement of computational power to make valid interpretations (Casci, 2012).

Voluminous data in genomics, proteomics, transcriptomic and metabolomics have been witnessed in the recent past. Therefore projects use computational techniques to glean biological knowledge from literature and from other public databases to address challenges of analysis and interpretation. To pin point, specify and identify the problems of the present day agriculture and to implement solutions at the right time, computing applications are the upcoming viable options for agricultural sciences (BPI, 2004).

The integration of information on the key biological processes allows us to achieve complete understanding of the biology of organisms as a long-term goal. The rationale behind strengthening the fast emerging field of Bioinformatics therefore, lies in.

- Bioinformatics and computational biology are parallel to the high-data generation in molecular biology at global scale.

- This field has grown as a cross-boundary research and development sector in the area of biology, agriculture, molecular biology, genetics, chemistry, health, environment and biostatistics.

- Bioinformatics is not only an area of primary research but its technological embedding makes this area critically applicable for work in all fields of biological, chemical and physical fields related to agriculture and environment.

Handling Massive Biological Data-Flow

Worldwide, big-data flow in the biological research coming out from the technologically-driven wet-lab experiments are accumulating at a phenomenal rate. India is obviously a part of this phenomenon because Indian researchers are taking integrative approaches to generate big-data in plant, microbe, fish and animal genomics, proteomics and phenomics. Currently, the data from the myriad of whole genome, transcriptome, proteome, metabolome and other interrelated projects are increasing with the fast pace from all the domains of biological, medical and agricultural sciences (Jimenez-Lopez et al., 2013). Since, the data are wide ranging and of diverse quality addressing signatures within the organisms or their networks, the process of extracting information lying within it becomes critical. The information dynamics, quantity and variety of information within a single experiment that includes the study of gene expression can be imagined from the fact that it involves analysis of genes, determination of protein structures encoded by the genes and details of how these products interact with each other. The ease with which computers can handle large amount of diverse data at a time and probe complex dynamics observed, makes computers indispensable to assist in biological research.

In the era of fast accumulating biological data coming out from the phenotypic analysis of many crop plants, live-stocks, microbes, climates, habitats and other interrelated entities and from the massive data generation from the - omics research (genomics, proteomics and metabolomics), bioinformatics has come forward across the globe to solve the problems of analysis, prediction, storage, management, pattern recognition, submission, retrieval and storage of the data to find out fruitful outcome. Complete whole genome sequencing projects on various crop plants along with more than 300 on-going projects, livestock (742 on-going projects) and microbes (4000 completed, > 8000 ongoing) have generated huge data resources at a large pace in many parts of the world (Source: NCBI). A core collection of 3,000 rice accessions from 89 countries were re-sequenced in China (http://www.gigasciencejournal.eom/content/3/1/7). Bioinformatics-driven analysis and interpretations of huge data can help in finding out novel alleles for important rice phenotypes and serve to understand genomic diversity within rice. Such big data in biological and agricultural sciences have attracted biologists to look at the cellular, molecular and metabolic levels to decipher agriculturally important traits (Bansal, 2005).

The size of data generated on each and every aspect of the organisms is too large to decipher, analyze, store, manage and retrieve. Therefore, the need of bioinformatics and its integration in the present days of agricultural research and education with high end computer applications, tools and software, database development and management, computational biology, biotechnology and bio-statistics are inevitable.

As there is an enormous increase of raw sequenced data, there is an increasing need of annotation that converts raw data into a significant form that is handy to biologists and can be interpreted into biological information (Fraser *et al.*, 2000). At present genome annotation is a combined approach of experimental and computational tools where computational methods represent a noteworthy fraction of the intensifying area of bioinformatics. Bioinformatics also facilitate the analysis of huge quantity of genome expression data generated by the technologies as microarrays and SAGE (Jones *et al.*, 2006).

Biological resources for data flow

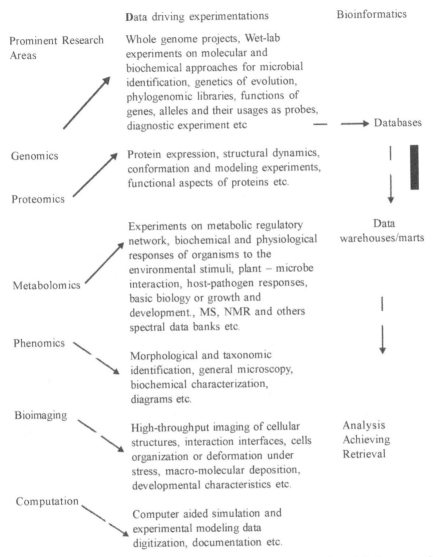

Fig. Experimental resources for data flow in biological research and their connectivity with bioinformatics

Research and Developmental Concerns

With the advent of widespread, fast growing, very complex arid often interdependent research data in the field of biological and agricultural sciences, the problem of non-concluding analysis and interpretation is increasingly felt. Since these data sets are different and regularly generated by different methods and sets of models, their integration at a common platform is not feasible. Responses of organisms with complex systems to the environment, pathogens and abiotic stresses and interactions with other organisms are among the few examples on which huge information have been gathered during the past several decades. This has posed a tough task of identifying, characterizing and integrating globally available research information to the field of agriculture, medical, pharmaceutical, biochemical and environmental etc. In reality, this would have been the greatest problem in biological research where computational systems, biostatistics and information technologies are finding their increasing applications (Ouzounis, 2012) However, these technologies developing parallel to the biological research enabled scientists in interpreting results. The assembling and integration of these technologies in solving the problems related to biological systems have given rise to "bioinformatics" (NABG, 2010).

Experimental biologists are more likely to create new technologies that are emerging with a very fast pace to enable high throughput collection of useful biological data. The complex cellular data will in turn be correlated with higher levels of phenotypic data based on the observations of the nature of cells, organs and organisms. This is why the establishment of connectivity between what is being observed at the molecular level with that of the phenotypic characters of the organisms under the environmental conditions is among the most emerging research areas. The amount and variety of biological data from recently developed methods enabled bioinformatics to perform the study of entire system at a single time (real time) rather than in individual components (genes, proteins, metabolites etc.). This has opened a new area of Systems Biology that is basically "Integrative Biology", which means for the integration of biological data for a better and meaningful understanding of the functioning of biological systems. Bioinformatics for Systems Biology links and connects many disciplines. It brings scientists from different areas like life sciences, computational biology, biomedical sciences and mathematics under a common framework (Jones *et al.*, 2006).

Our understanding of basic biology is being facilitated through comparison of organisms at different evolutionary distances, in order to reconstruct both the tree of life and the emergence of important phenotypic traits. Also, there is a growing expectation that bioinformatics will help fuel the creation of computational models (both qualitative and quantitative) which will allow us to capture, store and maintain biological models that help explain experimental observations. Algorithms in bioinformatics cover research in all aspects of computational biology. The emphasis is on discrete algorithms that address important problems in molecular biology, genomics, and proteomics that are computationally efficient, implemented and tested in simulations and on real datasets and provide new biological results and insights. Exact and approximate algorithms pertain to genomics, sequence analysis, gene and signal recognition, alignment, molecular evolution, phylogenetics, structure determination or prediction, gene expression and gene networks, proteomics, functional genomics, and drug design. In particular, bioinformatics tools include the BLAST program (homology searching), GENSCAN, GENIE (gene-finding), SAPS (statistical analysis of protein sequences), CLUSTAL, ITERALIGN, (multiple sequence alignment), r-SCAN STATISTICS (target array clustering, over-dispersion), etc. (Cochrane and Galperin, 2009). These programs are used by thousands of researchers every day in molecular biology and medicine. The BLAST protocol currently serves more than one lakh queries per day at the National Center for Biotechnology Information (NCBI) in Washington, DC.

Developments and Initiatives

Looking into the needs of Indian researchers certain landmark efforts have been taken-up in India in the past at various platforms to address the problems of biological data management, transfer and applications. India is credited in the whole world to establish a BioTechnology Information System (BTIS) network in 1987 for creating an infrastructure aimed to harness biotechnological applications through the use of bioinformatics. Department of Biotechnology (DBT) has successfully established a high-speed and high-bandwidth network in the form of Virtual Public Network (VPN) named as BIOGRID INDIA. Eleven nodes for this system have been established in the first phase and now being actively used for pursuing bioinformatics activities such as human resource development and R&D in bioinformatics and dissemination of biotechnology information to the researchers in the country. The nodes are interconnected through 2mbps dedicated leased circuit line at each location and 4Mbps Internet bandwidth shared from the central server by all the nodes. The BIOGRID allows exchange of database and software which have been created/acquired by the individual centers/nodes of BTIS. This resource- sharing helps in enhancing the value and usefulness of the BTIS, a true resource sharing network in India. The Department of Information Technology (D1T), India established Centre for Development of Advanced Computing (C-DAC) to deploy the nation-wide computational grid GARUDA which connected many cities across the country with an aim to bring Grid Network Computing for research laboratories and industry. Another network, GARUDA is a collaborative program of scientific and technological researchers on a nation-wide grid comprising of computational nodes, mass storage and scientific instruments. GARUDA aims to provide the technological advances required to enable data and compute intensive science. Department of Biotechnology, GOI has also provided support for the development of a highly curated international database entitled 'Manually Curated Database of Rice Proteins' (www.genomeindia. org/biocuration) based on semantic digitization of published experimental data on rice (BPI, 2004).

In Indian agricultural system, ICAR has taken a stepping-up initiative to establish National Agricultural Bioinformatics Grid (NABG) coupled with a high-end supercomputing facility and large data storage device to support agricultural research for all purposes of bioinformatics tasks right from sequence submission and retrieval to the annotation of whole genome sequencing projects, comparative genomics, transcriptomics, proteomics and metabolomics. This grid-based supercomputing facility has been established in 2013 at Indian Agricultural Statistics Research Institute, New Delhi with extensions at five National Bureaux of the ICAR including National Bureau of Plant Genetic Resources (NBPGR), New Delhi, National Bureau of Fish Genetic Resources (NBFGR), Lucknow, National Bureau of Animal Genetic Resources (NBAGR), Karnal, National Bureau of Agriculturally Important Microorganisms (NBAIM), Mau (UP) and National Bureau of Agriculturally Important Insects (NBAII), Bangalore for the integration and curation of biological data in the specific domains with an aim to provide bioinformatics services to the Indian scientists in a similar manner in which NCBI, EMBL, KEGG or other international public databases are providing. At the same time, to strengthen the capacity of Indian scientists, several National level training programs are regularly being organized at a mission mode. Masters level academic courses in Agricultural Bioinformatics was started in Indian Agricultural Statistics Research Institute, New Delhi and the Ph.D. program is underway (NABG, 2010).

Goals for Bioinformatics

Sequencing projects worldwide have resulted in a unique wealth of biological data for which bioinformatics has met the huge demand for analysis, storage and interpretation. The first row of genome and protein sequence information flow created demand for computational tools for search, comparison and analysis of nucleic acid, protein sequences and macromolecular structures (Allen, 2004). The second type of data flow related to the expression profiling (gene, protein, metabolite and ultra-structure) similarly generated demand for such tools that allowed the data to be minimized, understood and integrated. Future waves of data will support innovations in phenomic, genomic, proteomic and metabolomic research including large scale genomic comparisons, protein structures, simulation models, interactions, taxonomy, compartmentalization and overall turnover of metabolic pathways and regulatory networks (Al-Haggar, 2013).

For such integrated and cumulative efforts, research, education and services in bioinformatics need to be interdependent and mutually supportive. Future developments in the area need to have following goals-

1. Research goals should fulfill nationally and internationally recognized capabilities to strengthen multidisciplinary research tasks within and outside

NARS system in India. Multiplicity and plurality in the bioinformatics research and development are essential in order to establish this area as a mainstream of scientific research.

2. Educational goals should aim to develop an established system of educational offerings in the area of bioinformatics at different levels including Masters and Ph D programs as well as high end orientation programs for scientific and technical staff.

3. Service-oriented goals should be user-centric to support research efforts and develop bioinformatics tools and techniques as per the research needs and without any barrier of physical location where the actual research is being carried out. Participatory programs should be initiated with software developing companies, institutions, research departments and/or individuals that can support researchers in developing algorithms, databases and analytical as per the demand of the experiments to fetch maximum information.

4. Infrastructure resources should be aimed at maintaining efficient physical and technical infrastructure with full capability to allow fast and efficient computation, analysis, modeling, visualization and retrieval of biological data.

5. Organizational capabilities need to be created for the integration of research, education and service-oriented missions garner requisite institutional resources, leadership and back-up resources to realize these goals.

6. Development of data storage models and schemes with the capacity of digitizing every aspect of data generated by agricultural research including phenotypes, traits (molecular/biochemical), environmental conditions, plant development stage, tissue etc.

Priorities

National level priorities for bioinformatics should address the problems and perspectives of the current day biological and agricultural research. Major foreseen areas are as under:

1. Prioritize to cater the needs of the people who are engaged in generating data from biological experimentation related to agriculture.

2. Integration with the educational system should be given due priority to produce well-trained bioinformatics professionals and contribute to bioinformatics service.

3. Creation of bioinformatics research community involving extensive collaborations between the bioinformatics and life sciences researchers across the country.

4. Establishment of a high-end computational and statistical bioinformatics support system.

5. Establishing a nation-wide capacity building system for conveying bioinformatics techniques and information.

6. Increasing the practice of data digitization at institutional level for conversion of the same in databases with uniform platforms for end-use access.

7. Increase in the computational and data storage facilities and its upgradation should be addressed to accommodate bioinformatics and biological sciences researchers.

8. Recruiting and hiring support personnel working with the high-end clusters and high-performance computational environments, and large-scale storage systems commensurate with the infrastructure.

9. Gaining visibility by not only producing high quality research papers but also through developing novel bioinformatics tools, software and other integrated packages for data management.

10. Attracting appropriate funding from various National funding sources to cater the needs of infrastructure, man power and capacity building.

Strategies for Future Bioinformatics Infrastructures

- Extensive and wide spread competence development of man-power to handle challenges in data analysis.

- Serve the needs and priorities of a very complex community of users.

- Respond to the distinct demands of biological researchers.

- Provide a biological information environment with information about other biological (or biology-related) data and not simply genetic and molecular information.

- Comprehensive and multi-disciplinary data environment for integrative disciplines (e.g. systems biology, metagenomics, drug discovery etc.).

- Overcome several bottle-necks in bioinformatics resources exploitation such as lack of resource interoperability, programmatic access, input/output format standardization and user-friendly web interfaces.

Policy Paper 72 Bioinformatics in Agriculture: Way Forward

- Establish optimal community synergy between resource providers/developers and users.

- Involvement of future users during resource development phase, efficient capture of users' feedback information, development of resources documentation and tutorials.

Needs to Support Bioinformatics Services

Overarching needs

- *Recognition among leadership that bioinformatics is essential for agricultural research*

Such leadership supported envisioned programs are essentially required to establish bioinformatics resources to support high throughput-based Indian agricultural research

- *Strengthening of organizational framework support for bioinformatics.*

Initiatives are needed to equip NARS with world class bioinformatics centers with at least minimum of bioinformatics resources and capacity building.

Specific needs

Research

- Identifying truly national priorities in agricultural bioinformatics.

- Human resource development to strengthen human capital knowledge in bioinformatics.

- Parallel funding for collaborative research support for both wet- and dry-lab work.

- Support for partnering development in IT sector for new algorithms, tools, software and database to cater analytical needs of biological researchers.

- Centralized space to support bioinformatics research and cohesive research environment.

Education

- Mass awareness about the necessities, capabilities and services on bioinformatics.

- Masters, doctoral and post-doctoral programs and associated administrative support.

- Capacity building through orientation courses among NARS and university people.
- Bioinformatics courses:
  - Should be updated to include new aspects like systems biology, biocuration.
  - Emphasize the importance and functioning of various algorithms used in bioinformatics analysis. Competence development should be emphasized to inculcate the capacity to design and modify algorithms based on a biological problem.

Service

- Develop sustainable funding models to support public-private partnership
- Space for bioinformatics services especially connected with the biological/agricultural sciences
- Cloud-computing concept should be implemented where a centralized high-end infrastructure and highly qualified and motivated personnel may offer project/job wise assistance to various labs involved in wet-lab analysis over high speed internet.

Policy Options

- Developing a cohesive understanding in the wet-lab and dry-lab (bioinformatics) scientists.
- Partnering among the NARS scientists for data generation, sharing and management.
- Providing translational bioinformatics services for the NARS people.
- Maintenance of large-scale genomics and other biological databases.
- Audit and control of databases that are increasingly becoming larger and larger.
- Ensuring security in stretched and quickly changing computing environment.
- Proper internet network and computational infrastructure to work with high end servers.
- Management and design of large-scale genotype and phenotypic data.
- In-house curriculum development and pro-bioinformatics environment for institutions.

- Systematic government or inter-government funding to support upcoming programs.
- Closing-up the gaps between the intellectual haves and have-nots for bioinformatics and computational biology.
- Developing interface research coupling bioinformatics and systems biology.
- A dynamic interface (with active participation) with the international bioinformatics community to keep abreast with latest and path-defining bioinformatics implementation.

Recommendations

- Priority areas for biological tasks assisted by the bioinformatics in agricultural research with a better understanding of genomic, proteomic, metabolomic and phenomic data may be identified at national level. The work within these areas may be distributed across the country at a diverse level but, it need to be integrated.
- The importance of the application of the subject in routine research programs should be propagated among the Indian researchers especially the young researchers.
- To cater the need of Indian researchers, high-end infrastructure fully capable of handling processing of huge data at a time and storage of the processed data at the same time may be developed at National level and be maintained regularly.
- Targeted orientation and focused approach for database generation based on the work of Indian scientists may be initiated as a long-term goal to generate huge databases on Indian work. The concept of 'Biocuration' should be extensively explored to enable the development of highly intricate databases.
- Promotion of bioinformatics research and education in India in the specific context of agricultural system is a basic need these days. Looking into this need, focus on the Human Resource Development Programs, that too primarily based on sensitization about the subject and the latter related to specific issues in the subject should be designed and popularized.
- A greater understanding about how to digitize phenotypic data into a computer readable form is essentially required and this may be integrated. Extensive use of various ontologies as well as development of new ontologies to efficiently digitize all kinds of agricultural data such as phenotype, environmental conditions etc. are required.
- For targeted applications of bioinformatics in crop improvement programs, genomics of important crop plants, animals, fishes, insects and microbes,

systems biology of organisms and impact assessment of climatic changes on crops, biologists may be trained in different aspects of computational biology. This may help in fetching down the specific and targeted information.

- Service-oriented bioinformatics tasks may be started in the NARS system by making and implementing National level guidelines.

- Projects integrating data generation by laboratory experimentation and its integration with the bioinformatics need financial support to make a scope of the biological data generation at faster.

References

1. Al-Haggar M, Khair-Allaha BA, Islam MM and Mohamed A (2013). Bioinformatics in high throughput sequencing: application in evolving genetic diseases. *J Data Mining Genomics Proteomics* 4: 3.
2. Allen G (2004). Bioinformatics : New Technology Models for Research, Education and Service. *EDUCAUSE Research Bulletin* 8: 1-9.
3. Bansal AK (2005). Bioinformatics in microbial biotechnology - a mini review. *Microbial Cell Factories* 4: 19.
4. Bioinformatics policy of India (BPI-2004). Department of Biotechnology, www.dbtindia.nic.in/draftpolicy.doc.
5. Casci T (2012). Bioinformatics: Next-generation omics. *Nature Reviews Genetics* 13: 378-379.
6. Cochrane GR and Galperin MY (2009). The 2010 Nucleic acids research database issue and online database collection: a community of data resources. *Nucleic Acids Research* 38: D1-D4.
7. Fraser CM, Eisen J, Fleischmann RD, Ketchum KA, Peterson S (2000). Comparative Genomics and Understanding of Microbial Biology. *Emerging Infectious Diseases,* 6: 505-512
8. Jimenez-Lopez JC, Gachomo EW, Sharma S, Kotchoni SO (2013). Genome sequencing and next-generation sequence data analysis: A comprehensive compilation of bioinformatics tools and databases. *American Journal of Molecular Biology* 3: 115-130.
9. Jones MB, Schildhauer MP, Reichman OJ and Bowers S (2006). The new Bioinformatics: integrating ecological data from the gene to the biosphere. *Annual Review of Ecology, Evolution, and Systematics.* 37: 519-544.
10. Mochida K and Shinozaki K (2010). Genomics and bioinformatics resources for crop improvement. *Plant and Cell Physiology* 51: 497-523.
11. National Agricultural Bioinformatics Grid (NABG). 2010. http://cabgrid.res.in/cabin/.
12. Ouzounis CA (2012). Rise and Demise of Bioinformatics? Promise and Progress. *PLoS Comput Biol* 8: e1002487.
13. Singh DP, Prabha R, Rai A and Arora DK (2012). Bioinformatics-assisted microbiological research: tasks, developments and upcoming challenges. *American Journal of Bioinformatics* 1: 10-19.
14. Varshney RK, Graner A and Sorrells ME (2005). Genomics-assisted breeding for crop improvement *Trends in Plant Science* 10: 621-630.
15. Vassilev D, Leunissen J, Atanassov A, Nenov A, Dimov G (2005). Application of Bioinformatics in Plant Breeding. *Biotechnol and Biotechnol.* 18: 140-152.

Chapter – 1

Bioinformatics in Agriculture

H.P. Singh

Abstract

Indian agriculture made a rapid stride, converting the country from food scarcity to sufficiency. But challenges for the Indian agriculture in 21st century are much greater than before. The growing population has to be fed and surplus has to be produced with declining land, water and threat of climate change. Food and livelihood security of increasing population is a cause of concern and has received focused attention across the globe. Food security emphasizes access to sufficient, safe and nutritious food for an active and healthy life. Similarly, livelihood security has the emphasis on income and other resources to enable households to meet their fair needs.

The horticulture which includes fruits, vegetables, spices, flowers and medicinal and aromatic plants, has proved beyond doubt its potentiality for gainful diversification. Initiatives taken by Government and other stakeholders have impacted the development in terms of increased production, productivity and availability of horticultural crops. One of the significant developments is that horticulture has moved from rural confine to commercial production and this changing scenario has encouraged private sector investment in production system management. Now the horticulture sector contributes 29.65 per cent to agricultural GDP that has achieved a growth rate of 5-6 per cent during the decade. India has emerged as the second largest producer of fruits and vegetables and has first position in several horticultural crops. Production and export of flowers have increased manifold and the country has a major stake in global trade of spices and cashew nuts. Exports of medicinal plants, fruits and vegetables have also exhibited rising trend.

The emerging trend worldwide and also in the country is indicative of a paradigm shift in dietary needs of the people with rise in income, which demand for more horticultural produces. Fruits and vegetables add to food basket, while floriculture and plantation crops enhance the access to food as it provides better income per unit of land besides providing opportunity for employment. Since the growing of horticultural crops is rewarding to the farmers in terms of returns per unit area, the sector is expected to contribute significantly for food and nutritional security, employment opportunity and poverty alleviation. However, the challenges ahead are to have sustainability and competitiveness. Productivity of many horticultural crops continue to be low, quality of produce needs improvement, resources that improve efficiency require up-gradation, post-harvest losses continue to be high and there is growing regional disparity in production and use of technologies. The Government has addressed many of these constraints with initiatives like Technology Mission for Integrated Horticulture Development in North-East Region including Sikkim and National Horticulture Mission to achieve the transformation of horticulture utilizing the latest technologies.

Bioinformatics Holds the Key

In the recent past, we have witnessed how biotechnological tools can supplement various conventional approaches in conservation, characterization and utilization for increasing production and productivity of horticultural crops. The emerging field of bioinformatics is an integrated field arising from merging of biology and informatics. It is a conglomeration of various new frontiers of sciences such as genomics, proteomics, metabolomics etc. The rich warehouse of proteome and genome information which is nearly doubling every year, has significant implications and applications in various areas of science including agricultural sector.

The critical mission of bioinformatics involves gene discovery, prediction of structure and/or function of protein, metabolic pathway analysis, identification of gene to metabolite and understanding the system as a whole. The sequencing of the genomes of microbes, plants and animals have enormous benefits for the agricultural community. Utilization of bioinformatics in different areas like genomics, comparative genomics, proteomics, metabolomics, interactomics and synthetic biology would take the current research into accelerated mode which is needed for the present knowledge driven predictive science.

Bioinformatics will ultimately allow the evolution of biology from a descriptive science to a predictive science based on 'omics' data mining.

Genomics is the study of an organism's entire genome. Genomics appeared in the 1980s, and accelerated in 1990s with the initiation of genome projects for several species. A major branch of genomics is still concerned with sequencing the genomes of various organisms, although the knowledge of full genomes have created the possibility for the field of functional genomics, mainly concerned with patterns of gene expression during various conditions. More reasonable approach to gene discovery is through the development of Expressed Sequence Tags (ESTs), which provide a wealth of information in a relatively short time at a reasonable cost. These ESTs are an important source not only for the discovery of candidate genes and genetic markers, but also for the development of microarrays, until the whole genome sequence becomes available. EST data from a number of different genotypes are also a rich source for the discovery of Single Nucleotide Polymorphism (SNP) and Simple Sequence Repeat (SSR) markers, which are highly useful in Marker Assisted Selection (MAS) and linkage mapping.

The physical mapping of complex genomes is based on the construction of a genomic library and the determination of the overlaps between the inserts of the mapping clones to generate an ordered cloned representation of nearly all the sequences present in the target genome. Bacterial Artificial Chromosome (BAC) libraries have become the main vehicle for performing map-based gene cloning and physical mapping. These libraries represent a potentially useful resource for the study of comparative genome organisation and evolution, construction of genome wide physical maps and to develop chromosome-specific cytogenetic DNA markers for chromosome identification. BAC clones and physical maps are essential components in linking phenotypic traits to the responsible genetic variation, to integrate the genetic data, for the comparative analysis of genomes and to speed up and improve potential and effectiveness of MAS for breeding.

Proteomics is the study of the full set of proteins in a cell type or tissue, their structures and functions and the changes during various conditions. Most importantly, while the genome is a rather constant entity, the proteome differs from cell to cell and is constantly changing through its biochemical interactions with the genome and the environment. One organism has radically different protein expression in different parts of its body, in different stages of its life cycle and in different environmental conditions. The entirety of proteins in existence in an organism throughout its life cycle, or on a smaller scale the entirety of proteins found in a particular cell type under a particular type of stimulation, is referred to as the *proteome* of the organism or cell type, respectively. Since proteins play a central role in the life of an organism, proteomics is instrumental in discovery of biomarkers.

Phenomics is the study of *phenomes*— the physical and biochemical traits of organisms — as they change in response to genetic mutation and environmental influences. It is used in functional genomics, pharmaceutical research and metabolic engineering. Developments in computer technology, instrumentation and bioinformatics will help to meet the demands set by the molecular revolution. As a result, the field of 'phenomics' is still being born. This will integrate multidisciplinary research, with the goal of understanding the complex phenotypic consequences of genetic mutations at the level of the organism.

Transcriptomics deals with the study of messenger RNA molecules or the *transcriptome*, which is the set of all RNA molecules, including mRNA, rRNA, tRNA, and other non-coding RNA produced in one or a population of cells. The term can be applied to the total set of transcripts in a given organism, or to the specific subset of transcripts in a particular cell type. Unlike the genome, which is roughly fixed for a given cell line (excluding mutations), the transcriptome can vary with external environmental conditions. Because it includes all *mRNA* transcripts in the cell, the transcriptome reflects the genes that are being actively expressed at any given time, with the exception of mRNA degradation phenomena such as transcriptional attenuation. Transcriptomics, also referred to as expression profiling is often using high-throughput techniques based on DNA microarray technology. The use of next-generation sequencing technology to study the transcriptome at the nucleotide level is known as RNA-Seq.

Metabolomics is the systematic study of the metabolic profile of cells in tissues. The *metabolome* represents the collection of all metabolites in a biological cell, tissue, organ or organism, which are the end products of cellular processes. Use of high throughput technologies such as microarray and gene chips for expression profiling and improved tools of analytical organic chemistry (e.g. NMR, GC-MS, LC-MS and TOF-MS) allows the establishments of profiles of all metabolites present in specific plant cells and tissues. With major breakthroughs in genomics and analytical chemistry, plant biologists are now aiming to integrate transcriptomic, proteomic and metabolomic datasets using bioinformatics tools in sufficient detail that affords system-wide predictions of plant development in response to both genetic and environmental perturbations. Using the growing collection of available sequence data and combining bioinformatics and functional genomics should lead to a greater understanding of the genetic networks which are activated during plant responses. Systems biology or interactomics is the complete interlinking of various interactions of an organism and attempts to discover and understand biological properties which emerge from the interactions of many system elements. It aims at system-level understanding of biological systems and to draw the network of processes. System-level understanding of the plant has been a long-standing goal of biological sciences. With the progress of genomics and proteomics and range of other projects, in-depth knowledge of molecular nature of biological system is accumulated that leads to the possibilities for a system-level understanding at molecular-level. Understanding of dynamics of the system with the help of bioinformatics will allow the construction of theory/model with powerful prediction capability.

Cheminformatics (also known as chemoinformatics and chemical informatics) is the use of computer and informational techniques, applied to a range of problems in the field of chemistry. The primary application of cheminformatics is in the storage of information relating to compounds. The efficient search of such stored information includes topics that are dealt with in computer science as data mining and machine learning. In future, it should be possible to directly manipulate the content and composition of many chemical constituents in plants that may add their *medicinal or nutraceutical properties* by employing various bioinformatics techniques like gene identification, promoter prediction and primer design to amplify the beneficial genes and embedding in required crops of interest to make transgenic crops with the highest yield. Chemical pesticide reduction is also possible by adopting cheminformatics methods to identify naturally occurring chemical compounds in crops which act against pests.

Bioinformatics in Horticulture

Comparative genomic analysis is the cornerstone of *in silico*-based approaches to understand biological systems and processes across plant species and bioinformatics plays an underpinning role in it. Comparative genetics of the plant genomes has shown that the organization of their genes has remained more conserved over evolutionary time. These findings suggest that information obtained from the model crop systems can be used to suggest improvements to other food crops. It can be used as a tool for accelerating the normally laborious task of gene isolation. 'Orphan crops' which have not yet received the investment of research effort or funding required to develop significant public bioinformatics resources can exploit comparative genomics and bioinformatics to assist research and crop improvement provided they are related to a well-characterized model plant species.

The goal of plant genomics is to understand the genetic and molecular basis of all biological processes in plants that are relevant to the species. This understanding is fundamental to allow efficient exploitation of plants as biological resources in the development of new cultivars with improved quality and reduced economic and environmental costs. Traits considered of primary interest are pathogen and abiotic stress resistance, quality traits for plant and reproductive traits determining yield. Bioinformatics tools help in identification of specific markers which can amplify the gene expressed and then maps that area.

High-Throughput Sequencing (HTS) technologies enable application of global molecular approaches to horticultural crops. Sequencing of a gene or protein can lead to explore the information about the entire biochemical and physiological properties of the gene/protein. HTS is also applied for the generation of Expressed Sequence Tags (ESTs) which provide a wealth of information in a relatively short time at a reasonable cost. ESTs for many of the horticultural crops are currently available in dbEST. Several institutes had started to look at these data more critically and several markers are identified using *in silico* tools.

Diseases of crops are the major threat in production strategy. Molecular diagnostics kits can be developed with the knowledge of genomics. One of the very useful criteria in bioinformatics is the drug candidate identification. It also recognizes that the field of bioinformatics is crucial for validating these potential drug targets and for determining which ones are the most suitable for entering the drug development pipeline. Such tools of bioinformatics allow the successful identification of drugs against pathogenic bacterial, viral and fungal diseases of crops. Chemical pesticide reduction is also possible by adopting cheminformatics methods to identify naturally occurring chemical compounds in crops which act against pests. Biological control of pests is also possible by this approach. The design of agrochemicals based on an analysis of the components of signal perception and transduction pathways to select targets, and with cheminformatics, to identify potential compounds that can be used as herbicides, pesticides or insecticides.

Nutritional genomics takes advantage of the many genes that have been cloned for vitamin pathways and for the synthesis of many other "nonessential" compounds and macronutrients. In future, it should be possible to directly manipulate the content and composition of many chemical constituents in plants that may add their medicinal or nutraceutical properties by employing various bioinformatics techniques like gene identification, promoter prediction and primer design to amplify the beneficial genes and embedding in required crops of interest to make transgenic crops with the highest yield and superior quality.

Biodiversity, including genetic resources and species and ecosystem diversity, is of great importance to agriculture. As a natural-resource-based industry, agriculture depends on a healthy diversity of organisms and ecosystems which are its foundation. It is because of this close relationship that natural-resource-based industries, such as agriculture, have a direct impact on biodiversity. Information on all types of biological repositories including clones, cell lines, organisms and seeds have to be compiled, archived and linked. Most of the technologies being developed for MAS are also applicable in the evaluation, conservation and use of biodiversity, including microbial, animal and plant genetic resources. Bioinformatics can contribute to promoting databases that will result in preserving biodiversity details of important species and habitat digitally.

Excellent databases and tools are available free on the internet. Specific databases exclusively on certain crops are made available to the public by some international research organizations. We need to effectively utilize this rich information to reorient our research efforts than duplicating them. However, our immediate attention is warranted towards consolidating information available on our crops and to generate excellent information resources which are made available to the public.

Indian Bioinformatics Scenario

India, rich in skills and expertise in biological sciences, can definitely deliver and achieve marked success in bioinformatics too. In agriculture, the outcome of biotechnology and bioinformatics research must reach the resource poor farmers. Understanding the potential benefits of bioinformatics, ICAR is committed to develop strong expertise in bioinformatics to tackle problems in agriculture sector.

In India, the utilization of bioinformatics tools has enabled to develop cultivars through MAS. The genomic knowledge of rice aided in precise and quicker ways of identifying the markers which might have taken several years through conventional breeding methods. Utilization of bioinformatics tools and initiation of genomics made better understanding of pathogenic organisms like Oomycetes, Phytoplasmas, viruses and bacteria involved in several plant diseases. This knowledge is also vital for the development of new disease diagnostic tools and better control measures. The first draft sequence of the potato genome has been released by The Potato Genome Sequencing Consortium (PGSC), in which India is also a member. Tomato genomics certainly helped in deriving improved varieties.

High-Throughput Sequencing (HTS) technologies enable application of global molecular approaches to horticultural crops. Such technologies enable genome assisted selection, mapping of resistance genes etc. HTS has been installed at

Central Potato Research Institute (CPRI), Shimla which can be used for the genomics of potato, mango, banana, coconut and other horticultural crops.

A number of databases have been developed by IISR Calicut, Kasargod, CPCRI and other institutes. The Bioinformatics Centre at G.B. Pant University of Agriculture and Technology, Pantnagar is maintaining the database related to biotic and abiotic stresses and molecular diagnostics in crops. Recently ICAR launched a National Agricultural Bioinformatics Grid (NABG). It will serve as a national computational facility and will help in developing national databases and data warehouses. The grid will develop software, tools, algorithms, genome browsers etc. exclusively in the field of agricultural bioinformatics through systematic and integrated approaches. NABG will also help in capacity building for research and development in agricultural bioinformatics.

Conclusion

The new, budding field of agricultural genomics will pave the way towards a global understanding of plant and pathogen biology, and agriculture is likely to benefit from its application. The plant genomics is yet to gain momentum in our country. Indian horticulture has so far not made full advantage of the vast resources available across the globe in this sector. We are yet to initiate any studies on some of the emerging areas like systematic analysis of plant hormones, plant phenome analysis using 'loss-of-function' and 'gain-of-function' mutant lines, functional analysis of plant transcription factors (TFs) etc. We need to urgently prioritize and systematically plan our research endeavors in bioinformatics, develop proper infrastructure and establish sound collaborations.

To conclude, bioinformatics has transformed the discipline of biology (life science) from a purely lab-based science to an information science as well. Increasingly, biological studies begin with a scientist conducting vast numbers of database and website searches to formulate specific hypotheses or to design large-scale experiments. The implications behind this change for both science and medicine are staggering.

Chapter – 2

Bioinformatics Databases and Internet Resources

R. Amutha, R. Muthukumaran and B. Sangeetha

Abstract

Bioinformatics is a discipline combining biology, mathematics and the technology including computational tools and databases. It is one of the fast growing fields in the scientific community. This chapter is intended to provide an overview of the important databases of bioinformatics used for research in molecular biology, taxonomy, conservation biology, forensics and medicine and for the commercial applications in pharmaceutical industry and agricultural biotechnology. This chapter also describes various internet resources available in the field of Bioinformatics.

Introduction

Bioinformatics has emerged as one of the important scientific disciplines which revolutionized application of life sciences in our day to day life. A simple definition of bioinformatics is: "Bioinformatics is the use of computational techniques for the consolidation and analysis of large scale experimental biological data". Hence, Bioinformatics has transformed the life science discipline from a purely wet lab-based science to a computationally assisted information science.

Application of bioinformatics in life science has multi facets right from gene discovery to drug design where the knowledge on sequence, structure and function of genes and proteins take the lead. The objectives of biological database generation are to: (i) organize the large scale experimental biological data for further investigation, (ii) make the biological data available globally and (iii) avoid duplication of experiments.

Types of biological data

Biological databases are libraries of bioscience information including genomics, proteomics, metabolomics, microarray gene expression, phylogenetics etc. collected from various experiments, published literature, high throughput experiment technology and computational analyses. Hence, the biological data have then own inherited complexity. Various types of biological data at the nucleotide sequence level are: Genomic DNA, complementary DNA, recombinant DNA, Expressed sequence tags and genomic survey sequences.

Genomic DNA

Genomic DNA is the full complement of DNA contained in the genome of a cell or organism including both the coding and the non-coding sequences. The term "genomic DNA" (nuclear DNA) is used to distinguish from other non-nuclear or non-chromosomal DNA (eg. plasmid DNA, transposable elements, Mitochondrial and Chloroplast DNA). The genome of almost all organisms is DNA with an exception in some RNA viruses.

The genomic DNA of eukaryotes is well organized in the form of nucleosomes which fold through a series of higher order structures to form a chromosome. In contrast, the genomic DNA in prokaryotes may be circular, linear or irregularly organized in the form of nucleoid without a nuclear membrane.

The genome of an organism possesses discrete islands of genes which code for proteins and the associated regulatory elements to control the gene expression. The number of genes varies widely from organism to organism. The bulk of non-coding DNA is mainly made of repetitive sequences which seem to have structural and regulatory importance.

Complementary DNA

cDNA is a single-stranded DNA with a complementary sequence to an RNA molecule. cDNA is synthesized from an RNA template in the presence of the enzyme reverse transcriptase. The single stranded cDNA is converted to double-strand by the enzyme DNA polymerase.

The major hurdle in expressing eukaryotic genes in prokaryotic systems is lack of RNA processing splicing machinery in prokaryotes. To circumvent this problem, complementary DNA is produced from mature processed eukaryotic mRNA and cloned into vector for expression. cDNA is used as gene probes in identification of target chromosomal genes. It is also used in the creation of cDNA library. Partial sequences of cDNA are called ESTs.

Recombinant DNA (rDNA)

Recombinant DNA (rDNA) is a form of artificial DNA that is constructed outside of living cells by joining natural or synthetic DNA segments to DNA molecules that can replicate in a living cell, or molecules that result from their replication. It differs from naturally occurring genetic recombination as it is engineered artificially. A recombinant protein is a protein that is derived from the recombinant DNA.

Some of the important applications of recombinant DNA include: (i) Recombinant Vaccines (ii) Prevention and cure of sickle cell anaemia and cystic fibrosis, production of clotting factors, insulin, etc.

Expressed Sequence Tags (ESTs)

Expressed sequence tags (ESTs) represent informative source of expressed genes and are generated from cDNA. ESTs are short DNA molecules (300 - 500 bp) reverse-transcribed from a cellular mRNA population. Once cDNA representing an expressed gene has been isolated, we can sequence a few hundred nucleotides from either end of the molecule to create two different kinds of ESTs (5' and 3').

Based on sequencing, ESTs are classified into two types, 5'-EST and 3'-EST. Sequencing only the beginning portion of the cDNA produces 5' EST and sequencing the ending portion of the cDNA molecule produces what is called a 3' EST. A 5' EST is obtained from the coding portion of a transcript where as the 3' EST generated from the 3' end of a transcript fall within non-coding region. ESTs have proven to be efficient in identifying novel genes and are the powerful tools in the hunt for genes involved in hereditary diseases.

Genomic Survey Sequences (GSSs)

GSSs are nucleotide sequences similar to Expressed Sequence Tags. Unlike cDNA or mRNA, GSS are from genomic orgin and the name is derived from the homonym NCBI GenBank division. It should be noted that two classes (exon trapped products and gene trapped products) may be derived via a cDNA intermediate. Care should be taken when analyzing sequences from either of these classes, as a splicing event could have occurred and the sequence represented in the record may be interrupted when compared to genomic sequence.

The data stored in GSS contain:

- random "single pass read" genome survey sequences
- cosmid/BAC/YAC end sequences
- exon trapped genomic sequences
- Alu PCR sequences
- transposon-tagged sequences

Genome survey sequences are often used for the mapping of genome sequencing. NCBI maintains a GSS database. The GSS database is a collection of unannotated short single-read primarily genomic sequences from GenBank including random survey sequences, clone-end sequences and exon-trapped sequences.

Primary databases

There is a rapid increase in the amount of sequence data available due to advances in sequencing machines, which need careful classification and storage. This is facilitated by means of databases. Primary databases are archival and non-curated. Primary database consists of experimental results collectively stored and maintained by various resources like GenBank/EMBL/DDBJ. Protein sequence data are derived from translation of nucleotide sequences and also resulting from direct sequencing experiments submitted to Swiss-Prot, GenPept and PIR. The sequences submitted by authors to GenBank/EMBL/DDBJ are in standard format and possess unique numbers called as accession numbers. If there is a change or update in the sequence by the author, it is given a version number and released. The primary sequence record includes primary accession numbers, version numbers, protein ID numbers, gene identifier (GI) numbers and header records.

Composite databases

A database integrates different primary sources, based on a criteria to determine the inclusion of sources. In other ways, it is a database that amalgamates variety of different primary sources. They eliminate the need to interrogate multiple resources. The selection of sources and redundancy level lead to development of many composite databases. For example, NRDB combines sequences from GenBank, PDB, Swiss-Prot and PIR; UniProt is a collection of protein sequences from PIR-PSD, Swiss-Prot and TrEMBL.

Bioinformatics Databases and Internet Resources

Secondary databases

In addition to primary databases, there are many secondary or pattern databases. These databases contain results from analyses of primary database. Many software tools are available to analyze and interpret the biological significance of data available in the primary database. Swiss-Prot appears to be the primary source for all protein secondary databases. Some of the Secondary databases are (i) PROSITE, a database of regular expressions/patterns, (ii) PRINTS, a fingerprints database, (iii) BLOCKS, a database of motifs/BLOCKS, (iv) Pfam a database based on Hidden Markov Models. The development of secondary database includes: (i) reducing overlapping/redundancy from primary databases, (ii) access to vast amount of pre-processed biological data and (iii) availability of interconnectivity links.

Literature databases

PubMed

PubMed[1] is the most widely used free literature searching tool for biologists (*http://www.ncbi.nlm.nih.gov/pubmed*). It comprises more than 20 million citations for biomedical literature from MEDLINE, life science journals and online books. The MEDLINE database, the premier source for biomedical literature, is governed by NLM. The NLM Catalog provides access to NLM bibliographic data for journals, books, audiovisuals, computer software, electronic resources and other materials. Links to the library's holdings in LocatorPlus, NLM's online public access catalog are also provided. PubMed includes links to many sites providing full text articles and other related resources.

The interface allows the user to specify a search term (any alpha numeric string) and a search field (e.g. title, text word, journal or author). Queries retrieve abstracts from most of the major journals, although not all journals are indexed, particularly newer journals or journals with lower impact factors. Similar to other web search engines, Pubmed also uses the three main Boolean operators AND, OR and NOT. A wildcard function, is used to designate any character or combination of characters. One of the very effective features of PubMed is the option to retrieve related article, which helps to 'get into' the literature of a topic very easily. Description on Pubmed search guidelines can be found on NCBI website.

Books on research on the new topics are freely accessible through online biomedical textbooks. More than 16,000,000 biomedical journal abstracts can be viewed through PubMed. PubMed Central is a database that contains the whole research articles from over 300 research journals. Information on human genetic disorders is maintained in a separate website called Online Mendelian Inheritance in Man (OMIM). OMIM is a catalog of human genes and genetic disorders, or Gene Reviews, peer-reviewed articles on genetic testing, diagnosis, and management of inherited disorders. Online Mendelian Inheritance in Animals (OMIA) is a database of genes, inherited disorders and traits in animal species other than humans and mouse.

PLoS

The Public Library of Science (http://*www.plos.org*) is a nonprofit organization of scientists and physicians committed to making the world's scientific and medical literature a public resource [2, 3]. The main objectives of PLOS are: (i) to provide unlimited access to the world's library of scientific knowledge to everybody all over the world, (ii) to facilitate the educational and medical & scientific research communities with free search of every published full article with a view to inculcate new ideas and techniques and (iii) to develop innovative ways to explore and use the world's treasury of scientific ideas and discoveries for future developments.

Now PLOS has become a platform for the research communities to publish their most important work under the open access model. Hence the PLoS journals are immediately available online at free of cost and no restrictions on subsequent use or redistribution as long as the author(s) and source are cited.

Biomed central

BioMed Central [4] is an open access publisher (Springer Science and Business Media) committed to providing immediate open access to peer-reviewed biomedical research (http://*www.biomedcentral.com*). BioMed Central publishes over 200 online journals covering the whole of Science, Technology and Medicine (STM). All the original research articles in journals published by BioMed Central are open accessible and are immediately and permanently available online without charge. Only few of the journals require an institutional or a personal subscription to view other content, such as reviews or paper reports. BMC provide latest articles from various scientific fields under the heading, Gateways.

Classification of Bioinformatics Databases

Genome databases

Genomes of living organisms express profound diversity depending on the size and shape of the genetic material. The genetic material can be either the

Bioinformatics Databases and Internet Resources

DNA (single- or double- stranded) or RNA. Moreover, some genomes are linear (e.g., mammals) and others are closed and circular (e.g., most bacteria).

A genome database is a repository for complete information about a particular organism. It stores and curates data generated worldwide by researchers engaged in the mapping of the genome project. Main goal of development of genome database is to make available the genome information to the research community by providing them constantly revised and updated scientific knowledge.

Genome databases are used for the storage and analysis of genetic and physical maps. A genome database should define four data types such as sequence, physical, genetic and bibliographic data. Sequence data includes annotated molecular sequences. Physical data includes data such as sequence-tagged sites, coding regions, noncoding regions, control regions, telomeres, centromeres, repeats and metaphase chromosome bands. Genetic data includes locus name, location, recombination distance, polymorphisms, breakpoints, rearrangements and disease association. Bibliographic data consists of biological abstracts and links to other related literature and literature databases.

EBI Genomes: EBI Genomes[5] provides access and statistics for the completed genomes and information about ongoing projects (http://*www.ebi.ac.uk/genomes/*). The "Genome Reviews" database provides an up-to-date, standardized and comprehensively annotated view of the genomic sequence of organisms with completely deciphered genomes. Recently, Genome Reviews is updated to have the genome databases of archaea, bacteria, bacteriophages and selected eukaryota.

Genome Biology: NCBI[6] provides several genomic biology tools and resources, including organism-specific pages that include links to many web sites and databases relevant to that species and complete genomes (*www.ncbi.nlm.nih.gov/sites/genome, www.ncbi.nlm.nih.gov/guide/genomes-maps/*).

Ensembl: Ensembl[7] is a collaborative project between EMBL-EBI and the Sanger Centre to develop software for automatic annotation of eukaryotic genomes (http://*www.ensembl.org*). The Ensembl database is a repository of stable, automatically annotated human genome sequences, available as both an interactive website and standalone files. Ensembl annotates and predicts new genes, with annotation from the InterPro protein family databases and with additional annotations from databases of genetic disease [OMIM], expression [SAGE] and gene family. The Ensembl site is one of the leading sources of human genome sequence annotation and provides much of the analysis for publication by the international human genome project of the draft genome. It is an open source software project to develop a portable system to

16   Agriculture Bioinformatics

handle large genomes and additional features for sequence analysis and data storage.

Viral genome database

Viruses are intracellular parasites whose life cycle has two distinct phases: (i) an intracellular and (ii) an extracellular phase and have the ability to cause disease in other living organisms. The availability of both sequence and structural level information regarding the pathogenic viruses will be helpful in targeting these viruses. Some of the important viral resources available online are: (i) The Universal Virus Database of the International Committee on Taxonomy of Viruses (ICTVdb), (ii) HPV sequence database, (iii) Viral Bioinformatics Resource Centre (VBRC) (iv) ViperDB which is a database for icosahedral virus capsid structures, (v) Sub viral database, a partial and complete sequence data of the smallest known auto-replicable species and (vi) VGDB that stores genes and proteins of about 21 pox and related viruses.

ICTVdb

The International Committee on Taxonomy of Viruses (ICTV) is the "international court" of experts that rules on names and relationships of all viruses. ICTV authorizes and organizes the taxonomic classification of all viruses of animals (vertebrates, invertebrates and protozoa), plants (higher plants and algae), bacteria, fungi and archaea from the family level to strains and isolates. The objectives of ICTV are: (i) to develop an internationally acceptable taxonomy for viruses, (ii) to communicate taxonomic decisions to all users of virus names, in particular the international community of virologists (iii) to maintain an index of virus names and (iv) to maintain the ICTV database.

The universal virus database, ICTVdb[8] has been well developed by ICTV since 1991 (*http://www.ictvdb.org/*) and is open to public. The database classifies viruses based on their chemical characteristics, genomic type, nucleic acid replication, diseases, vectors, and geographical distribution, among other characteristics. It uses world standard DELTA (Description Language for Taxonomy) format for interactive identification and data retrieval, which is capable of storing a wide diversity of data and translate it into a language suitable for traditional reports and web publication.

VirGen

VirGen[9] comprises primary and derived data of viral genomes and is designed to serve as a single-stop solution for accession, retrieval and analysis of viral genome sequences (*http://bioinfo.ernet.in/virgen/virgen.html*). The main focus of VirGen is to provide an annotated and a curated database comprising complete genome sequences of viruses, value-added derived data and data mining tools. Unique features of VirGen include easy access to viral genome records, graphical representation of genome organization, a set of non-redundant genome records, pre-computed data on multiple alignments of genomes/proteomes and predicted antigenic determinants. It is the first database to curate viral genomes for 'alternate names' of proteins and to archive the results of whole-genome phylogeny. Comparative genome analysis data facilitate the study of genome organization and evolution of viruses, which would have implications in applied research to identify candidates for the design of vaccines and antiviral drugs.

Bacterial genome databases

Bacteria are ubiquitous in nature. The size of Bacterial chromosomes ranges from 0.6 Mbp to over 10 Mbp and archael chromosomes range from 0.5 Mbp to 5.8 Mbp. Consequently the number of genes found in bacteria is highly variable (between 575 and 5,500 genes) depending upon the species. Prokaryotes tend to have less "JUNK" DNA in compared to eukaryotes. The first bacterial genome to be sequenced is *Haemophilus influenzae* containing 1,181,000 base pairs. There are many organizations involved in sequencing and developing bacterial genome databases. Some of the major bacterial resources are listed below: NCBI's Genomes, JCVI's CMR (comprehensive microbial Resource), Welcome trust Sanger institute and EBI genome databases. They contain information completed & ongoing genome project and importance of the organisms to be sequenced with links to sequences, publications, name of the centre involved in sequencing. Databases include genome maps, genome assembly, gene prediction and functional annotation. It also offers tools for comparative genomics.

Genomes OnLine Database (GOLD)

Genomes OnLine Database (GOLD)[10] is a world wide web resource for comprehensive access to information regarding complete and ongoing genome projects around the world (*http://igweb.integratedgenomics.com/GOLD/*). As of September 2009, GOLD contains information for more than 5800 sequencing projects, of which 1100 have been completed and their sequence data deposited in a public repository. GOLD was created in 1997 and from April 2000, it has been licensed to Integrated Genomics. GOLD has evolved from a genome / metagenome project monitoring system into a universal genome project core catalog/indexer charged with the task of providing data interconnectivity, exchange and dissemination

18   Agriculture Bioinformatics

Microbial Genome Database (MBGD)

The Microbial Genome Database (MBGD)[11] for comparative analysis is a comprehensive platform for microbial comparative genomics (*http:// mbgd.genome.ad.jp*). To utilize the MBGD database as a comprehensive resource for investigating microbial genome diversity, the following advanced functionalities have been developed: (i) enhanced assignment of functional annotation, including external database links to each orthologous group, (ii) interface for choosing a set of genomes to compare based on phenotypic properties, (iii) the addition of more eukaryotic microbial genomes (fungi and protists) and some higher eukaryotes as references and (iv) enhancement of the MyMBGD mode, which allows users to add their own genomes to MBGD and now accepts raw genomic sequences without any annotation. MBGD is constantly updated and now contains almost 1000 genomes.

Organism specific genome databases

OMIM/OMIA

Online Mendelian Inheritance in Man (OMIM)[12, 13] is the database on genes and genetic disorders observed in humans (http://*www.ncbi.nlm.nih.gov/omim*). OMIM focuses on the relationship between phenotype and genotype. It is updated daily, and the entries are hyperlinked to other genetics resources like (i) OMIM Gene Map, and the NCBI Map Viewer entry for the chromosomal locus (ii) GenBank, RefSeq, dbSTS, and dbSNP records for related nucleotide sequence records, (iii) UniProt, TrEMBL for related protein sequence records, etc.

Entries for genes may contain sections on gene structure, gene function, mapping, molecular genetics, animal models, allelic variants. Entries of genetic disorders contain informations summarising the current literature on history, inheritance mapping, heterogeneity of the disorder, pathogenesis, molecular genetics, population genetics, clinical and biochemical features, clinical management, animal models.

Online Mendelian Inheritance in Animals (OMIA)[14] is a database of genes, inherited disorders and traits in animal species (http://*www.ncbi.nlm.nih.gov/ omia*). The database contains textual information and references, and links to other databases like OMIM, PubMed, Gene, etc.

SGD

The Saccharomyces Genome Database (SGD)[15] provides information for the complete *Saccharomyces cerevisiae* (baker's and brewer's yeast) genomic

Bioinformatics Databases and Internet Resources   19

sequence, along with its genes, gene products, the phenotypes of its mutants, and related literature (*http://genome-www.stanford.edu/Saccharomyces/*). The amount of information and the number of features provided by SGD have increased greatly after the release of the complete genomic sequence of *Saccharomyces cerevisiae*. SGD aids researchers by providing the basic information, tools to analyze the features of genome and relationships between genes. SGD presents information using a variety of user-friendly, dynamically created graphical displays illustrating physical, genetic and sequence feature maps.

WormBase

WormBase[16] is the repository of nematodes and is both manually and automatically curated (*http://www.wormbase.org/*). Initially worm base started consisting of *Caenorhabditis elegans*. *Caenorhabditis elegans* is a well established genetics model due its smaller size, rapid generation time and compact genome. Worm base provides new insights and views regarding *Caenorhabditis elegans*. It contains open source GBrowse software, BLAST and BLAT for similarity searches. Gene and protein summary gives details such as orthologs, paralog descriptions. Comparative studies are carried out using orthologs information derived from the Ensembl compara pipeline, homology from Inpranoid and inline tree pictures from TreeFam. WormBase has Textpresso text mining system. WormBase provides 5597 gene summaries (25% of protein-coding genes). WormBase facilitates Gene Ontology, phenotype ontology and identification transcription factor binding site information.

PlasmoDB

Plasmodium species are obligate intracellular protozoan parasites of humans and animals, and is the causative agents of malaria. The Plasmodium Genome database, PlasmoDB[17] (*http://PlasmoDB.org*) is a functional genomic database for Plasmodium spp. that provides a resource for a rapid and convenient access to the terabytes of genomic-scale data and visualization in a gene-by-gene or genome-wide scale. Combinatorial use of data analysis tools enables powerful data mining queries, such as combining gene and protein expression data to monitor changes through various life-cycle stages. Functional predictions can be used to explore potential targets for anti-malarial drug development.

FlyBase

The FlyBase[18] Drosophila genetics database, FlyBase (*http://flybase.org*) is the primary database of integrated genetic and genomic data about the organism *Drosophila melanogaster* and related *Drosophila dipterans*. FlyBase is one of

20

the organizations contributing to the Generic Model Organism Database (GMOD). FlyBase contains a complete annotation of the *Drosophila melanogaster* that is updated frequently. It also includes a searchable bibliography of research on Drosophila genetics in the last century. Information in FlyBase originates from a variety of sources ranging from large-scale genome projects to the primary research literature. Data-types include sequence-level gene models, molecular classification of gene product functions, mutant phenotypes, mutant lesions and chromosome aberrations, gene expression patterns, transgene insertions, and anatomical images. Links between FlyBase and external databases provide extensive opportunity for extending exploration into other model organism databases and resources of biological and molecular information.

TAIR

Arabidopsis thaliana is an experimental model organism for plants. The database of Arabidopsis thaliana provides genetic and molecular biology information and is maintained by TAIR (The Arabidopsis information resource) of the Carnegie institution for science Department of Plant Biology Stanford, California (*www.arabidopsis.org/about/index.jsp*)[19]. TAIR collects large amount of data from manual curation of literature and direct submissions from research community. TAIR focuses mainly on Arabidopsis genomic and genetic data, including genes, clones, ecotypes, markers, expression data, SNPs, mutations, proteins and germplasms. A variety of pipelines generate computational data. Gene structure pipelines generate exons, UTRs and add new genes. Functional pipeline updates gene ontology terms and functions. Mapping pipeline provides genome positions, ESTs, transposon insertions, makers, SNPs. The data retrieval and analysis is done by Gbrowse or interactive SeqViewer provided by TAIR. Interactive tools available at TAIR are MapViewer (for comparison of physical, genetic, sequence maps), AraCyc (pathways for biochemical pathways visualization), BLAST, Pattern Matching, Motif analysis, Vxlnsight (gene expression), Microarray elements search tool and restriction analysis tools. The TAIR10 release contains 27,416 protein coding genes, 4827 pseudogenes

FlyBase is an online bioinformatics database of the biology and genome of the model organism *Drosophila melanogaster* and related *Drosophilid dipterans*. The FlyBase project is carried out by a consortium of Drosophila researchers and computer scientists at Harvard University and Indiana University in the United States, and University of Cambridge in the United Kingdom. FlyBase is one of the organizations contributing to the Generic Model Organism.

Genome browsers

Genome browser is a graphical interface to display information from a biological database for genome analysis. Genome browsers enable researchers to visualize and browse entire genomes with annotated data including gene prediction and structure, proteins, expression, regulation, variation, comparative analysis, etc.

Ensembl

The Ensembl Genome Browser[7] of EBI is an extensive database browser for a large variety of species genomes from Human to yeast (http://www.ensembl.org). It can be used to browse an annotated graphical display of the genome and associated annotations for specific genomic regions and also to search the databases for additional information. A comparison of whole genome alignments and conserved regions across various species is possible with Ensembl using the tool, BLAST.

The project Ensembl started as a joint scientific project between the European Bioinformatics Institute and the Wellcome Trust Sanger Institute, in 1999. The Ensembl remains to provide a centralized resource for geneticists, molecular biologists and other researchers studying the genomes of our own species and other vertebrates and model organisms and is one of several well known genome browsers for the retrieval of genomic information. In Ensembl, sequence data is fed into a software called pipeline which creates a set of predicted gene locations and saves them in a MySQL database for subsequent analysis & display and makes them available for free access. The unique feature of Ensembl is the ability to automatically generate graphical views of the alignment of genes and other genomic data against a reference genome.

VEGA Genome Browser

The Vertebrate Genome Annotation (VEGA)[20] database of The Wellcome Trust *Sanger* Institute is a community resource for browsing manual annotation from a variety of vertebrate genomes of finished sequence (*http://vega.sanger.ac.uk*). Vega is distinct from other genome browsers as it has (i) a standardized gene classification including pseudogenes and non-coding transcripts, (ii) annotated PolyA sites/signals, (iii) manual curation of data and (iv) annotation of haplotypes, etc. Manual annotation is currently more accurate in identifying splice variants, pseudogenes poly (A) features, non-coding and complex gene structures and arrangements than current automated methods. The database also contains annotation from regions and displays multiple species annotation (human, mouse, dog and zebrafish) for comparative analysis.

NCBI Map Viewer

The NCBI Map Viewer[21] (*http://www.ncbi.nlm.nih.gov/projects/mapview/*) has been developed to facilitate genomic analysis using genetic mapping to find gene information, where genetic mapping refers to determine the relative position between two genes on a chromosome. The NCBI Map Viewer provides graphical displays of features on NCBI's assembly of human genomic sequence data as well as cytogenetic, genetic, physical, and radiation hybrid maps. eGenome provides a direct, element-specific link to the NCBI Map Viewer from each element record.

The Map Viewer provides special browsing capabilities for a subset of organisms in Entrez Genomes. Map Viewer allows us to view and search an organism's complete genome, display chromosome maps, and zoom into progressively greater levels of detail, down to the sequence data for a region of interest. The number and types of available maps vary by organism.

KEGG

Kyoto Encyclopedia of Genes and Genomes (KEGG)[22] is a knowledge base for systematic analysis of gene functions in terms of the networks of genes and molecules (*http://www.genome.jp/kegg/*). KEGG provides a computational prediction of higher-level complexity of cellular processes and organism behaviors from genomic and molecular information. The major component of KEGG is the PATHWAY database that consists of graphical diagrams of biochemical pathways including most of the known metabolic pathways and some of the known regulatory pathways. The pathway information is represented by the ortholog group tables that summarizes orthologous and paralogous gene groups among different organisms. KEGG maintains two databases; (i) GENES database for the gene catalogs of all organisms with complete genomes and selected organisms with partial genomes, which are continuously re-annotated, and (ii) LIGAND database for chemical compounds and enzymes. Each gene catalog is associated with the graphical genome map for chromosomal locations. KEGG also develops and provides various computational tools for (i) reconstructing biochemical pathways from the complete genome sequence, and (ii) predicting gene regulatory networks from the gene expression profiles.

MIPS

The Munich Information Center for Protein Sequences (MIPS)[23] is a research center hosted at the Institute for Bioinformatics and Systems Biology (IBIS) in Germany. It focuses on the systematic analysis of genome information including the development and application of bioinformatics methods in genome annotation, gene expression analysis and proteomics. MIPS supports and maintains a set of generic databases as well as the systematic comparative analysis of microbial, fungal, and plant genomes.

MIPS-GSF develops systematic classification schemes for the functional annotation of protein sequences, and provides tools for the comprehensive analysis of protein sequences (http://*www.helmholtz-muenchen.de/en/ibis*). MIPS-GSF updates the information on the Comprehensive Yeast Genome Database (CYGD), the *Neurospora crassa* genome (MNCDB), the databases for the comprehensive set of genomes (PEDANT genomes), the database of annotated human EST clusters (HIB), the database of complete cDNAs from the DHGP (German Human Genome Project), as well as the project specific databases for the GABI (Genome Analysis in Plants) and HNB (Helmholtz–Netzwerk Bioinformatik) networks.

UCSC Genome Browser

The University of California, Santa Cruz (UCSC) Genome Browser[24] website (*http://genome.ucsc.edu/*) provides a large database of publicly available sequence and annotation data along with an integrated tool set for examining and comparing the genomes of organisms, aligning sequence to genomes, and displaying and sharing users' own annotation data. The Browser is a graphical viewer optimized to support fast interactive performance and is an open-source, web-based tool suite built on top of a MySQL database for rapid visualization, examination, and querying of the data at many levels. The Genome Browser Database, browsing tools, downloadable data files, and documentation can all be found on the UCSC Genome Bioinformatics website.

The UCSC Browser has expanded to accommodate genome sequences of all vertebrate species and selected invertebrates for which high-coverage genomic sequences is available. As of September 2009, genomic sequence and a basic set of annotation 'tracks' are provided for 47 organisms, including 14 mammals, 10 non-mammal vertebrates, 3 invertebrate deuterostomes, 13 insects, 6 worms and a yeast. The UCSC Genome Browser provides a source of deep support for a wide range of biomedical molecular research.

Sequence Databases

Nucleotide Sequence Databases

The three primary nucleotide sequence databases that have been widely used are: (i) GenBank, the genomic information maintained by National Institutes of Health, USA, (ii) EMBL of the European Bioinformatics Institutes, Europe and (iii) the DDBJ of National Institute of Genetics DNA Databank, Japan.

Under the direction of the International Nucleotide Sequence Database Collaboration (INSDC), these databases gather, maintain, and share data. The entries in these three databases are synchronized on daily basis and accession numbers are ranged in a consistent manner between these three centers.

GenBank

The GenBank[25] sequence database is an open access, annotated collection of all publicly available nucleotide sequences and their protein translations (*http://www.ncbi.nlm.nih.gov/GenBank/index.html*) submitted by various research communities. Sequences are submitted via BanKlt or Sequin programs. There are 106,533,156,756 bases in 108,431,692 sequence records as of August 2009. GenBank is built by direct submissions from individual laboratories, as well as from bulk submissions from large-scale sequencing centers.

EMBL

The EMBL Nucleotide Sequence Database[5] incorporates, organizes and distributes nucleotide sequences from public sources (*http://www.ebi.ac.uk/embl/*). Data are exchanged between the collaborating databases on a daily basis. Sequences are submitted through WEBIN. The database contains 301,588,430,608 nucleotides in 199,575,971 entries until 2010. Database releases are produced quarterly. The latest data collection can be accessed via FTP, email and WWW interfaces. The EBI's Sequence Retrieval System (SRS) integrates and links the main nucleotide and protein databases as well as many other specialist molecular biology databases. A variety of tools (e.g. FASTA and BLAST) are available for sequence similarity searching.

Centre for Information Biology and DNA DataBank of Japan (CIB/DDBJ)

DDBJ[26] began the DNA databank activities in 1986 at the national institute of genetics (NIG) with the endorsement of the ministry of education, science, sport and culture (http://*www.ddbj.nig.ac.jp/*). DDBJ is the sole DNA databank in Japan, which is officially certified to collect DNA sequences from researchers and to issue the internationally recognized accession number to data submitters. Sequences are submitted through SAKURA and MSS(Mass survey system). As of September 2010 this database contains 120,919,931,265 bases in 128,607,782 entries. Similar to GenBank and EMBL, DDBJ also provides many tools for retrieval and analysis of the data which are developed at DDBJ.

Bioinformatics Databases and Internet Resources   25

Protein sequences databases

Swiss-Prot

SWISS-PROT [27] (http://*www.expasy.ch/sprot/*) is a curated protein sequence database which strives to provide a high level of annotation (i.e. description of protein function, its domains structure, post-translational modifications, variants, etc.). SWISS-PROT of Swiss Institute of Bioinformatics is well known for its minimal redundancy, high quality of annotation, use of standardized nomenclature, and links to specialized databases. Sequences are submitted through SPIN tool. SWISS-PROT Release 2010_12 of 30-Nov-10 comprises 184,678,199 amino acids from 193,537 entries. Recent developments of the database include format and content enhancements, cross-references to additional databases, new documentation files and improvements to TrEMBL, a computer-annotated supplement to SWISS-PROT.

TrEMBL

Due to the large number of sequences generated by different genome projects, EBI introduced another database, TrEMBL (translation of EMBL nucleotide sequence database)[27] in association with Swiss-Prot for computer-annotated entries derived from the translation of all coding sequences in the nucleotide databases GenBank, EMBL and DDBJ. This database is divided into two sections: SP-TrEMBL contains sequences that will eventually be transferred to Swiss-Prot and REM-TrEMBL contains patent application sequences, fragments of less than eight amino acids, and sequences that have proven not to code for a real protein. In order to avoid the redundancy in collecting all the nucleotide databases from GenBank, EMBL and DDBJ, the proteins have been classified based on the occurrence of functional domains, repeats and important sites. TrEMBL Release 2010_12 of 30-Nov-10 database contains 410, 901, 5043 entries in 12769092 entries.

UniProt

UniProt (Universal Protein Resource) database is the most reliable, comprehensive and widely used resource for protein sequence and annotation data[28] (http://www.uniprot.org/). The UniProt Consortium comprises the European Bioinformatics Institute (EBI), the Swiss Institute of Bioinformatics (SIB) and the Protein Information Resource (PIR).

The UniProt database is divided into several sub-databases:

- UniProt Knowledgebase (UniProtKB) consists of accurate protein sequences with functional annotation

- UniProt Archive (UniParc) stores the stable, non-redundant, corpus of publicly available protein sequence data

- UniProt Reference Clusters (UniRef) datasets provide non-redundant reference clusters based primarily on UniProtKB

- UniProt Metagenomic and Environmental Sequences (UniMES) is a database specifically developed for metagenomic and environmental data

UniProt Knowledgebase (UniProtKB)

Sequences submitted to UniProtKB[29] database come from the translations of coding sequences (CDS) submitted to INSDC collaborating databases, protein structures databases, biological experiments, literature and gene prediction. (*http://www.uniprot.org/help/uniprotkb*)

UniProtKB consists of two sections

- UniProtKB/Swiss-Prot section contains manually-annotated records with information extracted from literature and curator-evaluated computational analysis.

- UniProtKB/TrEMBL section contains computationally analyzed records that await full manual annotation.

Annotation of a single sequence consists of the description of the following: function(s), enzyme-specific information, biologically relevant domains and sites, post-translational modifications, sub-cellular location(s), tissue specificity, developmentally specific expression, structure, interactions, splice isoform(s), diseases associated with deficiencies or abnormalities, etc.

The annotation activities of the Swiss-Prot group can be divided into 2 major parts:

- Model organism-oriented annotation which is focused on specific group of organisms, for example: Mammals, Plants, Bacteria, etc.

- Transversal annotation which is focused on issues common to all organisms, for example: post-translational modifications, protein-protein interactions, etc.

UniProt Archive (UniParc)

UniParc[30] (*http://www.ebi.ac.uk/uniparc/*) database is a comprehensive and non-redundant database. It contains publicly available protein sequences from Swiss- Prot, TrEMBL, PIR-PSD, EMBL, Ensembl, International Protein Index (IPI), PDB, RefSeq, FlyBase, WormBase and the patent offices in Europe, the

Bioinformatics Databases and Internet Resources 27

United States and Japan. A sequence that exists in many copies in different databases is represented in UniParc database as a single entry which allows in identifying the same protein from different source databases. UniParc provides cross-references to the source databases (accession numbers), sequence versions, and status (active or obsolete). A UniParc sequence version is incremented each time the underlying sequence changes to reveal sequence changes in all source databases.

UniProt Reference Clusters (UniRef)

UniRef[31] (*http://www.ebi.ac.uk/uniref/*) database is the UniProt Reference Clusters. The two major objectives of UniRef are: (i) to facilitate sequence merging in UniProt, and (ii) to allow faster and more informative sequence similarity searches. UniProt Knowledgebase still contains a certain level of redundancy as it is not possible to use fully automatic merging without risking merging of similar sequences from different proteins. However, such automatic procedures are extremely useful in compiling the UniRef databases to obtain complete coverage of sequence space while hiding redundant sequences (but not their descriptions) from view.

UniRef database consists of three sub-databases:

- UniRef100 database combines identical sequences and sub-fragments with 11 or more residues (from any organism) into a single UniRef entry. The sequences are derived from UniProtKB and UniParc databases.

- UniRef90 is build by clustering UniRef100 sequences such that each cluster is composed of at least 90% sequence identity.

- UniRef50 is build by clustering UniRef100 sequences such that each cluster is composed of at least 50% sequence identity.

Such clustering allows reducing a size of UniRef100 database of approximately 40% (UniRef90) and 65% (UniRef50), thus the time needed for similarity searches is significantly reduced.

UniProt Metagenomic and Environmental Sequences (UniMES)

UniMES is a repository specifically for metagenomic and environmental data. The predicted proteins from this dataset are combined with automatic classification by InterPro to enhance the original information with further analysis. Data from UniMES is not included in UniProtKB or UniRef, but is included in UniParc. UniMES currently contains data from the Global Ocean Sampling Expedition (GOS) which was originally submitted to the International Nucleotide Sequence Databases.

Sequence motifs databases

Protein signature databases have become vital tools for identifying distant relationships in novel sequences and hence are used for the classification of protein sequences and for inferring their function.

PROSITE

PROSITE[32] consists of documentation entries describing protein domains, families and functional sites, as well as associated patterns and profiles to identify them (*www.expasy.org/prosite/*). It is complemented by ProRule, a collection of rules based on profiles and patterns, which increases the discriminatory power of these profiles and patterns by providing additional information about functionally and/or structurally critical amino acids. Among the various databases dedicated to the identification of protein families and domains, PROSITE is the first one created and is mainly used to annotate the domain features of UniProtKB/Swiss-Prot entries. PROSITE currently consists of a large collection of biologically meaningful motifs that are described as patterns or profiles, and linked to the brief description of protein family or domain. The close relationship of PROSITE with the SWISS-PROT protein database allows the evaluation of the sensitivity and specificity of the PROSITE motifs and their periodic reviewing. In return, PROSITE is used to help annotate SWISS-PROT entries.

Prosite Patterns: Protein sequence segments of length 10 to 20 amino acids which are important to the biological function of a group of proteins and are conserved in both structure and sequence during evolution are variously termed as a patterns, motifs, signatures or fingerprints. Some of the biologically significant regions or residues are: (i) enzyme catalytic sites, (ii) prostethic group attachment sites (heme, pyridoxal-phosphate, biotin, etc.), (iii) amino acids involved in binding a metal ion, (iv) cysteines involved in disulphide bonds, etc. It should be noticed that some families or domains are defined not only by one pattern but also by the co-occurrence of two or more patterns of low specificity. In that case, the simultaneous occurrence of linked patterns gives good confidence that the matched protein belongs to the set being considered. The advantages of patterns are; (i) their easy intelligibility for the user and the fact that patterns are directed against the most conserved residues which are more relevant for the biological function of the protein family or domain, and (ii) reasonably lesser time required for the scan of a protein database with patterns on any computer.

Prosite Profiles: Profiles are quantitative motif descriptors providing numerical weights for each possible match or mismatch between a sequence residue and a profile position. Unlike Pattern, a mismatch at a highly conserved position in a Profile is accepted provided that the rest of the sequence displays a sufficiently high level of similarity. In fact Patterns and Profiles have complementary qualities. Patterns confined to small regions with high sequence similarity are often powerful predictors of protein functions such as enzymatic activities. Profiles covering complete domains are more suitable for predicting protein structural properties.

The generalized profiles used in PROSITE are an extension of the sequence profiles introduced by Gribskov and coworkers[33]. The numerical weights of a profile serve to define a quality score for a profile-sequence alignment. A sequence region that can be aligned to a profile with a score higher than a threshold score is considered a match. Searching for multiple occurrences of a particular domain within the same sequence requires the execution of a dynamic programming algorithm that finds a maximal set of high-scoring profile sequence alignments above a threshold score. Different alignment modes (global or local) are defined by profile intrinsic parameters. The profile format used in PROSITE comprises fields for accessory parameters which define the search method to be used for a particular domain. They allow specification of appropriate cut-off values, different score normalization modes, and instructions as to how to treat partly overlapping matches.

ProDom

ProDom[34] is a comprehensive database of protein domain families generated from the global comparison of all available protein sequences (*http://prodom.prabi.fr*). The ProDom database consists of domain family entries and each entry provides a multiple sequence alignment of homologous domains and a family consensus sequence. The ProDom building procedure MKDOM2 is based on recursive PSI-BLAST searches [35]. The source protein sequences are non-fragmentary sequences derived from UniProtKB (Swiss-Prot and TrEMBL databases). ProDom was first established in 1993[36]. It is now maintained by the PRABI (bioinformatics center of Rhone-Alpes). Recent improvements include the use of three-dimensional (3D) information from the SCOP database; a completely redesigned web interface called MOLSCRIPT for visualization of ProDom domains on 3D structures; coupling of ProDom analysis with the Geno3D homology modelling server; Bayesian inference of evolutionary scenarios 7 for ProDom families. In addition ProDom-SG, a ProDom-based server is dedicated to the selection of candidate proteins for structural genomics.

Pfam

Pfam[37, 38] is a widely used database of protein families and domains that include their annotations and multiple sequence alignments generated using hidden Markov models. It is an open access resource established by Dr Alex Bateman at the Wellcome Trust Sanger Institute in 1998. Pfam-A is the manually curated portion of the database having more than 10,000 entries and for each entry, a protein sequence alignment and a hidden Markov model is stored. As the entries in Pfam-A does not cover all known proteins, an automatically generated supplement, Pfam-B is created. Pfam-B contains a large number of small families derived from clusters produced by an algorithm called ADDA. The database iPfam builds on the domain description of Pfam.

InterPro

InterPro [39] (*www.ebi.ac.uk/interpro/*) is an integrated documentation resource for protein families, domains, regions and sites. InterPro combines a number of databases; ProDom, PROSITE, PRINTS, PANTHER, PIRSF, Pfam, SMART, TIGRFAMs, Gene3D and SUPERFAMILY by uniting these member databases, InterPro capitalizes on their individual strengths, producing a powerful integrated database and diagnostic tool (InterProScan). InterPro release 30.0 contains 21178 entries and is given in the Table shown below.

Signature Database	Version	Signatures	Integrated Signatures**
GENE3D	3.3.0	2386	1253
HAMAP	180510	1656	1180
PANTHER	6.1	30128	2267
PIRSF	2.73	3233	2765
PRINTS	41.1	2050	2012
PROSITE patterns	20.66	1308	1292
PROSITE profiles	20.66	901	839
Pfam	24.0	11912	11462
PfamB	24.0	142303	0
ProDom	2006.1	1894	976
SMART	6.1	895	882
SUPERFAMILY	1.73	1774	1141
TIGRFAMs	9.0	3808	3796

**Not all signatures of member databases are integrated at the time of InterPro release

Sequence file formats

Sequence formats are simply the way in which the amino acid or DNA sequence is recorded in a computer file. Different programs expect different formats.

Bioinformatics Databases and Internet Resources

GenBank format

GenBank is the NIH genetic sequence database, an annotated collection of all publicly available DNA sequences. Each GenBank entry includes (i) a concise description of the sequence, (ii) the scientific name and taxonomy of the source organism, and (iii) a table of features that identifies coding regions and other sites of biological significance, such as transcription units, sites of mutations or modifications, and repeats. The feature table also lists the protein translations for coding regions. Bibliographic references are linked to the Medline for all published sequences.

The entries in GenBank file are:

- LOCUS:- is a short name for this sequence (a maximum of 32 characters)
- DEFINITION: Definition of sequence (Maximum of 80 characters).
- ACCESSION: accession number of the entry.
- VERSION: Version of the entry.
- DBSOURCE: The source, the date of creation / modification of the data entry.
- KEYWORDS: Keywords for the entry.
- AUTHORS: Authors for the work.
- TITLE: Title of the publication.
- JOURNAL: Journal reference for the entry.
- MEDLINE: Medline ID.
- COMMENT: Lines of comments.
- SOURCE ORGANISM: The organism from which the sequence was derived.
- ORGANISM: Full name of organism (Maximum of 80 characters).
- AUTHORS: Authors of this sequence (Maximum of 80 characters).
- ACCESSION: ID Number for this sequence (Maximum of 80 characters).
- FEATURES: Features of the sequence.
- ORIGIN: Beginning of sequence data.
- // End of sequence data.

EXAMPLE: GenBank format (Accession No: X14897)

LOCUS MMFOSB 4145 bp mRNA linear ROD 12-SEP-1993

DEFINITION Mouse fosB mRNA.

ACCESSION X14897

VERSION X14897.1 GI:50991

KEYWORDS fos cellular oncogene; fosB oncogene; oncogene.

SOURCE Mus musculus.

ORGANISM Mus musculus

Eukaryota; Metazoa; Chordata; Craniata; Vertebrata; Euteleostomi;

Mammalia; Eutheria; Rodentia; Sciurognathi; Muridae; Murinae; Mus.

REFERENCE 1 (bases 1 to 4145)

AUTHORS Zerial, M., Toschi, L., Ryseck, R.P., Schuermann, M., Muller, R. and Bravo, R.

TITLE The product of a novel growth factor activated gene, fos B, interacts with JUN proteins enhancing their DNA binding activity

JOURNAL EMBO J. 8 (3), 805-813 (1989)

MEDLINE 89251612

PUBMED 2498083

COMMENT clone = AC113-1; cell line = NIH3T3.

FEATURES Location/Qualifiers

source 1.4145

/organism = "Mus musculus"

/db_xref = "taxon:10090"

CDS1202.2218

/note = "fosB protein (AA 1-338)"

/codon_start = 1

/protein_id = "CAA33026.1"

/db_xref = "GI:50992"

/db_xref = "MGD:95575"

Bioinformatics Databases and Internet Resources

/db_xref="SWISS-PROT:P13346"

translation="MFQAFPGDYDSGSRCSSSPSAESQYLSSVDSFGSPPTAAASQEC
AGLGEMPGSFVPTVTAITTSQDLQWLVQPTLISSMAQSQGQPLASQPPAVDPYDMPGT
SYSTPGLSAYSTGGASGSGGPSTSTTTSGPVSARPARARPRRPREETLTP
EEEEKRRV
RRERNKLAAAKCRNRRRELTDRLQAETDQLEEEKAELESEIAELQKE
KERLEFVLVAH
KPGCKIPYEEGPGPGPLAEVRDLPGSTSAKEDGFGWLLPPPPPPPL
PFQSSRDAPPNL
TASLFTHSEVQVLGDPFPVVSPSYTSSFVLTCPEVSAFAGAQRTSGSEQPS
DPLNSPS
LLAL"

BASE COUNT   960 a 1186 c 1007 g 991 t   1 others

ORIGIN

    1 ataaattctt attttgacac tcaccaaaat agtcacctgg aaaacccgct ttttgtgaca
   61 aagtacagaa ggcttggtca catttaaatc actgagaact agagagaaat actatcgcaa
  121 actgtaatag acattacatc cataaaagtt tccccagtcc ttattgtaat attgcacagt
  181 gcaattgcta catggcaaac tagtgtagca tagaagtcaa agcaaaaaca aaccaaagaa
  241 aggagccaca agagtaaaac tgttcaacag ttaatagttc aaactaagcc attgaatcta
  301 tcattgggat cgttaaaatg aatcttccta caccttgcag tgtatgattt aactttttaca
  361 gaacacaagc caagtttaaa atcagcagta gagatattaa aatgaaaagg tttgctaata
  421 gagtaacatt aaataccctg aaggaaaaaa aacctaaata tcaaaataac tgattaaaat
  481 tcacttgcaa attagcacac gaatatgcaa cttggaaatc atgcagtgtt ttatttaaga
  541 aaacataaaa caaaactatt aaaatagttt tagagggggt aaaatccagg tcctctgcca
  601 ggatgctaaa attagacttc aggggaattt tgaagtcttc aattttgaaa cctattaaaa
  661 agcccatgat tacagttaat taagagcagt gcacgcaaca gtgacacgcc tttagagagc
  721 attactgtgt atgaacatgt tggctgctac cagccacagt caatttaaca aggctgctca
  781 gtcatgaact taatacagag agagcacgcc taggcagcaa gcacagcttg ctgggccact
  841 ttcctccctg tcgtgacaca atcaatccgt gtacttggtg tatctgaagc gcacgctgca
  901 ccgcggcact gcccggcggg tttctgggcg gggagcgatc cccgcgtcgc cccccgtgaa

961 accgacagag cctggacttt caggaggtac agcggcggtc tgaaggggat ctgggatctt

1021 gcagagggaa cttgcatcga aacttgggca gttctccgaa ccggagacta agcttccccg

1081 agcagcgcac tttggagacg tgtccggtct actccggact cgcatctcat tccactcggc

1141 catagccttg gcttcccggc gacctcagcg tggtcacagg ggccccctg tgcccaggga

1201 aatgtttcaa gcttttcccg gagactacga ctccggctcc cggtgtagct catcaccctc

1261 cgccgagtct cagtacctgt cttcggtgga ctccttcggc agtccaccca ccgccgccgc

1321 ctcccaggag tgcgccggtc tcggggaaat gcccggctcc ttcgtgccaa cggtcaccgc

1381 aatcacaacc agccaggatc ttcagtggct cgtgcaaccc accctcatct cttccatggc

1441 c

FASTA Format

A sequence in FASTA format begins with a single-line description, followed by lines of sequence data. The description line starts with a greater-than (">") symbol. It is recommended that all lines of text be shorter than 80 characters in length.

Example: FASTA format

>gi|532319|pir|TVFV2E|TVFV2E envelope protein

ELRLRYCAPAGFALLKCNDADYDGFKTNCSNVSVVHCTNLMN
TTVTTGLLLNGSYSENRTQIWQKHRTSNDSALILLNKHYNLTVT
CKRPGNKTVLPVTIMAGLVFHSQKYNLRLRQAWCHFPSNWKG
AWKEVKEEIVNLPKERYRGTNDPKRIFFQRQWGDPETANLW
FNCHGEFFYCK....

PIR Format

The first line in PIR format starts with a greater-than (">") sign, followed by a two-letter code describing the sequence type (P1, F1, DL, DC, RL, RC, or XX), which is followed by a semicolon, and then by the sequence identification code (the database ID-code). The second provides a textual description of the sequence. The sequence begins from the third line and extends until the end which is marked by an asterisk (*) character. A file in PIR format may comprise more than one sequence. The PIR format is also often referred to as the NBRF format.

Bioinformatics Databases and Internet Resources

Example PIR Format

>P1; CRAB_ANAPL

ALPHA CRYSTALLIN B CHAIN (ALPHA (B)-CRYSTALLIN).

MDITIHNPLI RRPLFSWLAP SRIFDQIFGE HLQESELLPA SPSLSPFLMR

SPIFRMPSWLETGLSEMRLEKDKFSVNLDVKHFSPEELKV KVLGDMVEIH

GKHEERQDEHGFIAREFNRKYRIPADVDPLTITSSLSLDG VLTVSAPRKQ

SDVPERSIPI TREEKPAIAG AQRK*

ALN/ClustalW2 Format

In ALN format, the first line starts with the words "CLUSTAL W" followed by one or more empty lines. After one or more empty lines, the multiple sequence alignment follows as one or more blocks of sequence data. Each block consists of: (i) one line for each sequence in alignment, and the no. of lines depend on the no. of sequences being compared (ii) line showing the degree of conservation for the amino acid alignment columns and (iii) one or more empty lines. In each block, the column represents (i) the sequence name, (ii) space, (iii) up to 60 amino acid sequence symbols, (iv) a cumulative count of residues for the sequences respectively.

The characters used to represent the degree of conservation are: (i) asterisk (*) for identical residues or nucleotides (ii) colon (:) for conserved substitutions, (iii) dot (.) for semi-conserved substitutions and (iv) a blank indicates no match.

GCG/MSF Format

The GCG / MSF format file may begin with as many lines of comment or description as required. But, the first line of description would start with "MSF:", and also includes the sequence length, type and date plus an internal check sum value. The next line is a mandatory blank line inserted before the sequence names. Then a one line sequence description (on sequence name, length, checksum and a weight value) of all the sequences follows. Another blank line is added followed by a line starting with two slashes "//", which denotes the end of the name list. Following another blank line, each sequence line starts with the sequence name which is separated from the aligned sequence residues by white space. In each line, 50 amino acids are aligned and each 10 amino acids are separated by space.

Example: GCG/MSF Format

MSF: 510 Type: P Check: 7736 ...

Name: ACHE_BOVIN oo Len: 510 Check: 7842 Weight: 16.0

Name: ACHE_HUMAN oo Len: 510 Check: 8553 Weight: 17.8

Name: ACHE_MOUSE oo Len: 510 Check: 229 Weight: 12.5

Name: ACHE_RAT oo Len: 510 Check: 8410 Weight: 14.2

Name: ACHE_XENLA oo Len: 510 Check: 2702 Weight: 39.2

//

ACHE_BOVIN MAGALLCALL LLQLLGRGEG KNEELRLYHY LFDTYDPGRR
PVQEPEDTVT

ACHE_HUMAN MARAPLGVLL LLGLLGRGVG KNEELRLYHH LFNNYDPGSR
PVREPEDTVT

ACHE_MOUSE MAGALLGALL LLTLFGRSQG KNEELSLYHH LFDNYDPECR
PVRRPEDTVT

ACHE_RAT MTMALLGTLL LLALFGRSQG KNEELSLYHH LFDNYDPECR
PVRRPEDTVT

ACHE_XENLA MESGVRILSL LILLHNSLAS ESEESRLIKH LFTSYDQKAR
PSKGLDDVVP

ACHE_BOVIN ISLKVTLTNL ISLNEKEETL TTSVWIGIDW QDYRLNYSKG
DFGGVETLRV

ACHE_HUMAN ISLKVTLTNL ISLNEKEETL TTSVWIGIDW QDYRLNYSKD
DFGGIETLRV

ACHE_MOUSE ITLKVTLTNL ISLNEKEETL TTSVWIGIDW HDYRLNYSKD
DFAGVGILRV

ACHE_RAT ITLKVTLTNL ISLNEKEETL TTSVWIGIEW QDYRLNFSKD
DFAGVEILRV

ACHE_XENLA VTLKLTLTNL IDLNEKEETL TTNVWVQIAW NDDRLVWNVT
DYGGIGFVPV

Structure and derived databases

The primary structure databases

Protein Data Bank (PDB)

Structure databases archive, annotate and distribute sets of atomic coordinates.
The Protein Data Bank (PDB)[40] (http://www.rcsb.org/) is a repository for the

3-D structural data of large biological molecules, such as proteins, nucleic acids, carbohydrates, etc. obtained mainly from X-ray crystallography and NMR spectroscopy (www.rcsb.org/pdb). The PDB is overseen by an organization called the Worldwide Protein Data Bank, wwPDB and freely accessible on the Internet via the websites of its member organizations (PDBe, PDBj, and RCSB). PDB provides information on: (i) experimental details about the structure determination, (ii) the amino acid sequences, (iii) structure details (atomic coordinates, details on secondary structures, disulphide bridges) including cofactors, inhibitors, and water molecules and (iv) literatures and references about the respective biomolecule. PDB contains 64995 proteins, 2208 nucleic acid, and 2990 protein/nucleic acid complexes as on December, 2010.

Cambridge Structural Database (CSD)

The Cambridge Structural Database (CSD)[41] is a vast and ever growing compendium of accurate three-dimensional structures of organic and metal–organic compounds created by the Cambridge Crystallographic Data Centre (CCDC) (*http://www.ccdc.cam.ac.uk/products/csd/*). CSD also provides a teaching environment by illustrating the key chemical concepts of the structures deposited, and a number of teaching modules. All of this material is freely available from the CCDC website, and the subset can be freely viewed and interrogated using WebCSD, an internet application for searching and displaying CSD information content. The Cambridge Structural Database (CSD) also provides software for: (i) search and information retrieval (ConQuest), (ii) structure visualization (Mercury), (iii) numerical analysis (Vista) and (iv) database creation (PreQuest). The CSD System also incorporates IsoStar, a knowledge base of intermolecular interactions, containing data derived from both the CSD and the PDB. It contains 501,857 structures as on Jan 2010.

Molecular Modeling DataBase (MMDB)

The Molecular Modeling DataBase (MMDB)[42], also known as "Entrez Structure," is a database of experimentally determined structures obtained from PDB (*http://www.ncbi.nlm.nih.gov/sites/entrez?db=structure*). The MMDB, developed by NCBI Computational Biology Branch has a number of useful features that facilitate computation on the data and link them to many other data types in the Entrez system. The Author annotated features provided by PDB are recorded in MMDB. The agreement between atomic coordinate and sequence data is verified, and sequence data are obtained from PDB coordinate records, if necessary, to resolve ambiguities. Data are mapped into a computer friendly format and transferred between applications using Abstract Syntax Notation 1 (ASN.1). This validation and encoding supports the interoperable display of sequence, structure and alignment. Uniformly defined secondary-structure and 3D-domain features are added to support structure neighbor calculations. As of Dec 2010, it contains 69,964 total structure records. It includes 17,590 proteins, 605 DNA only, 435 RNA only structures.

The Secondary structure databases

Structural Classification of Proteins (SCOP)

The Structural Classification of Proteins (SCOP)[43] (*http://scop.berkeley.edu*) database is a comprehensive ordering of all proteins of known structures according to their evolutionary and structural relationships based on the fundamental evolutionary unit, protein domain. Protein domains in SCOP are grouped according to species and hierarchically classified into families, superfamilies, folds and classes. Family contains proteins with related sequences but typically distinct functions. Superfamily bridges protein families with common functional and structural features inferred to be from a common evolutionary ancestor. Structurally similar superfamilies with different characteristic features are grouped into Folds, which are further arranged into Classes based mainly on their secondary structure content and organization.

SCOP release 1.75 contains the following entries

Class	Number of folds	Number of superfamilies	Number of families
All alpha proteins	284	507	871
All beta proteins	174	354	742
Alpha and beta proteins (a/b)	147	244	803
Alpha and beta proteins (a+b)	376	552	1055
Multi-domain proteins	66	66	89
Membrane and cell surface proteins	58	110	123
Small proteins	90	129	219
Total	1195	1962	3902

Class Architecture Topology Homology (CATH)

The CATH [44] (*http://www.cathdb.info/*) Protein Structure Classification is a semi-automatic, hierarchical classification of protein domains of protein structures available in the Protein Data Bank. Each protein has been chopped into structural domains and assigned into homologous superfamilies (groups of evolutionarily related domains). This classification procedure uses a combination of automated and manual techniques which include computational algorithms, empirical and statistical evidence, literature review and expert analysis. There are four major levels in this hierarchy: Class, Architecture,

Topology (fold family) and Homologous superfamily. Class describes the overall secondary-structure content of the domain. Architecture is a large-scale grouping of topologies which share particular structural features. Topology refers high structural similarity and Homologous superfamily indicates a demonstrable evolutionary relationship. CATH v3.3 release database built from 97,625 PDB chains and contains 1,288 chains, 2,593 superfamilies, 10,019 families, 14,473 domains.

Families of Structurally Similar Proteins (FSSP)

Protein families are known to retain the shape of the fold even when sequences have diverged below the limit of detection of significant similarities at the sequence level. Structural comparisons merge protein families of known 3D structure into structural classes, the members of which may or may not be evolutionarily related. The FSSP database[45] (*http://swift.cmbi.kun.nl/swift/fssp*) of structural alignments provides a rich source of information for the study of both divergent and convergent aspects of the evolution of protein folds. FSSP is a database of structural alignments of proteins available in PDB. FSSP data sets can be used to derive generalized sequence patterns specific for a given family of divergently related proteins. The alignments of remote homologs are the result of pairwise all-against-all structural comparisons in the representative set of protein chains. All alignments are based purely on the 3-D coordinates of the proteins and are derived by an automatic structure comparison program (Dali).

Catalytic Site Atlas (CSA)

CSA, the Catalytic Site Atlas[46] provides catalytic residue annotation for enzymes in the Protein Data Bank (http://*www.ebi.ac.uk/thornton srv/databases/ CSA*). The database consists of two types of annotated sites: (i) an original hand annotated set containing information extracted from the primary literature, using defined criteria to assign catalytic residues, and (ii) an additional homologous set, containing annotations inferred by PSI BLAST and sequence alignment to one of the original set. The CSA can be queried via Swiss Prot identifier and EC number, as well as by PDB code. The CSA is updated on a monthly basis to include homologous sites found in new PDBs.

Molecular functions - Enzymatic catalysis

KEGG ENZYME Database

KEGG ENZYME[22, 47] contains the information about Enzyme Nomenclature obtained from the ExplorEnz database (*www.genome.jp/kegg/*). Additional information is included both computationally and manually. Manually added information includes the KEGG reaction data with parent-child relationship and the source organism and protein sequence information in each reference. Links to various data in KEGG and other databases are computationally generated. The accession number of an ENZYME entry is the EC (Enzyme Commission) number given by the Nomenclature Committee of the IUBMB and the IUPAC. The role of this database within KEGG has diminished, but the EC number is still the simplest way to link to KEGG from outside resources. KEGG ENZYME has 5,296 entries until 2010.

BRENDA

The BRENDA (BRaunschweig ENzyme Database) enzyme information system [48] is a manually annotated repository for enzyme data (*www.brenda-enzymes.org*). Its contents are not restricted to specific groups of enzymes, but include information on all enzymes that have been classified in the EC scheme of the IUBMB (International Union of Biochemistry and Molecular Biology) irrespective of the enzyme's source. The range of data includes the catalyzed reaction, detailed description of the substrate, cofactor and inhibitor specificity, kinetic data, structure properties, information on purification and crystallization, properties of mutant enzymes, participation in diseases and amino acid sequences. Each single entry is linked to the enzyme source and to the literature reference. Data queries can be performed by a number of different ways, including an EC-tree browser, a taxonomy-tree browser, an ontology browser and a combination query of up to 20 parameters. The newly implemented web-service provides instant access to the data for programmers via a SOAP interface (Simple Object Access Protocol).

The BRENDA Genome Explorer is an enzyme-centered genome visualization tool for browsing and comparing enzyme annotations in full genomes. It closes the gap between genomic and enzymatic data and allows the alignment of genomes at a given enzyme-coding gene and its orthologs, thus allowing to visually compare the genomic environment of the gene in different organisms. It maintains a record of 5117 different enzymes as of release 2010.2

Bioinformatics Database Search Engines

Text-based search engines

Entrez

Entrez[1] is an integrated database retrieval system of NCBI and allows the user to search and browse all the NCBI databases through a single gateway (http://*www.ncbi.nlm.nih.gov/Entrez/*). Entrez provides access to DNA and protein sequences derived from many sources, including genome maps, population sets and biomedical literatures via PubMed. Entrez grows with addition of new search features and the most recently added search engines are: (i) ProbeSet that allows searches for DNA data from gene-expression experiments and (ii) the Molecular Modelling Database of 3D structures (MMDB) for proteins by molecular weight range, by protein domain or by structure.

Sequence retrieval server (SRS)

The sequence retrieval server (SRS) serves a similar role to Entrez, for the major European sequence databases [49] (http://srs.ebi.ac.uk/). SRS is a flexible sequence query tool which allows the user to search a defined set of sequence databases and knowledge-bases by accession number, keyword or sequence similarity. SRS encompasses a very wide range of data, including all the major EMBL sequence divisions. SRS goes one step ahead than Entrez by enabling the user to create analysis pipelines by selecting retrieved data for processing by a range of analysis tools, including ClustalW, BLAST and InterProScan. In addition to the number and variety of databases to which it offers access, SRS offers tight links among the databases, and fluency in launching applications. A search in a single database component can be extended to a search in the complete network. Similarity searches and alignments can be launched directly, without saving the responses in an intermediate file. In SRS, the databases are grouped by categories such as nucleotide sequence-related, protein-related, etc. Similar to Entrez, we can search one or more databases either in all fields, or assign terms to categories. The search results page would allow creating and downloading reports on the selected matches.

DBGET / LinkDB

The integrated database retrieval system DBGET/LinkDB[50] (http://www.genome.jp/dbget/) is the backbone of the Japanese GenomeNet service. DBGET is used to search and extract entries from a wide range of molecular biology databases, while LinkDB is used to search and compute links between entries in different databases. DBGET/LinkDB is designed to be a network distributed database system with an open architecture, which is suitable for incorporating local databases or establishing a specialized server environment. It also has an advantage of simple architecture allowing rapid daily updates of all the major databases. The WWW version of DBGET/LinkDB at GenomeNet is integrated with other search tools, such as BLAST, FASTA and MOTIF, and with local helper applications, such as RasMol.

Sequence-based (sequence similarity) search engines

Sequence similarity searches use alignments to determine a match. Alignment of two sequences is matching of the two sequences, except that they allow the most common mutations: insertions, deletions and single-character substitutions. The basic considerations in using a sequence similarity search are global sequence alignment and local sequence alignment. Global alignment forces complete alignment of the input sequences, whereas local alignment will align the most similar segments. The choice of global Vs local depends on the assumptions made by the user as to whether the sequences are related over the entire length or presumed to share only isolated regions of homology. As similarity will span segments rather than entire sequences, local alignment is the most popular database similarity search.

Needleman-Wunsch (1970) proposed an algorithm for optimal global alignment and in 1981, the Smith-Waterman algorithm for optimal local alignment of two sequences was introduced. These two methods were developed prior to whole genome sequencing. Today, the special purpose parallel machines and the massive computational time required by these algorithms have rendered them almost obsolete. Programs like BLAST[51] and FASTA[52] make use of the heuristic methods to align sequences locally. Most users prefer BLAST to speed up alignment searches.

BLAST and FASTA

The BLAST programs (Basic Local Alignment Search Tools) are a set of sequence comparison algorithms introduced in 1990 that are used to search sequence databases for optimal local alignments to a query. The BLAST analysis tools are available in the public domain via the BLAST web pages (*www.ncbi.nlm.nih.gov/BLAST/*).

FASTA is a DNA and protein sequence alignment software package first described (as FASTP) by David J. Lipman and William R. Pearson in 1985. The FASTA package contains various programs for analyzing protein-protein, DNA-DNA, protein-translated DNA (with frameshifts) and ordered or unordered peptide searches and is available online at (*www.ebi.ac.uk/Tools/sss/*).

The given query sequence is filtered to remove low-complexity regions. A list of words of certain length is made which are evaluated for matches with any database sequence using substitution scoring matrices (PAM and BLOSUM) and these scores are added along with the penalties for mismatches and for including gaps in the alignment (to obtain optimal alignment). To obtain only the most significant matches, a cutoff score is selected. This procedure is repeated for each word in the query sequence. If a good match is found then an alignment is extended from the match area in both directions as far as the score continue to grow. The significance of an alignment is evaluated statistically with the help of values like E-value, P-value, and Z-score. This approach permits FASTA or BLAST to run 100 times faster than conventional Smith-waterman, at the cost of missing few alignments. Some of the adjustable parameters provide the user the flexibility to trade off between speed and accuracy. BLAST, in general, tends to be faster whereas FASTA returns fewer false hits.

Motif-based search engines

ScanProsite

ScanProsite[53] is a new and improved version of the web-based tool for detecting PROSITE signature matches in protein sequences (*www.expasy.org/tools/scanprosite/*). For a number of PROSITE profiles, the tool now makes use of ProRules (context-dependent annotation templates) to detect functional and structural intra-domain residues. The detection of those features enhances the power of function prediction based on profiles. Both user-defined sequences and sequences from the UniProt Knowledgebase can be matched against custom patterns, or against PROSITE signatures. Several output modes are available including simple text views and a rich mode providing an interactive match and feature viewer with a graphical representation of results.

eMOTIF

The eMOTIF[54] database is a collection of highly specific and sensitive protein sequence motifs representing conserved biochemical properties and biological functions (*http://motif.stanford.edu/*). As the groups of amino acids in motifs represent critical positions conserved in evolution, search algorithms employing the eMOTIF patterns can identify and classify more widely divergent sequences than the methods based on global sequence similarity. eMOTIFS are able to classify novel proteins into protein subfamilies. In addition eMOTIFS focus on just those highly conserved residues which improve the signal-to-noise ratios of the eMOTIFS and enhance the specificity of the automated sequence classification.

Structure-based (structure similarity) search engines

VAST

VAST[55] Search is NCBI's structure-structure similarity search service, which compares 3D coordinates of a newly determined protein structure to those, exists in the MMDB/PDB database (*www.ncbi.nlm.nih.gov/Structure/VAST/*). Using VAST algorithm, it computes a list of structure neighbors, viewing superposition and alignments by molecular graphics. From the MMDB Structure summary pages, which are retrieved via Entrez, structure neighbors are available for protein chains and individual structural domains. This service is meant to be used with newly determined protein structures that are not yet part of MMDB.

Dali

The Dali server[56] (*http://ekhidna.biocenter.helsinki.fi/dali_server*) is a computational service for structure comparison using the same machinery used in deriving the FSSP database. The web site consists of: (i) a server that compares newly solved structures against structures in the Protein Data Bank (PDB), (ii) a database that allows browsing precomputed structural neighbourhoods and (iii) a pairwise comparison engine that generates suboptimal alignments for a pair of structures. Each part has its own query form and results page. The key purpose of such interactive analysis is to check whether conserved residues line up in multiple structural alignments and how conserved residues and ligands cluster together in multiple structure superimpositions and therefore leads to evolutionary discoveries, which is not detected by sequence analysis.

References

1. Schuler, G.D. *et al.* (1996) Entrez: molecular biology database and retrieval system. Methods Enzymol., 266:141-62.
2. Bloom, T., *et al.* (2008) PLoS Biology at 5: the future is open access. PLoS Biol., 6(10):e267.
3. Campbell, A.M. (2004) Open access: a PLoS for education. PLoS Biol., 2(5): p. E145.
4. Butler, D. (2000). BioMed Central boosted by editorial board. Nature, 405(6785): 384.
5. Brooksbank, C., Cameron, G. and Thornton, J. (2010) The European Bioinformatics Institute's data resources. Nucleic Acids Res., 38(Database issue): D17-25.
6. Sayers, E.W., *et al.* (2010) Database resources of the National Center for Biotechnology Information. Nucleic Acids Res., 38(Database issue): D5-16.
7. Hubbard, T., *et al.* (2002). The ensembl genome database project. Nucleic Acids Res., 30(1): 38-41.
8. Buechen-Osmond, C. and Dallwitz, M. (1996) Towards a universal virus database - progress in the ICTVdB. Arch. Virol., 1141(2): 392-9.
9. Kulkarni-Kale, U., *et al.* (2004) VirGen: a comprehensive viral genome resource. Nucleic Acids Res., 2004. 32 (Database issue): D289-92.
10. Liolios, K., *et al.* (2006) The genomes on line database (GOLD) v.2: a monitor of genome projects worldwide. Nucleic Acids Res., 34 (Database issue): D332-4.
11. Uchiyama, I. (2003) MBGD: microbial genome database for comparative analysis. Nucleic Acids Res., 31(1): 58-62.
12. Hamosh, A., *et al.* (2000) Online mendelian inheritance in man (OMIM). Hum Mutat., 15(1):57-61.

Bioinformatics Databases and Internet Resources   45

13. Hamosh, A., *et al.* (2005) Online mendelian inheritance in man (OMIM), a knowledgebase of human genes and genetic disorders. Nucleic Acids Res., 33(Database issue): D514-7.

14. Nicholas, F.W. (2003) Online mendelian inheritance in animals (OMIA): a comparative knowledgebase of genetic disorders and other familial traits in non-laboratory animals. Nucleic Acids Res., 31(1):275-7.

15. Cherry, J.M., *et al.* (1998) SGD: Saccharomyces genome database. Nucleic Acids Res., 26(1): 73-9.

16. Harris, T.W., *et al.*, (2010) WormBase: a comprehensive resource for nematode research. Nucleic Acids Res., 38(Database issue): D463-7.

17. Plasmo, D.B. (2001) An integrative database of the plasmodium falciparum genome. Tools for accessing and analyzing finished and unfinished sequence data. The Plasmodium Genome Database Collaborative. Nucleic Acids Res., 29(1):66-9.

18. The FlyBase database of the drosophila genome projects and community literature. (1999) The FlyBase Consortium. Nucleic Acids Res., 27(1): 85-8.

19. Swarbreck, D., *et al.* (2008) The arabidopsis information resource (TAIR): gene structure and function annotation. Nucleic Acids Res., 36(Database issue): D1009-14.

20. Loveland, J. (2005) VEGA, the genome browser with a difference. Brief Bioinform, 6(2): 189-93.

21. Wolfsberg, T.G. Using the NCBI map viewer to browse genomic sequence data. Curr. Protoc Bioinformatics, Chapter 1: p. Unit 15: 1-25.

22. Kanehisa, M. and Goto, S. (2000) KEGG: Kyoto encyclopedia of genes and genomes. Nucleic Acids Res., 28(1): 27-30.

23. Mewes, H.W., *et al.* (2004) MIPS: analysis and annotation of proteins from whole genomes. Nucleic Acids Res., 32(Database issue): D41-4.

24. Kuhn, R.M., *et al.* (2009) The UCSC genome browser database: update 2009. Nucleic Acids Res., 37(Database issue): D755-61.

25. Benson, D.A., *et al.* (2009) GenBank. Nucleic Acids Res., 2009. 37 (Database issue): D26-31.

26. Okubo, K., *et al.* (2006) DDBJ in preparation for overview of research activities behind data submissions. Nucleic Acids Res., 34(Database issue): D6-9.

27. Bairoch, A. and Apweiler, R. (1997) The SWISS-PROT protein sequence data bank and its supplement TrEMBL. Nucleic Acids Res., 25(1): 31-6.

28. Wu, C.H., *et al.* (2006) The Universal Protein Resource (UniProt): an expanding universe of protein information. Nucleic Acids Res., 34(Database issue): D187-91.

29. Apweiler, R., *et al.* (2004) UniProt: the universal protein knowledgebase. Nucleic Acids Res., 32(Database issue): D115-9.

30. Leinonen, R., *et al.* (2004) UniProt archive. Bioinformatics, 20(17): 3236-7.

31. Suzek, B.E., *et al.* (2007) UniRef: comprehensive and non-redundant UniProt reference clusters. Bioinformatics, 23(10): 1282-8.

32. Sigrist, C.J., *et al.* (2002) PROSITE: a documented database using patterns and profiles as motif descriptors. Brief Bioinform., 3(3): 265-74.

33. Gribskov, M., McLachlan, A.D. and Eisenberg, D. (1987) Profile analysis: detection of distantly related proteins. Proc. Natl. Acad. Sci., USA, 84(13): 4355-8.

34. Bru, C., *et al.* (2005) The ProDom database of protein domain families: more emphasis on 3D. Nucleic Acids Res., 33(Database issue): D212-5.

35. Altschul, S.F., *et al.* (1997) Gapped BLAST and PSI-BLAST: a new generation of protein database search programs. Nucleic Acids Res., 25(17): 3389-402.

36. Sonnhammer, E.L. and Kahn, D. (1994) Modular arrangement of proteins as inferred from analysis of homology. Protein Sci., 3(3):482-92.

37. Sonnhammer, E.L., Eddy, S.R. and Durbin, R. (1997) Pfam: a comprehensive database of protein domain families based on seed alignments. Proteins, 28(3): 405-420.

38. Finn, R.D., *et al.* (2008) The Pfam protein families database. Nucleic Acids Res., 36(Database issue): D281-8.
39. Mulder, N.J., *et al.* (2005) InterPro, progress and status in 2005. Nucleic Acids Res., 33(Database issue): D201-5.
40. Berman, H.M., *et al.* (2000) The protein data bank. Nucleic Acids Res., 28(1): 235-42.
41. Allen, F.H. and Taylor, R. (2004) Research applications of the Cambridge Structural Database (CSD). Chem. Soc. Rev., 33(8): 463-75.
42. Wang, Y., *et al.* (2007) MMDB: annotating protein sequences with Entrez's 3D-structure database. Nucleic Acids Res., 35(Database issue): D298-300.
43. Murzin, A.G., *et al.* (1995) SCOP: a structural classification of proteins database for the investigation of sequences and structures. J. Mol. Biol., 247(4): 536-40.
44. Orengo, C.A., *et al.* (1997) CATH—a hierarchic classification of protein domain structures. Structure, 5(8): 1093-108.
45. Holm, L., *et al.* (1992) A database of protein structure families with common folding motifs. Protein Sci., 1(12): 1691-8.
46. Porter, C.T., Bartlett, G.J. and Thornton, J.M. (2004) The Catalytic Site Atlas: a resource of catalytic sites and residues identified in enzymes using structural data. Nucleic Acids Res., 32 (Database issue): D129-33.
47. Kanehisa, M., *et al.* (2004) The KEGG resource for deciphering the genome. Nucleic Acids Res., 32(Database issue): D277-80.
48. Barthelmes, J., *et al.* (2007) BRENDA, AMENDA and FRENDA: the enzyme information system in 2007. Nucleic Acids Res., 35(Database issue): D511-4.
49. Etzold, T. and Argos, P. (1993) SRS—an indexing and retrieval tool for flat file data libraries. Comput. Appl. Biosci., 9(1): 49-57.
50. Fujibuchi, W., *et al.* (1998) DBGET/LinkDB: an integrated database retrieval system. Pac Symp Biocomput., 683-94.
51. Altschul, S.F., *et al.* (1990) Basic local alignment search tool. J. Mol. Biol., 215(3): 403-10.
52. Pearson, W.R. and Lipman, D.J. (1988) Improved tools for biological sequence comparison. Proc. Natl. Acad. Sci., USA, 85(8): 2444-8.
53. de Castro, E., *et al.* (2006) ScanProsite: detection of PROSITE signature matches and ProRule-associated functional and structural residues in proteins. Nucleic Acids Res., 34 (Web Server issue): W362-5.
54. Huang, J.Y. and Brutlag, D.L. (2001) The EMOTIF database. Nucleic Acids Res., 29(1): 202-4.
55. Gibrat, J.F., Madej, T. and Bryant, S.H. (1996) Surprising similarities in structure comparison. Curr. Opin. Struct. Biol., 6(3): 377-85.
56. Holm, L. and Rosenstrom P., (2010) Dali server: conservation mapping in 3D. Nucleic Acids Res., 38 Suppl: W545-9.

Bioinformatics Databases and Internet Resources

Annexure – I

Important Internet Resources of Bioinformatics

Literature Databases
PubMed — www.ncbi.nlm.nih.gov/pubmed/
PLoS — www.plos.org/
Biomed Central — www.biomedcentral.com/

Genome Databases
Genome Databases at NCBI — www.ncbi.nlm.nih.gov/Genomes/
Organelle Genome Sequences — www.ncbi.nlm.nih.gov/genomes/ GenomesHome.cgi?taxid=2759&hopt=html
Parasites Genome Database and Genome Research — www.ebi.ac.uk/parasites/parasite-genome.html

Viral Genome Database
Retroviral Genotyping and Analysis — www.ncbi.nlm.nih.gov/retroviruses/
ICTVdB — www.ictvdb.org/
VirGen — http://202.41.70.249/virgen/virgen.asp

Bacterial Genomes Database
Genomes OnLine Database (GOLD) — www.genomesonline.org/
TIGR:The Comprehensive Microbal Resource — www.jcvi.org/
Microbial Genome Database (MBGD) — http://mbgd.genome.ad.jp/

Organism Specific Genome Database
OMIM/OMIA — www.ncbi.nlm.nih.gov/omim/www. ncbi.nlm.nih.gov/omia/
SGD — http://genome-www.stanford.edu/ Saccharomyces/
WormBase — www.wormbase.org/
PlasmoDB — http://PlasmoDB.org/
FlyBase — http://flybase.org/
A. thaliana Information Resource TAIR — www.arabidopsis.org/
C. elagans Genome Project — www.sanger.ac.uk/Projects/C_elegans/
Dictyostelium discoideum Genome Information — www.biology.ucsd.edu/others/dsmith/ dictydb.html/
Drosophila melanogaster Berkeley Drosophila Genome Project — www.fruitfly.org/
E.coli Genome and Proteome Database (GenProtEC) — http://genprotec.mbl.edu/
E.coli Genome Project — www.genome.wisc.edu/
S. cerevisiae, YPD Yeast Proteome Database — www.biobase-international.com/index.php?id=ypd
Mouse *(Musmusculus)* genome informatics — www.informatics.jax.org/
Rice *(Oryza sativa)* Genome Project — http://rgp.dna.affrc.go.jp/
Human Genome Resources at NCBI — www.ncbi.nlm.nih.gov/genome/guide/ human/index.shtml

Genome Browsers

Ensembl — www.ensembl.org/
VEGA genome browser — http://vega.sanger.ac.uk/
NCBI-NCBI map viewer — *www.ncbi.nlm.nih.gov/projects/mapview/*
KEGG — www.genome.jp/kegg/
MIPS — http://mips.helmholtz-muenchen.de/proj/ppi/
UCSC Genome Browser — http://genome.ucsc.edu/

Gene and Genome Relationship and Proteome Analysis

COG:Cluster of Orthologous group — www.ncbi.nlm.nih.gov/COG/
DOGS:Databases of Genome sizes — www.cbs.dtu.dk/databases/DOGS/
GeneQuiz — http://swift.cmbi.kun.nl/swift/genequiz/
Gene and Disease — www.ncbi.nlm.nih.gov/disease/
SEQUEST — http://fields.scripps.edu/sequest/
Taxonomy Browser at NCBI — www.ncbi.nlm.nih.gov/Taxonomy/taxonomyhome.html
UniGene — www.ncbi.nlm.nih.gov/UniGene/

SEQUENCE DATABASES

Nucleotide Sequence Databases

GenBank — www.ncbi.nlm.nih.gov/genbank/
EMBL — www.ebi.ac.uk/embl/
DDBJ — *www.ddbj.nig.ac.jp/*

Protein Sequences Databases

Swiss-Prot — www.expasy.ch/sprot/
UniProt — www.uniprot.org/
UniProt Knowledgebase (UniProtKB) — www.uniprot.org/help/uniprotkb
UniProt Archive (UniParc) — www.ebi.ac.uk/uniparc/
UniProt Reference Clusters (UniRef) — www.ebi.ac.uk/uniref/

Sequence Motifs Databases

PROSITE — www.expasy.org/prosite/
ProDom — http://prodom.prabi.fr/
Pfam — www.sanger.ac.uk/Pfam/
InterPro — www.ebi.ac.uk/interpro/
PIR — http://pir.georgetown.edu/
PRINTS — www.bioinf.manchester.ac.uk/dbbrowser/PRINTS/index.php

The Primary Structure Databases

Protein Data Bank (PDB) — www.rcsb.org/pdb/
Cambridge Structural Database (CSD) — www.ccdc.cam.ac.uk/products/csd/
Molecular Modeling Database (MMDB) — www.ncbi.nlm.nih.gov/sites/entrez?db=structure

The Secondary Structure Databases

Structural Classification of Proteins (SCOP) — http://scop.berkeley.edu/
Class Architecture Topology Homology (CATH) — www.cathdb.info/
Families of Structurally Similar Proteins (FSSP) — http://swift.cmbi.kun.nl/swift/fssp/
Catalytic Site Atlas (CSA) — www.ebi.ac.uk/thornton-srv/databases/CSA/

Bioinformatics Databases and Internet Resources

Database of Patterns and Sequence of Protein Families

3D-PSSM	www.sbg.bio.ic.ac.uk/~3dpssm/
BLOCKS	http://blocks.fhcrc.org/
DIP:Database of Interacting Protein	http://dip.doe-mbi.ucla.edu/
HOMSTRAD	http://tardis.nibio.go.jp/homstrad/
HSSP	http://swift.embl-heidelberg.de/hssp/

Molecular Functions - Enzymatic Catalysis

KEGG ENZYME database	www.genome.jp/kegg/
BRENDA	www.brenda-enzymes.org/
Array Express Database	www.ebi.ac.uk/arrayexpress/
Ecocyc	http://ecocyc.pangeasystems.com/ecocyc/
SMART	http://smart.embl-heidelberg.de/
SWISS-2DPAGE	www.expasy.ch/ch2d/
Yeast Transcriptome	http://bioinfo.mbb.yale.edu/genome/

RNA Databases

RDP	http://rdp.cme.msu.edu/
RNA modification database	http://medlib.med.utah.edu/RNAmods/
5S Ribosomal RNA databank	http://rose.man.poznan.pl/5SData/
Comparative RNA	www.rna.icmb.utexas.edu/
Nucleic acid database and structure resource	http://ndbserver.rutgers.edu/
Ribosomal RNA mutation database	http://ribosome.fandm.edu/
RNA modification database	http://medlib.med.utah.edu/RNAmods/
RNA structure database	www.rnabase.org/
tmRNA website	www.indiana.edu/~tmrna/
Vienna RNA package	www.tbi.univie.ac.at/~ivo/RNA/

Immunogenic and Receptor Databases

IMGT: ImMunoGeneTics Database	www.ebi.ac.uk/imgt/hla/http://imgt.cines.fr/
Glucocorticoid receptor resource	http://nrr.georgetown.edu/GRR/GRR.html/
Thyroid hormone receptor resources	http://nrr.georgetown.edu/NRR/TRR/trrfront.html

Text-based Search Engines

Entrez	www.ncbi.nlm.nih.gov/Entrez/
SRS	http://srs.ebi.ac.uk/
DBGET / LinkDB	www.genome.jp/dbget/

Sequence-based (Sequence Similarity) Search Engines

BLAST and FASTA	www.ncbi.nlm.nih.gov/BLAST/www.ebi.ac.uk/Tools/sss/

Motif-based search engines

ScanProsite	www.expasy.org/tools/scanprosite/
eMOTIF	http://motif.stanford.edu/

Structure-based (Structure Similarity) Search Engines

VAST	www.ncbi.nlm.nih.gov/Structure/VAST/
DALI	http://ekhidna.biocenter.helsinki.fi/dali_server

Chapter – 3

Analysis of Genetic Diversity in Crop Plants Using Molecular Marker Data

M.K. Rajesh and S. Jayasekhar

Abstract

Plant genetic resources constitute the chief component of agro-biodiversity and comprise of land races, modern cultivars and obsolete varieties, breeding lines, genetic stocks and wild species. They provide the basic materials to the plant breeders to utilize genetic variability for the development of high yielding cultivars with a broad genetic base. The utilization of these genetic resources, however, depends upon their efficient and adequate characterization and evaluation, which in turn entails efficient characterization standards and appropriate strategies.

Introduction

Analysis of trait data generated from characterization and evaluation of the genetic resources is used to understand and use diversity. Currently, a large number of distance measures are available for analyzing similarity/dissimilarity among accessions based on different traits representing different types of variables. The selection of the most appropriate distance measure for each trait is the prerequisite for diversity analysis. One of the approaches is to form clusters where accessions between clusters would be more diverse than the accessions within a cluster. The clustering algorithms require a distance/

similarity matrix between the accessions which can be calculated depending upon the nature or type of traits such as morphological and agronomic traits and/or molecular markers.

The availability of cost-efficient large scale genotyping techniques has greatly facilitated the assessment of genetic diversity within populations. Various computational tools have also been developed concurrently to analyze the genetic data derived from the genotyping experiments. In this review, the basics of population genetics, important parameters in genetic diversity analysis and the most widely used computer programs in population genetic studies have been described.

Basics of population genetics

Variation in alleles allows organisms to adapt to ever-changing environments. Alleles are different forms of the same gene that are expressed as different phenotypes. All of the alleles shared by all of the individuals in a population make up the population's gene pool. In diploid organisms, every gene is represented by two alleles, one inherited from each parent. The pair of alleles may differ from one another, in which case it is said that the individual is "heterozygous" for that gene. If the two alleles are identical, it is said that the individual is "homozygous" for that gene.

Population genetics is the study of allele frequency distribution and change under the influence of the four main evolutionary processes: natural selection, genetic drift, mutation and gene flow. It also takes into account the factors of population subdivision and population structure and attempts to explain such phenomena as adaptation and speciation.

Based on mendelian genetics, it is possible to predict the probability of the appearance of a particular allele in an offspring when the alleles of each parent are known. Similar predictions can be made about the frequencies of alleles in the next generation of an entire population. By comparing the predicted or "expected" frequencies with the actual or "observed" frequencies in a real population, one can infer a number of possible external factors that may be influencing the genetic structure of the population (such as inbreeding or selection).

A population is defined as a group of interbreeding individuals that exist together at the same time. A population may either be considered as a single unit or it can be subdivided into smaller units. Subdivisions of a population may be the result of ecological factors or behavioural factors. If a population is subdivided, the genetic links among its parts may differ, depending on the real degree of gene flow taking place.

A population is considered structured if:

(i) Genetic drift is occurring in some of its subpopulations,

(ii) Migration does not happen uniformly throughout the population, or

(iii) Mating is not random throughout the population.

A population's structure affects the extent of genetic variation and its patterns of distribution.

Genetic drift

'Genetic drift' refers to fluctuations in allele frequencies that occur by chance (particularly in small populations) as a result of random sampling among gametes, i.e. random changes in gene frequency which are not due to selection, gene mutation or migration. Genetic drift decreases diversity within a population because it tends to cause the loss of rare alleles, reducing the overall number of alleles. Because of genetic drift, small isolated populations often have unusual frequencies of a few alleles.

Gene flow

'Gene flow' is the passage and establishment of genes typical of one population in the gene pool to another by natural or artificial hybridization and backcrossing. 'Non-random mating' occurs when individuals those are more closely (inbreeding) or less closely related mate more often than would be expected by chance for the population. Self-pollination or inbreeding is similar to mating between relatives. It increases the homozygosity of a population and its effect is generalized for all alleles. Inbreeding *per se* does not change the allelic frequencies but, over time, it leads to homozygosity by slowly increasing the two homozygous classes.

Mutations could lead to occurrence of new alleles, which may be favourable or deleterious to the individual's ability to survive. If changes are advantageous, then the new alleles will tend to prevail by being selected in the population. The effect of selection on diversity may be:

- 'Directional', where it decreases diversity;

- 'Balancing', where it increases diversity. Heterozygotes have the highest fitness, so selection favours the maintenance of multiple alleles; and

- 'Frequency dependent', where it increases diversity. Fitness is a function of allele or genotype frequency and changes over time.

Migration

'Migration' implies not only the movement of individuals into new populations but that this movement introduces new alleles into the population (gene flow). Changes in gene frequencies will occur through migration either because more copies of an allele already present will be brought in or because a new allele arrives. Various factors which affect migration in crop species include breeding system, sympatry with wild and/or weedy relatives, pollinators, and seed dispersal. The immediate effect of migration is to increase a population's genetic variability and, as such, helps increase the possibilities of that population to withstand environmental changes. Migration also helps blend populations and prevent their divergence.

Hardy-Weinberg Principle / Law

The foundation for population genetics was laid in 1908, when Godfrey Hardy and Wilhelm Weinberg independently published which is known as the 'Hardy-Weinberg Equilibrium' or 'Hardy-Weinberg Principle', which states: "In a large, randomly breeding (diploid) population, allelic frequencies will remain the same from generation to generation; assuming no unbalanced mutation, gene migration, selection or genetic drift." When a population meets all of the Hardy-Weinberg conditions, it is said to be in Hardy-Weinberg equilibrium. The "equilibrium" is a simple prediction of genotype frequencies in any given generation, and the observation that the genotype frequencies are expected to remain constant from generation to generation as long as several simple assumptions are met. This description oprovides a counterpoint to studies on how populations change over time.

Testing for Hardy-Weinberg Equilibrium

The deviation of a population from Hardy-Weinberg equilibrium is an indication of the intensity of external factors and can be determined by a statistical formula called a chi-square, which is used to compare observed versus expected outcomes. The statistical test follows this formula:

$$HWT = \Sigma (O_i - E_i)^2/E_i$$

Where HWT = Statistical test for Hardy-Weinberg Equilibrium; O_i = Observed frequencies and Ei = Expected frequencies

If $X^2_{cal} \, d" \, X^2_{tab}$, then H_0 hypothesis is accepted and it follows that allele frequencies for loci in a given population are HWT equilibrium. If $X^2_{cal} \geq X^2_{cal}$, then H_0 hypothesis is rejected.

Important Parameters in Genetic Diversity Analysis

(a) Polymorphism or rate of polymorphism: A polymorphic gene is usually defined as one for which the most common alleles has a frequency of less than 0.95.

$$Pj = q \text{ d } 0.95$$

Where, Pj = rate of polymorphism and q = allele frequency

For a correct estimation of genetic distance, the genetic loci used in genetic distance analysis should be informative, *i.e.*, they should display sufficient polymorphism. The limit of allele frequency, which is set at 0.95, is arbitrary, its objective being to help identify those genes in which allelic variation is common. Rare alleles are defined as those with frequencies of less than 0.005.

This index is the best applied with codominant markers. It can also be used with dominant markers too, but restrictively, as the estimate based on dominant markers would be biased below the real number.

(b) Average number of alleles per locus: It is the sum of all the detected alleles in all loci, divided by the total number of loci. This parameter, which provides complementary information to that polymorphism, is given by:

$$N = (1/k) \sum_{i=1}^{k} n_i$$

Where: k = Number of loci and n_i = Number of alleles detected by locus

This parameter is the best applied in the case of codominant markers as dominant markers do not permit the detection of all alleles.

(c) Effective number of alleles: This measure, which explains about the number of alleles that would be expected in a locus in each population, is given by:

$$A_e = 1/(1 - h) = 1/\Sigma p_i^2$$

Where, pi = frequency of the ith allele in a locus and $h = 1 - \Sigma p_i^2$ = heterozygosity in a locus.

It ranges from 0 to 1. It can be calculated for both dominant and co-dominant markers.

By taking allele frequencies into account, this descriptor of allelic richness is less sensitive to rare alleles. This parameter plays a fundamental role in verification of sampling strategies. However, its calculation is affected by the sample size.

(d) Observed heterozygosity: A population's heterozygosity is measured by first determining the proportion of genes that are heterozygous and the number of individuals that are heterozygous for each particular gene. For a single gene locus with two alleles, the Observed Heterozygosity (H_o) is calculated as follows:

$$H_o = \frac{\text{Number of heterozygotes at a locus}}{\text{Total number of individuals surveyed}}$$

Derivations of the above formula are used to calculate the H_o when there are more than two alleles for a particular locus, which is particularly common when microsatellite or simple sequence repeat (SSR) markers are applied for analysis of populations.

(e) Expected heterozygosity: The Expected Heterozygosity (H_e) is defined as the estimated fraction of all individuals that would be heterozygous for any randomly chosen locus. It is the probability that, at a single locus, any two alleles, chosen at random from the population, are different to each other. For a locus j with I alleles, It is calculated as:

$$h_j = 1 - \Sigma p_i^2$$

Where, h_j = heterozygosity per locus and p = allele frequencies

H_e differs from the H_o because it is a prediction based on the known allele frequency from a sample of individuals. Deviation of the observed from the expected can be used as an indicator of important population dynamics.

(f) Effective population Size: One of the many variables of population dynamics that can influence the rate and size of fluctuation in allele frequencies is population size. Genetic drift, the random increase or decrease of an allele's frequency, affects small populations more severely than large ones, since alleles are drawn from a smaller parental gene pool. The rate of change in allele frequencies in a population is determined by the population's effective population size. The effective population size is the number of individuals that evenly contribute to the gene pool.

The actual number of individuals in a population is rarely the effective population size. This is because some individuals reproduce at a higher rate than others (have a higher fitness), the distribution of males and females may result in some individuals being unable to secure a mate, or inbreeding reduces the unique contribution of an individual. The effective population size is a theoretical measure that compares a population's genetic behavior to the behavior of an "ideal" population. As the effective population size becomes smaller, the chance that allele frequencies will shift due to chance (drift) alone becomes greater.

(g) Shannon index: Shannon's Information Index is a measure of gene diversity. It is based on information theory and is a measure of the average degree of "uncertainty" in predicting to what species an individual chosen at random from a collection of S species and N individuals will belong. This average uncertainty increases as the number of species increases and as the distribution of individuals among the species becomes even. The proportion of species i relative to the total number of species (p_i) is calculated, and then multiplied by the natural logarithm of this proportion ($\ln p_i$) in order to obtain the Shannon's Index (H').

$$H' = -\sum_{i=1}^{S} (p_i \ln p_i)$$

It can be shown that for any given number of species, there is a maximum possible H', $H_{max} = \ln S$ which occurs when all species are present in equal numbers. When Shannon index is near 1, it can be concluded that the population is highly heterozygous.

(h) Inbreeding and relatedness: Small effective population size can result in a high occurrence of inbreeding, or mating between close relatives. One of the effects of inbreeding is a decrease in the heterozygosity (increase in homozygosity) of the population as a whole, which means a decrease in the number of heterozygous genes in the individuals. This effect places individuals and the population at a greater risk from homozygous recessive diseases that result from inheriting a copy of the same recessive allele from both parents. The impact of accumulating deleterious homozygous traits is called 'inbreeding depression' - the loss in population vigor due to loss in genetic variability.

A set of parameters called F-statistics was developed[15]. The inbreeding coefficient (F_{IS}) defined as the probability that two homologous (same) alleles present in the same individual are identical by descent. F_{IS} is calculated by comparing the expected heterozygosity (H_e) with observed heterozygosity (H_o), and ranges from -1 (no inbreeding) to +1 (complete identity). If the values for both observed and expected heterozygosity are the same, F_{IS} will be zero. A positive value indicates that there is an increased number of homozygotes, and population may be inbred - the larger the number, the greater the extent of inbreeding. A negative value indicates that there are more heterozygous individuals than would be expected; this might happen for the first a few generations after two previously isolated populations become one.

The relationships among the F statistics can be deduced through the following:

$$(1 - F_{IT}) = (1 - F_{IS})(1 - F_{ST})$$

$$F_{IT} = 1 - (H_I/H_T)$$

$$F_{IS} = 1 - (HI/HS)$$

$$F_{ST} = 1 - (H_S/H_T)$$

Where, H_T = total gene diversity or expected heterozygosity in the total population as estimated from the pooled allele frequencies, HI = intrapopulation gene diversity or average observed heterozygosity in a group of populations, and H_S = average expected heterozygosity estimated from each subpopulation.

These statistical indices measure:

F_{IS} = the deficiency or excess of average heterozygotes in each population

F_{ST} = the degree of gene differentiation among populations in terms of allele frequencies

F_{IT} = the deficiency or excess of average heterozygotes in a group of populations

The chi-square test can be used to statistically analyze whether the difference between the observed and expected is not likely due to chance. If there is a significant increase in the expected number of heterozygotes, inbreeding can be ruled out as a possible population dynamic that is influencing the genotype frequencies.

Corrections for Sampling Error

There are two sources of allele frequency difference among subpopulations in a sample:

(i) Real differences in the allele frequencies among our sampled subpopulations
(ii) Differences which arise because allele frequencies in our samples differ from those in the subpopulations from which they were taken.

The G_{ST} approach was described to account for the sampling error. G_{ST} is an interpopulation differentiation measure when multiple loci are used for analysis. It measures the proportion of gene diversity that is measured among populations, when a large number of loci are sampled.

$$G_{ST} = D_{ST} / H_{T,}$$

where, D_{ST} = interpopulation diversity,

H_T = total diversity ($H_S + D_{ST}$),

Hs = intrapopulation genic diversity, and

$D_{ST} = H_T - H_S$.

Because of the complexity of its components, calculation of G_{ST} requires specialized computer software. It can be used with codominant markers and restrictedly with dominant markers, since it is a measure of heterozygosity. Another statistic, q, which incorporates an important source of sampling error ignored by G_{ST} was described [14].

Measurement of Genetic Distance

Various genetic distance measures have been proposed for analysis of molecular marker data, depending on whether the markers are dominant or co-dominant. For dominate markers, the total number of bands is conventionally set as the number of analyzed loci. For co-dominant markers, genetic similarity between two individuals number of alleles per locus determined for total collection, is in general higher than two, Opposite to the 1- and 0- allele for dominant markers. Generally, genetic distance in codominant markers are based on allele frequencies.

If we assume that a = 3, b = 1, c = 3 and d = 2 then:

(i)   Dice and Nei and Li: a/[a + (b + c)/2] =0.6

(ii)   Jaccard: a/(a + b + c) = 0.49

(iii)   Sokal and Sneath: a/[a +2(b + c)] = 0.273

(iv)   Roger and Tanimoto : (a + d)/[a + d + 2(b + c)]= 0.385

The Jaccard coefficient only counts bands present for either individual and treats double absences as missing data. If false-positive or false negative data occur, the index estimate tends to be biased. It can be applied with co-dominant marker data. Nei and Li coefficient counts the percentage of shard bands among two individuals and gives more weight to those bands they are present in both. It considers that absence has less biological significance, and so this coefficient has complete meaning in terms of DNA similarity. It can be applied with codominant marker data (RFLP, SSR).

Multivariate Analysis

One of the main concerns of plant breeders is to quantify the degree of dissimilarity in genetic resources, since knowledge concerning genetic distances is necessary for optimum organization of gene banks and for identifying parental combinations that produce progenies with maximum genetic variability, thereby increasing the chances of obtaining superior individuals [5]. Use of multivariate statistical algorithms is considered an important strategy to quantify genetic similarity. Multivariate analysis is based on the statistical principle of multivariate statistics, which involves observation and analysis of more than one statistical variable at a time. Multivariate techniques permit standardization of multiple types of information of a set of characteristics. The most widely used algorithms are principal component and canonical variable analysis, as well as clustering methods.

The principle of clustering methods is to join genotypes into groups, so that there is uniformity within and heterogeneity among groups. These methods depend on previous estimates of dissimilarity measures derived from discrete and continuous (or categorical) variables. These categorical variables can be defined as binary, nominal or ordinal. Among grouping methods, hierarchical clustering has been used most frequently, particularly the Single Linkage (SL) and Unweighted Pair Group Method using Arithmetic averages (UPGMA) methods. The reliability of clustering methods depends on the magnitude of the cophenetic correlation, which is the association between the genetic distance matrix and the matrix based on genotype grouping. SL considers absence which corresponds to homozygous loci and it can be used with dominate marker (RAPD, AFLP) because absence could correspond to homozygous recessives. UPGMA is the most common method for cluster analysis UPGMA can only be used when the evolutionary rate is nearly same for all groups included in the study, when studying the genetic diversity of germplasm collection, SL method should be preferred above the UPGMA clustering method, because genetic differences among accessions in germplasm are dominantly determined by selection and breeding rather than by evolutionary forces.

Resampling is a term used in statistics for bootstrapping and permutation. These procedures can be used in genetic diversity studies to assign confidence to the presence of clusters in a dendrogram. Bootstrapping is a statistical method for estimating the sampling distribution of an estimator by sampling with replacement from the original sample, major purpose of bootstrapping is deriving robust estimates of standard errors and confidence intervals of population parameters. A permutation test is a type of statistical significant test in which a reference distribution is obtained by calculating all possible values of the test statistics by rearranging the tables of the observed data.

Steps Involved in Analysis of Molecular Marker Data

Three main steps are involved in the statistical analysis of molecular data in diversity studies:

a) **Data collection:** The data on molecular markers are recorded in the following two forms:

i) Binary data: presence or absence of molecular marker bands

ii) Allelic data (based on allele size)

b) **Data analysis** using univariate and multivariate statistical approaches

c) **Interpretation of the data:**

Each step in the process should follow a standardized format if the output of one diversity study is to be compared to other studies and inferences drawn in this manner.

Software programs for analyzing genetic diversity

Many software programs for molecular population genetics studies have been developed; the important ones are given below:

i) CONVERT (http://www.agriculture.purdue.edu/fnr/html/faculty/rhodes/students%20and%20staff/glaubitz/software.htm)

CONVERT is a user-friendly, 32-bit Windows program that aids conversion of diploid genotypic data files into formats that can be directly read by a number of commonly used population genetics computer programs: GDA, GENEPOP, ARLEQUIN, POPGENE, MICROSAT, PHYLIP and STRUCTURE [1]. In addition, CONVERT can be used to produce a table of allele frequencies in a convenient format, allowing the visual comparison of allele frequencies across populations. The input file for CONVERT follows a 'standard' format that can be easily obtained via an EXCEL file containing the genotypic data. CONVERT can also read in input data files in GENEPOP format. CONVERT works on Windows 95/98/NT/2000/XP platforms.

ii) ARLEQUIN (http://cmpg.unibe.ch/software/arlequin/)

Released first in 1997, Arlequin is a freely available integrated population genetics software environment [12]. It is able to handle both large samples of molecular data (RFLPs, DNA sequences, microsatellites) and also conventional genetic data (standard multi-locus data or allele frequency data). Molecular data can be entered as DNA sequences, RFLP haplotypes, microsatellite profiles, or multilocus haplotypes. The graphical interface is designed to allow users to rapidly select the different analyses they want to perform on their data.

The data format is specified in an input file. The user can create a data file from scratch, using a text editor and appropriate keywords, or use the 'Project Outline Wizard'. Data can be imported from files created for other programs, including MEGA, BIOSYS, GENEPOP, and PHYLIP. Missing or ambiguous data can be included. A very detailed user manual is available, which includes a large amount of theoretical information, formulae and references. A large number of data can be analysed, and a Batch Files option is also available

iii) POWERMARKER (http://statgen.ncsu.edu/powermarker/)

PowerMarker was designed specifically for the use of SSR/SNP data in population genetics analyses [2]. Data can be imported from Excel or other formats, making data set-up very easy. Data can also be exported to NEXUS and Arlequin formats. It includes a '2D viewer' for linkage disequilibrium visualization. The user can edit graphics within PowerMarker or export them for publication. The program has been tested extensively for accuracy and efficiency. Full documentation is included. Several new modules for association study are included in the package. Several demonstration datasets are available to get started. The program is free, but requires having PHYLIP, TreeView and the Microsoft.net framework system (all freely available) and Excel 2000 (not free). Another disadvantage is that it is available only for Windows 98 and above (not for Macintosh or other systems).

iv) PAUP (http://paup.csit.fsu.edu/)

PAUP is widely used for inferring and interpreting evolutionary trees [13]. It originally meant Phylogenetic Analysis Using Parsimony, but now has many other options. Although not free, it is relatively inexpensive and available from Sinauer Associates, Sunderland, MA. A new version, 4.0 beta, has been released as a provisional version. Macintosh, PowerMac, Windows and Unix/OpenVMS versions are available; the Mac version has some extra features. The Windows version runs as a GUI application, however, unlike the Macintosh version, most options are command-line-driven. The advantage to running PAUP under Windows is that a scrollback display buffer is built into the program, an editor is provided, and commands are remembered between sessions (they can be recalled, edited, etc.). It is closely compatible with MacClade (another program available from Sinauer), since they use a common data format (NEXUS).

v) MEGA (http://www.megasoftware.net/)

MEGA (Molecular Evolutionary Genetics Analysis) software has been widely used since its creation in 1993. It uses DNA sequence, protein sequence, evolutionary distance or phylogenetic tree data. It is an integrated tool for conducting automatic and manual sequence alignment, inferring phylogenetic

Analysis of Genetic Diversity in Crop Plants Using Molecular Marker Data 63

trees, mining web-based databases, estimating rates of molecular evolution, and testing evolutionary hypotheses[3]. Although it was designed for the Windows platform, it runs well on Macintosh with a Windows emulator, Sun workstation (with SoftWindows95) or Linux (with Windows by VMWare). Online, a thorough manual is available, together with a bulletin board to interact with other users.

vi) GENEPOP (http://genepop.curtin.edu.au/)

Genepop is a population genetics software package, which has options for the following analysis: Hardy-Weinberg equilibrium, linkage disequilibrium, population differentiation, effective number of migrants, F_{st} or other correlations[11] (Raymond and Rousset, 1995). Genepop can be used either as a DOS-version or a Web-version. The web-version is easy to use: after choosing an option for the analysis, the data are typed or pasted into the text window provided and the results are obtained either by email or by viewing the output via the Web.

vii) POPGENE (http://www.ualberta.ca/~fyeh/popgene_download.html)

POPGENE is a user-friendly window-based computer package for the analysis of genetic variation among and within natural populations using co-dominant and dominant markers and quantitative traits [16]. This package provides the Windows graphical user interface that makes population genetics analysis more accessible for the casual computer user and more convenient for the experienced computer user. The current version is designed specifically for the analysis of co-dominant and dominant markers using haploid and diploid data. It performs most types of data analysis encountered in population genetics and related fields. It can be used to compute summary statistics (*e.g.*, allele frequency, gene diversity, genetic distance, *F*-statistics, multilocus structure, *etc.*) for (a) single-locus, single populations; (b) single-locus, multiple populations; (c) multilocus, single populations and (d) multilocus, multiple populations. The latest version also includes the module for quantitative traits.

viii) GDA (http://hydrodictyon.eeb.uconn.edu/people/plewis/software.php)

GDA (Genetic Data Analysis) is a programme which computes linkage and Hardy-Weinberg disequilibrium, some genetic distances, and provides method-of-moments estimators for hierarchical F-statistics.

ix) GenAlEx (http://www.anu.edu.au/BoZo/GenAlEx/)

GenAlEx ('**Gen**etic **A**nalysis in **Ex**cel') is a user-friendly cross-platform package for population genetic analysis that runs within Microsoft Excel [9]. GenAlEx enables population genetic data analysis of codominant, haploid and binary genetic data providing analysis tools applicable to plants, animals and microorganisms. It has tools for importing, editing and manipulating raw genotype and sequence data from automated sequencing or genotyping software. New 2D spatial autocorrelation procedures have been incorporated in addition to the existing wide range of spatial analysis options. Pairwise relatedness among individuals can be estimated. There are tools for genetic tagging applications, including location of matching genotypes and calculation of probabilities of identity. Data export options to a host of other population genetic software packages are also available.

x) TFGPA (http://www.marksgeneticsoftware.net/tfpga.htm)

TFGPA (Tools for Population Genetic Analyses) is a Windows program for the analysis of allozyme and molecular population genetic data [14]. This program calculates descriptive statistics, genetic distances, and F-statistics. It also performs tests for Hardy-Weinberg equilibrium, exact tests for genetic differentiation, Mantel tests and UPGMA cluster analyses. Additional features include the ability to analyze hierarchical data sets as well as data from either codominant markers such as allozymes or dominant markers such as AFLPs or RAPDs.

xi) STRUCTURE (http://pritch.bsd.uchicago.edu/structure.html)

The program *structure* is a free software package for using multi-locus genotype data to investigate population structure. Its uses include inferring the presence of distinct populations, assigning individuals to populations, studying hybrid zones, identifying migrants and admixed individuals, and estimating population allele frequencies in situations where many individuals are migrants or admixed. It can be applied to most of the commonly-used genetic markers, including SNPs, microsatellites, RFLPs and AFLPs. The basic algorithm was described by Pritchard *et al.* (2000).

Useful Internet Resources

The following are a list of Internet resources containing links to useful information pertaining to genetic diversity analysis, population genetics and other softwares available:

i) 'An alphabetical list of genetic analysis software' from the North Shore LIJ Research Institute (http://linkage.rockefeller.edu/soft/list1.html) contains a list of 520 programmes. Computer software on genetic linkage analysis for human pedigree data, QTL analysis for animal/plant breeding data, genetic marker ordering, genetic association analysis, haplotype construction, pedigree drawing, and population genetics are included here.

ii) 'Phylogeny Programs' (http://evolution.genetics.washington.edu/ phylip/ software.html) contains links to 365 phylogeny packages and 51 free web servers. Updates to these pages are made monthly. Many of the programs in these pages are available on the web, and some of the older ones are also available from ftp server machines. The programs listed below include both free and non-free ones. The packages are sorted in various ways (e.g. by methods, system used, analyzing particular kind of data, most recent etc.).

iii) Maize Genetics site (http://www.maizegenetics.net/bioinformatics) from Cornell's Institute of Genomic Diversity contains freely available software programme to evaluate linkage disequilibrium, nucleotide diversity and trait associations

iv) The European Molecular Biology Laboratory–European Bioinformatics Institute (EBI) site (http://www.ebi.ac.uk/) contains links to many useful programs and other sites.

v) Mathematical Genetics and Bioinformatics Site, University of Chicago (http://mathgen.stats.ox.ac.uk/software.html)

vi) Statistical genetics and Bioinformatics Site, North Carolina State University (http://statgen.ncsu.edu/brcwebsite/software_BRC.php) contains softwares for genetic data analysis developed and made available by researchers at or affiliated with the Bioinformatics Research Centers.

Conclusion

The analysis of genetic diversity within a species is imperative for gaining an insight into the process of evolution of the species at the population level. Many statistical packages and computer programmes are currently available for analyzing molecular data for assessment of genetic diversity. Most programs perform similar tasks and many of them are freely downloadable from the internet. The programmes, however, differ from each other in the type of marker they can handle, the manner in which the raw data are formatted and also in how the users select the details of the computations to be performed. Many of these programmes use a specific data-file format, but several of these programmes offer the possibility to read or write data from, or to other file formats. Many of these programs possess user-friendly and sophisticated graphical interfaces which help the users to easily select the type of analyses to be performed and to set up computational parameters. Currently, researchers are directing their efforts on development of newer programmes using more specialized methodologies.

References

1. Glaubitz, J.C. (2004) Convert: A user-friendly program to reformat diploid genotypic data for commonly used population genetic software packages. Molecular Ecology Notes, 4: 309-310.
2. Liu, J. (2003) Powermarker: New genetic data analysis software, Version 3.0. Free program distributed by author over internet.
3. Kumar, S., Dudley, J., Nei, M. and Tamura, K. (2008) MEGA: A biologist-centric software for evolutionary analysis of DNA and protein sequences. Briefings in Bioinformatics, 9: 299-306.
4. Miller, M.P. (1997) Tools for population genetic analysis (TFPGA) 1.3:A Windows program for the analysis of allozyme and molecular population genetic data. Distributed by the author.
5. Mohammadi, S.A. and Prasanna, B.M. (2003) Analysis of genetic diversity in crop plants-salient statistical tools and considerations. Crop Science, 43:1235-1248.
6. Nei, M. and Chesser, R.K. (1983) Estimation of fixation indices and gene diversities. Annals of Human Genetics, 47:253-259.
7. Nei, M. and Li, W. (1979) Mathematical model for studying genetic variation in terms of restriction endonucleases. Proceedings of National Academy of Sciences (USA), 76:5269-5273.
8. Nei, M. and Chesser, R.K. (1983) Estimation of fixation indices and gene diversities. Annals of Human Genetics, 47:253-259.
9. Peakall, R. and Smouse, P. E. (2006) GENALEX 6: Genetic analysis in excel. Population genetic software for teaching and research. Molecular Ecology Notes, 6: 288-295.
10. Pritchard, J.K., Stephens, M. and Donnelly, P. (2000) Inference of population structure using multilocus genotype data. Genetics, 155:945-959.
11. Raymond, M., and Rousset, F. (1995) GENEPOP (version 1.2): Population genetics software for exact tests and ecumenicism. Journal of Heredity, 86:248-249.
12. Schneider, S., Roessli, D. and Excoffier, L. (2000) ARLEQUIN, version 2.00-software for population genetics data analysis. Genetics and Biometry Laboratory, University of Geneva, Switzerland.
13. Swofford, D.L. (2002) PAUP: Phylogenetic analysis using parsimony (and Other Methods), Version 4. Sinauer Associates, Sutherland, MA. USA.
14. Weir, B.S. and Cockerham, C.C. (1984) Estimating F-statistics for the analysis of population structure. Evolution, 38:1358-1370.
15. Wright, S. (1951) The genetical structure of populations. Annals of Eugenics, 15: 323-354.
16. Yeh, F.C. and Boyle, T.J.B. (1997) Population genetic analysis of co-dominant and dominant markers and quantitative traits. Belgian Journal of Botany, 129:157.

Chapter – 4

Plant Genomics - A Bioinformatics Perspective

S. Raji Radhakrishnan and R. Keshavachandran

Abstract

Plant Bioinformatics is a new and rapidly evolving field driven by the advances in 'Omics' technologies. This chapter is intended to help both beginners and experienced researchers to develop and apply bioinformatics tools to specific areas of plant genomics research. This chapter will also help plant biologists to access and implement comparative genomics studies of the plant genes of interest. Also discussed are the databases which house the data on plant genes and genomes. The chapter ends with a review of basic Perl Programming for sequence motif analysis beneficial for plant biologists.

Introduction

Ever since the discovery of DNA structure, the field of plant genomics has experienced a dramatic change in the ways scientific problems are approached. With the recent advances in High Throughput analysis clubbed with methods like Physical mapping of genes, RFLP, RAPD, large scale EST sequencing and mRNA protein profiling, there is an enormous amount of biological data available for various analysis. As the amount of data grows tremendously, there

68   Agriculture Bioinformatics

is a demand for developing tools and methods to effectively access these assembled data for analysis, modeling, visualization and prediction. In this chapter, we emphasize on some of the key concepts in bioinformatics, tools and databases relevant to plant genomics.

In a biological system DNA, RNA and Protein sequences are the most fundamental factors at the molecular level. Several genomes have been sequenced to high quality in plants including *Arabidopsis thaliana* and rice [6]. Draft genome sequences are available for the tree *Poplar trichocarpa* (http://genome.jgi-psf.org/Poptr1/) and lotus (http://www.kazusa.or.jp/lotus/) and sequence efforts are in progress for several others including tomato, maize, *Medicago truncatula* and sorghum.

Two types of approaches can be considered for studying the plant genome:

- The comparative genome studies which present opportunities to study the evolution of plant genome structure and the dynamics of molecular evolutionary processes.

- The single gene study which identifies genes and other functional elements provide vital data for annotation of completed plant genomes. [17]

Bioinformatics renders the computational methods and tools required to interpret and assign biological significance to a raw DNA sequence. The branch of bioinformatics which deals with giving biological meaning to DNA sequence is referred to as Genome Annotation. A few attributes overlooked are factors such as regulatory elements, transposons, repetitive elements, motifs etc. and predicting the biological functions for each of these elements and the biological processes they take part in.

The Structure of Plant Genome

A genome can be divided based on two attributes: one comprising the proteins along with RNA encoding genes and the other, the non coding DNA. The term is often used to refer to those segments of DNA that are involved in the production of proteins, excluding other transcribed segments which encode functional or structural RNA molecules. The amount of DNA thought to be "non-coding" currently shrinks as ongoing research discovers more and more of the DNA to be transcribed [16].

Non Coding DNA makes up the largest fraction of eukaryotic genomes. For example, in *Arabidopsis*, which is considered to have a relatively compact genome, about 70% of the DNA is non-coding [7]. This fraction is much larger in many other eukaryotes, as an indication in the human genome less than 2% is believed to be protein encoding. [1]

Regulatory elements in plant genome

The development of plants depends not only on the structural genes encoding for proteins, but also on regulatory genes that coordinate and direct the activities of hundreds of structural genes. Regulatory genes encode for proteins that either activate or repress the ability of other genes to express themselves. Regulatory and structural genes therefore operate in vast, complex feedback systems that receive and respond to physiological cues generated from within and without the organism.[11] Networks of regulatory genes help to directly coordinate the activities of the genome, so that they directly or indirectly influence cell metabolism, growth and development.

Expressed sequence tags [EST's]

EST is a short sub-sequence of a transcribed cDNA sequence. EST is produced by sequencing of cloned mRNA. They are used to identify gene transcripts in gene discovery. Because these clones consist of DNA that is complementary to mRNA, the EST's represent portions of expressed genes. They may be present in the databases as either cDNA/mRNA sequence or as the reverse complement of the mRNA, the template strand. EST's provide a snapshot of the population of genes expressed from a specific tissue such as a leaf or root. EST's are useful for identifying landmarks on the genetic map. EST's from genomics databases can provide further clues about the gene function.

Resource: PlantGBD

PlantGDB intends to assemble all the major plant species on a regular basis, providing a frequent estimation of plant gene space. The assembly results and methods are freely available to the plant research community. EST assemblies are composed of GenBank EST's + cDNA sequences. Plant GDB builds the output as shown in Figure 4.1, enlisting the set of putative unique transcripts [PUT], set of similar proteins, alignment and blast output.

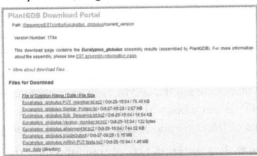

Figure 4.1: From PlantGDB snapshot for the query *Eucalyptus globulus*. It gives the automated downloadable version of the Blast output, alignment files, EST's and a list of similar proteins.

Patterns/regular expressions in plant genome

The conserved regions present in plant genome sequences are structural features like repetitive elements, transposons, promoter sequences and regulatory elements. Repetitive sequences/patterns have well defined functions. These regions in plant genome encode proteins necessary for their own reverse transcription and integration. Genes with similar repetitive elements may have similar biological functions. The knowledge of repetitive DNA in the genome is also important for evolutionary, genetic, taxonomic and applied studies.

Commonly Observed Features

Satellite DNA: Satellite DNA is also referred to as tandem repeats. Satellite DNA's vary from 140-180bp in length. They consist of highly repetitive DNA. These repetitions of DNA tend to produce a different frequency of the nucleotides adenine, cytosine, guanine and thymine and thus have a different density from the normal DNA and they form a 'satellite' band when genomic DNA is separated on a density gradient.

Microsatellites: Microsatellites are simple sequence repeats [with motifs 1-5bp long] while mini-satellites have longer or more repeating units [upto 40 bp]. They are used as molecular markers in genetics and also can be used to study gene duplication and deletion.

Telomeric DNA: This is a repetitive DNA at the end of the chromosome and protects the end of the chromosome from deterioration. Telomeric DNA consists of conserved 7bp repeats [CCCTAAA] and is added to terminal end by the enzyme telomerase.

LINE [Long Interspersed Nuclear Elements]: They are long DNA sequences of >5kb. Interspersed repeats are repeated DNA sequences located at dispersed regions in a genome. A LINE encodes a reverse transcriptase and other proteins.

SINE [Short Interspersed Nuclear Elements]: They are short DNA sequences. SINE's do not encode a functional reverse transcriptase protein and rely on other mobile elements for transposition. Previously confused for junk DNA's, recent research suggests that both LINE's and SINE's have significant role in transcription process.

MITE [Miniature Inverted-repeat Transposable Elements]: are short length about 400-600bp that occurs at the end of each element in an inverted fashion. Thousands of MITE's have been identified in the genomes of *Oryza sativa*. Unlike some types of transposons, MITE's do not appear to encode proteins.

Promoter Region: Promoter is the region of DNA that facilitates the transcription of a particular gene. They are located near the genes they regulate on the same strand and upstream [towards the 5' strand of the sense strand].

Many eukaryotic promoters between 10 and 20% of all genes, contain a TATA box [seq TATAAA], which in turn binds a TATA binding protein assisting in the formation of the RNA polymerase transcriptional complex.

Structural rRNA genes: Ribosomal RNA gene loci 18S-5.8S-28S and 5S rRNA in the genome encode the structural RNA components of ribosome. Other conserved regions include transposons and retrotransposons regions.

Transposons: They are full length autonomous elements encoding a protein called transposase, by which an element can be removed from one position and inserted into another. They typically have short inverted repeats at each end.

Retrotransposons: They are mobile genetic elements and are ubiquitous in eukaryotic genome, particularly abundant in plants. In maize 50%-80% and in wheat up to 90% of the genome are made up of retrotransposons.

Resource: PlantCARE: A database of plant cis-acting regulatory elements.

PlantCARE is a referential database for plant promoter elements. It houses 435 different names of plant transcription sites defining more than 159 plant promoters. PlantCARE features 417 cis-acting regulatory elements of which 150 are from monocotyledonous species, 263 from dicotyledonous species and four from conifers. The database can be queried on the basis of transcription factor [TF] sites, motif sequences, function, species, cell types, gene and literature references. The result page is linked to other databases such as EMBL, GenBank, TRANSFAC and MEDLINE[12]. The output of PlantCARE for the query *Solanum tuberosum* is shown below in Figure 4.2.

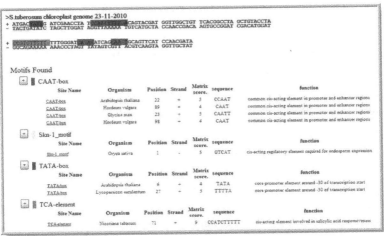

Figure 4.2: Output of analysis of *Solanum tuberosum* whole Chloroplast Genome for CIS-acting regulatory elements using PlantCARE. Also shown are the similar regulatory elements in other related plant species and their designated functions. Uniqueness of Plant Mitochondrial and Ribosomal Genome

Genes in Mitochondrial and Ribosomal membranes are said to be highly conserved from an evolutionary point of view. The mitochondrial genome of plants distinguishes itself from those of other higher eukaryotes in its unusual organization and gene-processing mechanisms.

Mitochondrial Genome: The plant mitochondrial genome differs from that of other higher eukaryotes in the vast diversity demonstrated in genome size and structure (reviewed by Hanson and Folkerts, 1992)[5]. Even within one plant family, a 10-fold difference in mitochondrial genome size can be observed. The mitochondrial genome encodes only a fraction [estimated at 20-30 proteins] of the gene products required for its function; the vast majority are encoded by the nucleus. The plant mitochondrial genome is known to encode three rRNA's [26S, 18S and a 5S rRNA unique to plants], ribosomal proteins and some of the necessary tRNA's. Those tRNA's necessary for mitochondrial translation that are not encoded by the mitochondrion are located in the nucleus (Dietrich et al., 1992) [2].

Ribosomal Genome: The gene encoding the 18S, 5.8S and 25S ribosomal RNA's are present in tandem arrays of unit repeats in a recognizable chromosome structure, the nucleolar organizer region [NOR]. The repeat unit consists of the coding sequences for each of these three RNA's as well as an internal transcribed spacer region and an intergenic region [Figure 4.3]. The number of repeating units varies between several hundred and over 20,000.[4]

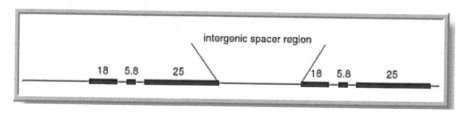

Figure 4.3: The repeat unit for the large ribosomal RNA genes [Courtesy to Plant genomics and proteomics by Christopher A Cullis]

Much of the variability in genome size in plants is accounted by the complex organizations observed among different plant species. Mitochondrial DNA recombination events that occur among homologous repeated sequences as well as among apparently non-homologous sequences result in a highly variable genome configuration in higher plants relative to other eukaryotic systems. These events contribute not only to the complexity of genome structure but to mitochondrial mutation frequency as well.

Plant Genomics - A Bioinformatics Perspective

The genetic code used to define mitochondria differs from the universal genetic code used by nuclear and chloroplast genes. Yeast and animal mitochondria use triplet TGC [or UGA] instead of the formal TGG to code for Tryptophan. Mitochondria in higher plants also tend to use CGG to code for Tryptophan. TGA which encodes Tryptophan in the mitochondria of other species appears to be used as a stop codon in plant mitochondrial genes. Like chloroplast, mitochondrial genome codes for a small but important number of mitochondrial polypeptides.

Predicting Genes in Genome Sequences - A Computational Approach

Gene prediction refers to identifying regions of sequences usually genomic DNA, that are biologically functional. This includes protein-coding genes and other functional elements such as RNA genes and regulatory regions. Methods to identify genes in a newly sequenced genome can be divided into two: [I] Similarity based methods, which utilize sequence similarity to known genes and [II] *Ab initio* or *de novo* methods, which predict genes using a statistical approach. Comparative methods employ sequence comparison between multiple related genomes to identify conserved genes. Given a known gene and an unannotated genome sequence, the main objective of gene prediction is to find a set of gene sequence which matches with the query / target sequence.

Sequence alignment reveals function, structure and evolutionary information in DNA and protein sequences. It is important to obtain the best possible optimal alignment to discover its information.

When two genomes have only recently diverged, the order of many genes, gene numbers, gene positions and even gene structures (exon–intron organization, splice site usage, and so on) remain highly conserved. New genes can also be identified from direct genome comparisons. By comparing the genomes of several closely related species, conserved regulatory regions can also be easily identified. For these reasons, making use of comparative genomic data will be a key challenge for the gene-prediction field.

a) Similarity Based Methods

This method includes searching the databases for similar sequences. Sequence similarity is done to trace out the biological significance of protein \ nucleotide sequences. The basic step in any similarity search is an alignment of two or more sequences. It is possible to compare new genes to known ones and to compare genes from different species. Sequence comparison is the most powerful and reliable method to determine evolutionary relationships between genes. Popular tools for sequence alignment are BLAST and FASTA. These tools also help to identify regions in a single sequence that have an unusual composition suggestive of an interesting function.

Scoring Matrix forms the basis of similarity search programs. The BLOSUM scoring matrices are based on the substitutions found in alignments of a large number of protein families of variable sequence similarity. The alignments are found in BLOCKS database. Overrepresented sequences in these alignments are grouped together to different extent to reduce their distribution. The BLOSUM62 matrix which has 62% of the alike sequences grouped, is the most commonly used matrix for scoring protein sequence alignments. [3]

b) Ab Initio Methods

Sequencing of DNA needs sophisticated *in vivo* experiments like gene knockout and other assays and the frontiers of bioinformatics are making it increasingly possible to predict the gene function based on sequence alone. In *ab initio* finding the genomic DNA is searched for certain landmarked protein coding genes. These could be specific sequences which indicate the presence of a gene nearby. Promoter region, ORF [Open Reading Frames], the continuous DNA sequence without a stop codon in the frame, 3'UTR untranslated regions [AATAAA for polyadenylation signal] etc serve as landmark in the analysis of a given DNA sequence.

In Eukaryotes, the finding of such landmark sequences is a challenging task for several reasons. The promoter and other regulatory signals in these genomes are more complex and less understood than in prokaryotes.

Hidden Markov Modes [HMM]: This method applies a statistical approach to the way similarity analyses are carried out. It can carry a variety of calculations. HMM models can predict the probability of finding a "pattern" in a given DNA. Using HMM model, family of protein sequences can be compressed into a profile called as probabilistic profile. The program takes a few sequences and align them one below the other and see for matches and variations; convert those matches and variations into a set of statistical values thereby creating an HMM profile. Every tool is trained with already annotated DNA.

Once the HMM profile is created, the program predicts the probability of finding the regular patterns in the gene sequence. The drawback in the tool is that they are trained over a certain groups of genes only. A gene finding tool must be sensitive and specific. Using HMM, the gene predicting tools can be designed such that it is specific and sensitive.

Plant Genomics - A Bioinformatics Perspective

Some of the commonly used gene predicting softwares which use HMM approach are HMMer, GeneMark, FrameD, ORF Finder, Genehacker, GenScan, mGENE, mSplicer etc. Comparison of closely related species gives better quality [sensitivity and specificity] for the analysis. Most of these programs operate on a single continuous sequence at a time and the results are generated in a diverse array of readable formats that must be translated to a standardized file format. These translated results must then be concatenated into a single source and then presented in an integrated form for human curation. [9]

A change in a single codon/base [mutation] affects the sensitivity of the input data. The reading frame specific codon bias provides a valuable tool for evolutionary and mutational studies.

Genome sizes

Genome sizes are typically given as gametic nuclear DNA contents ['C-values'] either in units of mass or in number of base pairs [most often in mega-bases, where 1 Mb = 106 bases]. The majority of modern genome size estimates are based on Feulgen densitometry or Flow Cytometry. Data from all such measurements are compiled into the databases along with updated taxonomy, analytical results and other relevant information including chromosome number, genetic maps, taxonomic data, methods used etc. The data collected from peer reviewed journals are shown in Table 4.1.

Table 4.1: Genome sizes of selected plants

Organisms	Genome Size	Number of genes predicted
Arabidopsis thaliana	120 Mb	25,498
Oryza sativa ssp. *indica*	420 Mb	32-50,000
Oryza sativa ssp. *japonica*	466 Mb	46,022-55,615
Physcomitrella patens	500 Mb	39,458
Populus trichocarpa	550 Mb	45,555
Vitis vinifera	490 Mb	30,434
Carica papaya	372 Mb	28,629
Cucumis sativus	367 Mb	26,682
Zea mays	2800 Mb	32,000
Sorghum bicolor	730 Mb	27,640
Glycine max	1,100 Mb	46,430
Brachypodium distachyon	272 Mb	-
Malus domestica	927 Mb	57,000

The estimated size of plant genome ranges from 130 million base pairs for *Arabidopsis thaliana*, 430 for rice, 550 for *Medicago truncatula*, 770 for apple, 950 for tomato to 5000 for barley, 16000 for wheat and 1800 for onion.

Whole genome sequencing determines the complete DNA sequence of an organism's genome. This includes sequencing all of an organism's chromosomal DNA as well as DNA contained in the mitochondria and for plants the chloroplast as well. The growth of advanced technologies like whole genome sequencing methods and high throughput methods have enabled the annotation of more genomes thereby more data entries in biological databases as shown in Figure 4.4.

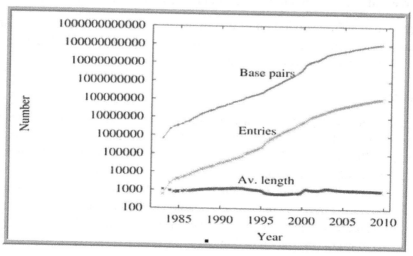

Figure 4.4: Biological data bases as recorded by the Center of Biological Sequence Analysis, The Technical University of Denmark. [http://www.cbs.dtu.dk/databases/DOGS/GBgrowth.php]

Model Plant Species in Genomics Research

Many aspects of biology are similar in most organisms, but it is much easier to study a particular aspect in one organism than in others. Some characteristics required for model plant species are rapid development with short life cycles, small adult size, ready availability and tractability. Many valuable data can be retrieved from model plant databases which facilitate comparative studies of the plant of interest. Model species are often diploids, with a few chromosomes and small genome. Some model species include *Arabidopsis*, sorghum, rice and *Brachypodium* which represent both dicotyledon and monocotyledon plant species as well as different plant development and flowering structures.

Over the last century, research on a small number of organisms has played a pivotal role in advancing our understanding of numerous biological processes. This is because many aspects of biology are similar in most organisms, but it is frequently much easier to study a particular aspect in one organism than in

Plant Genomics - A Bioinformatics Perspective

others. These much studied organisms are commonly referred to as **model organisms**, because each has one or more characteristics that make it suitable for laboratory study. A large amount of information can then be derived from these species which provide valuable data on the normal plant development; gene regulation and evolutionary processes.

Organism	Type	Relevance	Genome size	Number of genes predicted	Organization	Year of completion
Arabidopsis thaliana Ecotype:Columbia	Wild mustard Thale Cress	Model plant	120 Mb	25,498[22]	Arabidopsis Genome Initiative[23]	2000[22]
Oryza sativa ssp indica	Rice	Crop and model organism	420 Mb	32-50,000[24]	Beijing Genomics Institute, Zhejiang University and the Chinese Academy of Sciences	2002[24]
Oryza sativa ssp japonica	Rice	Crop and model organism	466 Mb	46,022-55,615[25]	Syngenta and Mynad Genetics	2002[25]
Physcomitrella patens	Bryophyte	Model organism early diverging land plant	500 Mb	39,458[26]	US Department of Energy Office of Science Joint Genome Institute	2008[26]
Populus trichocarpa	Balsam poplar or Black Cottonwood	Carbon sequestration, model tree, commercial use (timber), and comparison to A. thaliana	550 Mb	45,555[27]	The International Poplar Genome Consortium	2006[27]
Vitis vinifera	Grapevine PN40024	Fruit crop	490 Mb[28]	30,434[28]	The French-Italian Public Consortium for Grapevine Genome Characterization	2007[28]
Carica papaya	Papaya 'SunUp'	Fruit crop	372 Mb[29]	28,629[29]	Hawaii Agricultural Research Center and others	2008[29]
Cucumis sativus	Cucumber 'Chinese long' inbred line 9930	Vegetable crop	367 Mb[30]	26,682[30]	Chinese Academy of Agricultural Sciences, Beijing	2009[30]
Zea mays ssp mays	Corn (maize) B73	Cereal crop	2,800 Mb[31]	32,000[31]	NSF	2009[31]
Sorghum bicolor		Crop plant	730Mb[32]	27,640[32]	Multiple institutions	2009[32]

Figure 4.5: List of few of the completely sequenced plant species

Arabidopsis as a model plant: Mapping the genomic sequences of plants had its first major breakthrough in 2000 and was the result of 4 years of study by the Arabidopsis Genome Initiative. Arabidopsis is a small plant in the mustard family and is widely used as the model system for comparative genomics research in plant biology. This project includes over 100 scientists from around the world and the total cost was estimated around $78 million. The Thale Cress serves as the model plant for genomic sequencing because of its small genome, its ability to reproduce quickly in about 6 weeks and easy growth. The genome is organized into 5 chromosomes, mitochondrial and plastid genome and contains an estimated 20.000 genes [Figure 4.6]. NCBI's Map viewer tool facilitates the whole genetic map which has data on the location of a gene on a specific chromosome as shown in Figure 4.7.

The first sequenced genome of a plant, *Arabidopsis thaliana*, was published less than 6 years ago. Since that time, the complete rice genome [*Oryza sativa*; Goff *et al.*, 2002, Yu *et al.*, 2002] International Rice Genome Sequencing Project [2005], and a draft sequence of the poplar genome [*Populus trichocarpa*] have also been completed. In addition, the National Centre for Biotechnology Information [NCBI] Entrez Genome projects website reports that sequencing of several more plant genomes are in progress. Many of the obvious candidates for genome sequencing model species are those with economic importance. The sequencing studies have either already been completed or are underway. A list of a few of the completely sequenced plant species is shown in Figure 4.5.

Quantitative Trait Loci [QTL]: The position on a chromosome occupied by a particular gene that determines a quantitative trait expression is referred to as QTL. QTL are identified via statistical procedures that integrate genotype [molecular markers] and phenotype [physical observation data]. QTL are assigned chromosome locations based on the positions of markers on linkage maps.

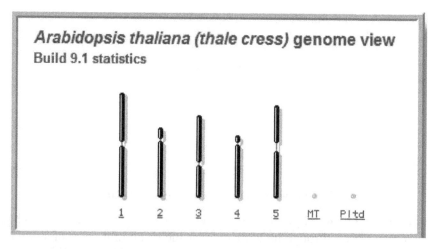

Figure 4.6: Chromosome Map of *A.thaliana* using MapViewer [www.ncbi.nlm.nih.gov/mapview]

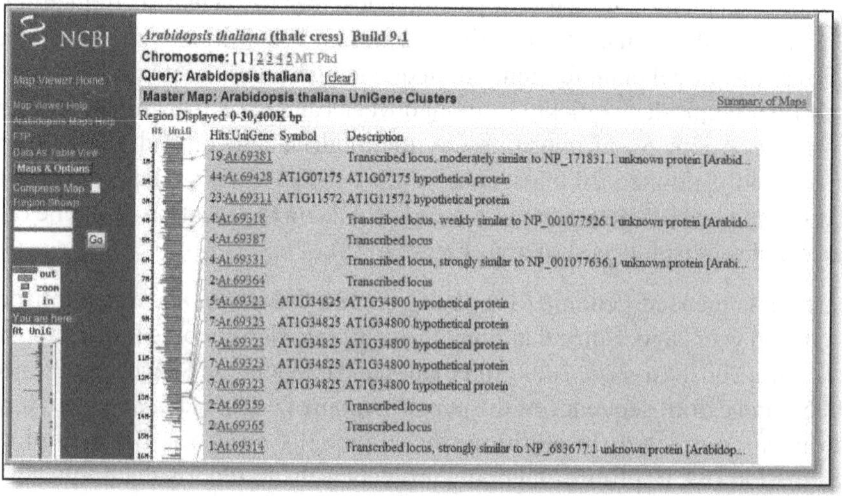

Figure 4.7: NCBI detailed MapView with gene information

Rice as a model plant: After the success of the Thale Cress, rice became an important target for sequencing studies since it serves as a staple in many countries. Rice is also used as a model plant species. As the number of plants with sequenced genomes increased, scientists set their sights on more complex plants in the hope of understanding plant biology better and provide new insights into genetic engineering.

Below is a flowchart [Figure 4.8] showing the role of Genomics in Product Development. Use of molecular markers in combination with novel physical mapping techniques has led to availability of genetic maps for many crop species.

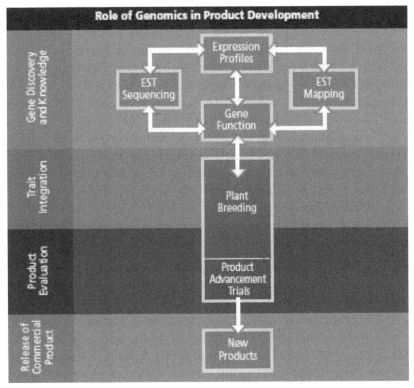

Figure 4.8: Showing the Role of plant genomics in product development
[*Courtesy* to Pioneer Hi Bred International Inc.,]

Resources for plant genomics

The goal of a novel model plant genomics resource database is to provide a community resource that allows researchers to build their research models by making use of data derived from one or more approaches, for example the forward and reverse genetics, comparative genomics, *in silico* predictions etc.[8]

AgBase

AgBase is a curated, open-source, web-accessible resource for functional analysis of agricultural plant and animal gene products. The database provides gene associated files of fully sequenced plant species. They use the vocabularies developed by the Gene Ontology (GO) Consortium to describe molecular function, biological process and cellular component for genes and gene products in agricultural species [14]. AgBase develops freely available tools for functional analysis, including tools for using Gene Ontology.

TIGR plant repeat databases

It is a collective resource for the identifying repetitive sequences in plants. The database is constructed of repetitive sequences for 12 plant genera: *Arabidopsis, Brassica, Glycine, Hordeum, Lotus, Lycopersicon, Medicago, Oryza, Solanum, Sorghum, Triticum* and *Zea*. The repetitive sequences within each database have been coded into super-classes, classes and sub-classes based on sequence and structure similarity. These databases are available for sequence similarity searches as well as downloadable files either as entire databases or subsets of each database. In order to ease the further comparative studies, they provide a resource for repetitive sequences in other genera within these families. Repetitive sequences have been combined into four databases to represent the Brassicaceae, Fabaceae, Gramineae and Solanaceae families as shown in Figure 4.9. Collectively, these databases provide a resource for the identification, classification and analysis of repetitive sequences in plant. (www.tigr.org/tdb/e2k1/plant.repeats/index.shtml)[15].

In plants, ploidy levels and repetitive sequences contribute significantly to the genome size. A number of different repetitive sequences have been reported in the plant genome and these can be classified into classes, subclasses and super-classes based on structure and sequence composition. The transposable element (TEs) super-class includes retrotransposons, transposons and Miniature Inverted-repeat Transposable Elements (MITEs). Other repetitive sequences associated are the centromere and telomere. Another super-class of repetitive sequences is rDNA which encodes the structural RNA components of ribosome. [13]

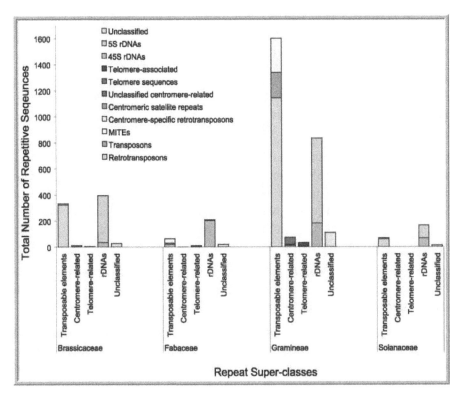

Figure 4.9: Repetitive elements/sequences of plants from four different families: Brassicaceae, Fabaceae, Gramineae and Solanaceae.

Repetitive DNA sequences of selected plant genera were queried from GenBank and other published records. After eliminating the duplicated sequences and vector sequences, the repeats were coded into five superclasses: Transposable elements, centromere related, telomere related, rDNA and Unclassified. The superclasses were then broken into major classes of repeats. The collected repetitive sequences within the same plant family were combined into a repeat database for the plant family.

Cereal Genomic Databases

Gramene: A comparative resource for cereal genomics. With a focus on rice, Gramene has marker, maps, gene, proteins, QTL, literature, diversity, pathway and ontology information on rice, maize, wheat, sorghum, barley, millet, rye, oats and wild rice.

International Rice Information System (IRIS) is the rice information of the International Crop Information System (ICIS) which is a database system that provides integrated management of global information on genetic resources and crop cultivars. This includes germplasm pedigrees, field evaluations, structural and functional genomic data (including links to external plant databases) and environmental (GIS) data.

Oryzabase, Japan: The Oryzabase consists of five parts (1) genetic resource stock information, (2) gene dictionary, (3) chromosome maps, (4) mutant images, and (5) fundamental knowledge of rice science.

PLEXdb (plant expression database) is a unified public resource for gene expression for plants and plant pathogens. PLEXdb serves as a bridge to integrate new and rapidly expanding gene expression profile data sets with traditional structural genomics and phenotypic data. The integrated tools of PLEXdb allow investigators to use commonalities in plant biology for a comparative approach to functional genomics through use of large-scale expression profiling data sets.

RIS - rice information system (BGI-RIS) is targeted to be the most up-to-date integrated information resource for rice genomes as well as a workbench for comparative genomic analysis among cereal crops.

Wheat genetic and genomic resources center: The WGGRC has established a national and international network to conduct and coordinate genetic studies in wheat. The WGGRC maintains a gene bank, along with evaluation and passport data, on 2,500 wheat species accessions. In addition, the WGGRC houses 2,200 cytogenetic stocks, the genetic treasures produced by a lifetime of work by wheat scientists.

Phytozome is a joint project of the Department Of Energy's Joint Genome Institute and the Center for Integrative Genomics to facilitate comparative genomic studies amongst land plants. Clusters of orthologous and paralogous genes which represent the modern descendents of ancestral gene sets are constructed at key phylogenetic nodes. These clusters allow easy access to clade specific orthology / paralogy relationships as well as clade specific genes and gene expansions. As of version 2.0, Phytozome provides access to five sequenced and annotated land plant genomes, clustered at six evolutionarily significant nodes. Where possible, each gene has been annotated with PFAM, KOG, KEGG, and PANTHER assignments, and publicly available annotations from RefSeq, UniProt, TAIR, JGI are hyper-linked and searchable, including *Sorghum bicolor* and *Oryza sativa*.

General Databases

National centre for biotechnology information: As the DNA sequencing methods progressed, so did the development of bioinformatics tools for storing and retrieving those data. First and foremost was the NCBI database founded in 1988 and the other International Centers for managing biological data like EBI [European Bioinformatics Institute]. NCBI processes the data submitted by the individual laboratories around the world in collaboration with two other databases EMBL and DDBJ. These three databases exchange data on a daily basis. NCBI conducts basic and applied research in computational, mathematical and theoretical problems in molecular biology and genetics, including genome analysis, sequence comparisons, sequence search methodologies, macromolecular structure, dynamics and interaction, and structure/function prediction

Plant oriented resources at NCBI: Customized plant genome BLAST http://www.ncbi.nlm.nih.gov/mapview/static/MVPlantBlast.shtml is available which searches the entire database and has options for multispecies plant genome map. The program selections are shown in Figure (4.10) which includes the search between highly similar sequences, most dissimilar sequences and translated database using a protein query. The Plant BLAST includes three datasets:

a) GenBank sequences derived from DNA used to identify mapped genetic loci in genetic maps of *Avena sativa* [oat], *Glycine max* [soybean], *Hordeum vulgare* [barley], *Lycopersicum* [tomato], *Triticum aestivum* [wheat] and *Zea mays* [corn].

b) The two collections of contigs developed by the Chinese WGS endeavors

c) The whole genome material for *Arabidopsis thaliana* and the *Oryza sativa* genome.

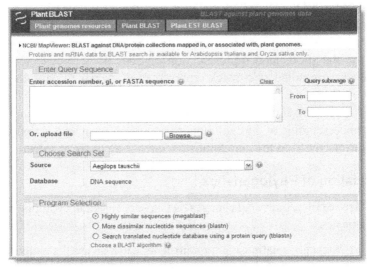

Figure 4.10: Snapshot of NCBI's Plant BLAST

Some of the research driven new technologies include production of plant based edible vaccines, RNA silencing to study genome's response in adverse conditions like drought, and application of knowledge of plant genomics for health protection and food production. A list of plant derived genome databases is shown in Table 4.2.

Table 4.2: Commonly used plant derived databases

Title	URL
Algae Base	http://www.algaebase.org/
Araliales Resource Center	http://rbg-web2.rbge.org.uk/URC/arc.htm
PLANTS	http://www.info.usda.gov/
Protein Data Bank	http://www.rcsb.org/pdb/
Protein Information Resource	http://pir.georgetown.edu/
European Bioinformatics Institute	http://www.ebi.ac.uk/
PlantSat [Motif search] [13]	http://w3lamc.umbr.cas.cz/PlantSat/ motifs.php

Status of On-going Plant Genome Sequencing Projects

The NCBI's Entrez Genome Projects give an outline of the plant genome sequencing projects with the group information, TaxID, Genome size, chromosome no:, sequencing methods and the center/consortium information.

Molecular Phylogenetics

Plastid, Nuclear and Mitochondrial genome can be used to study the phylogenetic relationship between a set of unique plants. It deals with the use of molecular [DNA, RNA, protein] information to gain information on an organism's evolutionary relationships through phylogenetic trees. Phylogenetic trees are the short hand representation of trees. It is represented as nested parentheses or called as Newick representation. A pair of parentheses which encloses all the descendents of that node represents each internal node. This format is used by computer programs to store tree files. Significant advances in our understanding of the evolutionary relationships of major groups of plants come primarily from the comparative sequence analysis of selected genes.

Application of Phylogenetics

1. Tree of life: Analyzing changes that have occurred in evolution of different species.

2. Phylogenetic relationships among genes can help predict which ones might have similar functions (e.g., homolog detection)

3. Follow changes occurring in rapidly changing species

4. Trace the Horizontal transfer of genes

Boot strapping

Bootstrapping: It is a statistical method to resample the phylogenetic trees. New datasets are created using the existing data to improve the tree branches i.e., the values used in building the tree are resampled to get a better relationship.

Choice of Macromolecular Sequences

Mitochondrial DNA: Since Mitochondrial DNA has faster mutation rate than nuclear DNA sequences, mitochondrial DNA is preferred more. This faster rate of mutation produces more variance between sequences and is an advantage while studying closely related species.

rRNA: For the study of divergent data, a gene which is highly conserved in all living organisms is preferred. For the study of more divergent species, we need a ubiquitous molecule that is highly conserved in all living organisms. Such a molecule is constrained at almost all residues. It would therefore have a low tolerance for mutations. rRNA is one such molecules.

As more genomes are sequenced we are becoming more interested in learning about protein and gene evolution. This can aid our understanding of the function of proteins and genes. Studies of protein and gene evolution involve the comparison of Homolog sequences having common origin but may or may not have common function.

New Themes in Plant Bioinformatics

Biodiversity informatics

Objectives

The goal of Biodiversity initiatives are to make the world's biodiversity data freely and universally available via the internet, share primary scientific biodiversity data and promote the development of biodiversity informatics around the world. The major objective of creating Biodiversity information database is not only to collate information, but also to add value to this information with added or supplementary data and have this information in uniform format.

Challenges faced by biodiversity informatics

The Biodiversity information is in the form of scattered, distributed and non-interoperable information sources. Lack of understanding between domain experts and IT managers is one of the major problems faced by this field. Proper management of Biodiversity data may prevent duplication and waste of intellectual ability.

Many of the Biodiversity programs failed even before they are launched. The major reason for this is the lack of understanding and sense of mistrust between the subject experts and information managers. Subject experts have to overcome the fear that collation of data would replace them. On the other hand, the information managers have to realize that they can never replace taxonomists and that they are only the care takers of the information and not the owner.

A few initiatives in the field of biodiversity informatics

Global Biodiversity Information Facility: [GBIF]: GBIF is an international organization that focuses on making biodiversity data available publicly. The GBIF makes the data accessible and searchable through a single portal. They include various data on plants, animals, fungi and microbes. The GBIF portal can be accessed at http://data.gbif.org/welcome.htm .

Integrated taxonomic information systems (ITIS)

The ITIS is the result of a partnership of US federal agencies formed to satisfy their mutual needs for scientifically credible taxonomic information. Since its inception, ITIS has gained valuable new partners. The goal of ITIS is to create an easily accessible database with reliable information on species name and their hierarchial classification. The database is reviewed periodically to ensure high quality with valid classification, revisions and additions of newly described species. The ITIS includes documented taxonomic information of flora and fauna from both aquatic and terrestrial habitats. Web site: http://www.itis.gov/

Species 2000

The species 2000 program was established by International Union of Biological Sciences (IUBS), in co-operation with the Committee on Data for Science and Technology (CODATA) and the International Union of Microbiological Societies (IUMS) in September 1994. Species 2000 has the objective of enumerating all known species of organisms on earth (animals, plants, fungi and microbes) as the baseline dataset for studies on global biodiversity. Web site: http://www.sp2000.org/

Plant Genomics - A Bioinformatics Perspective

Systems biology

We understand that the metabolic processes collectively make up an organism. There are a large number of interacting proteins and genes which facilitate these metabolic pathways. This paves way to an exciting cutting edge field in computational science namely Systems biology. Systems biology approach facilitates the scientists to study the gene/protein in a different dimension. Such analysis will include data on gene sequences, expression profiles, protein-protein interactions and even the reaction kinetics of a large number of metabolic intermediates.[10]

KEGG Bioinformatics for Plant Genomics Research

KEGG [Kyoto Encyclopedia for Gene and Genomes] is a bioinformatics resource to understand biological function from a genomic perspective. It has got cross reference to other major biological databases with pathways to represent cellular functions. KEGG consists of a suite of tools which include PATHWAY, GENES/Sequence Similarity database [SSDB], Bio-molecular Relations in Information Transmissions and Expression [BRITE] and LIGAND which is a composite database of compounds, drug, glycan, reaction, repair and enzymes.

KEGG EDRUG Database: It is a collection of crude drugs, essential oils etc., which are mostly natural products of plants. Each KEGG EDRUG entry is identified by the E number and is linked-up with the chemical component, efficacy information, and source species information.

KEGG PLANT Resource

Plants are known to produce diverse chemical substances including those with medicinal and nutritional values. KEGG EDRUG and other KEGG resources are being organized for understanding relationships between genomics and chemical information of natural products from plants. These genes can serve as markers to further study the complex metabolic pathways of economically important compounds and to improve the quality of plants.

Perl Programming in the 'Omics' Era

Why use perl in bioinformatics?

PERL stands for Practical Extraction and Reporting Language. Perl is one of the popular Bio-programming languages. It is well suited for various Bioinformatics tasks. It is available freely and runs on any platform [windows/Linux/Unix]. It needs an interpreter file for all platforms except for Linux. It is a scripting language. The scripts and executable files are interchangeable.

Usually Perl is run from command line. Once Perl program is installed in the system, the next step is to type the program in the Text editor. A text editor is used to type documents, such as programs and to save those contents into files. *Vi* and *emacs* are the commonly used editors. There are different Perl scripts used to retrieve data from databases.

The advantages of using Perl programming are:

- Ease of programming
- High portability, speed and program maintenance
- Flexible [Can do same things in many ways]
- Enormous quantity of code available

Types of analysis that can be carried out:

- Determine frequency of nucleotides
- Find DNA mutation
- Codon Bias Data
- To Retrieve and store DNA sequences

The following is a simple Perl program to print a set of DNA sequences.

A sample Perl Script:

```
#!/usr/bin/perl
print"Content-type:text/html\n\n";
print"<html><head><title>My First Sequence </title></head>\n\n";
print"<body>\n";
print"<p>ATGCATGCATGCATGC!</p>\n";
print"</body></html>";
```

Conclusion

The field of bioinformatics thus facilitates the analysis of genomics and post genomics data. It also integrates information from the related fields of transcriptomics, proteomics, metabolomics and phenomics. The 'omics' field in science altogether enables the identification of genes and genome products and can illuminate the functional importance of each gene present in the plant genome. The scope of this multidisciplinary field is vast and the availability of ever increasing data persuades researchers to develop new tools for analysis.

References

1. Brent, (2005) Genome annotation past, present, and future: How to define an ORF at each locus. Genome Research, 15(12):1777–86.
2. Christopher A cullis., Plant Genomics and Proteomics. John Wiley & Sons Inc. Publication. p 13.
3. David, W. Mount., (2004) Bioinformatics sequence and genome analysis., 2nd edition, p. 68.
4. Dietrich, A.J., Weil, J.H. and Marechal-Drouard, L. (1992) Nuclear-encoded transfer RNAs in plant mitochondria., Annual Review of Cell Biology, 8:115-131.
5. Folkerts, O. and Hanson, M.R. (1991) The male sterility-associated pcf gene and the normal atg9-1 gene in petunia are located on different mitochondrial DNA molecules. Genetics, 129: 885-895.
6. Goff, S.A. *et al.* (2002) A draft sequence of the rice genome (*Oryza sativa* L. ssp. *Japonica*). Science, 296: 92–100.
7. Gutirrez *et al.* (2004) Phylogenetic profiling of the Arabidopsis thaliana proteome: what proteins distinguish plants from other organisms? Genome Biology, 5(8):R53.
8. Jaiswal, P., Dharmawardhana, P. and Naithani, S. (2010) A model plant genome resource and comparative genomics. Acta. Hort. (ISHS), 859:31-41.
9. James C Estill. and Jeffrey L Bennetzen., (2009) Plant Methods, 5:8.
10. Jeff Augen (2005) Bioinformatics in the post genomics era., Addison-Wseley Publication.
11. Karl J. Niklas (1997) The Evolutionary Biology of Plants., The University of Chicago Press, p 101.
12. Lescot, M., Déhais, P., Moreau, Y., De Moor, B., Rouzé, P. and Rombauts, S. (2002) PlantCARE: a database of plant cis-acting regulatory elements and a portal to tools for *In silico* analysis of promoter sequences. Nucleic Acids Research., Database issue, 30(1): 325-327.
13. Macas, J., Meszaros, T. and Nouzova, M. (2002) PlantSat: a specialized database for plant satellite repeats. Bioinformatics, 18: 28-35.
14. McCarthy F.M., Bridges S.M., Wang N., Magee G.B., Williams W.P., Luthe D.S. and Burgess S.C. (2007) "AgBase: a unified resource for functional analysis in agriculture." Nucleic Acids Research. Database issue, D599-603.
15. Ouyang S. and Buell C.R. (2004) The TIGR Plant Repeat Databases: a collective resource for the identification of repetitive sequences in plants. Nucleic Acids Research, 32 (Database issue): D360-3.
16. Pearson., (2006) Genetics: What is a gene? Nature, 441(7092):398–401.
17. Scott Jackson., *et al.* (2006) Comparative sequencing of plant genomes: Choices to Make The Plant Cell 18:1100-1104 © 2006 American Society of Plant Biologists.

Chapter – 5

Integrating Knowledge of Bioinformatics in Medicinal Plant Research

Gurpreet Kaur, Pritika Singh and Pratap Kumar Pati

Abstract

In recent years there has been a world-wide shift in consumer choice and preferences for herbal drugs. Herbal drugs are crude preparations of various kinds of medicinal plants. It involves dried plant or any part such as leaf, stem, root, flower, or seed. According to WHO survey, about 70-80% of the world population particularly in the developing countries rely on non-conventional medicines mainly of herbal sources for their primary health care. [1] Herbal medicines are being widely accepted for their safety, efficacy, cultural acceptability, better compatibility with the human body and lesser side effects. With the legal acceptance in many countries of the world as an alternative system of medicine, the growth rate of ayurvedic and herbal industry is estimated to be more than 30% for the last 25 years. [2]

Introduction

With an ever-increasing global inclination towards the consumption of herbal medicines, there is a growing need of raw materials and to identify the active principles that should be available in optimum quantities at the requisite time. Currently, the herbal medicine sector faces numerous challenges such as narrow genetic base, identification and authentication of species, chemical characterization of active compounds and identification of the target molecules. The pharmacologically active metabolites are produced in very low amounts by the plants and their chemical synthesis is expensive, hence there is a growing pressure of procuring these active metabolites from the wild leading to diminishing the populations, loss of genetic diversity and local extinctions. Domestic cultivation of medicinal plants offers an attractive alternative but not much information is available regarding the cultivation practices for many plants and also, there are issues related to storage practices, quality, safety and stability assessment that need to be addressed. Futher, the study of secondary metabolites and their enhancement is a bottleneck due to lack of knowledge about characterized biosynthetic pathways and enzymes in many plants and how their synthesis is regulated at molecular level. To address these challenges, new molecular tools and biotechnological approaches are much warranted. However, the rapidly changing pace of technology and development of high-throughput research, the field of plant biotechnology is beginning to suffer from data overload. All biotechnological efforts have involved empirical, labor-intensive and time-consuming methods. Bioinformatics is one such approach in which biology and information technology converge. It is an interdisciplinary scientific tool that facilitates both the analysis and integration of information from 'omics' technologies including genomics, transcriptomics, proteomics, metabolomics and phenomics. The present chapter highlights some of the important areas in bioinformatics which play a significant role and could also have immense impact in medicinal plants research.

Online Bioinformatics Resources on Medicinal Plants

Activities in the field of medicinal plants have been enhanced significantly during the past couple of decades and a huge volume of data is being generated out of these works. About 85,000 plant species world-wide are reported to have medicinal properties.[3] Therefore, for maintaining the records in a consolidated form, there is need to store and manage all information. In accordance with this, various databases on medicinal plants have been developed that facilitate a wide variety of information retrieval including variation in number of vernacular names, geographical distribution, family, part of the plant investigated, propagation, agro-technique, seed storage, chemical profile (active constituents), mode of action, model organism (human, dog, rat etc. and their quantity) on which the clinical /experimental studies have been done, digitalization and bibliographic information. Various medicinal plants resources are available online (Table 5.1). Linking with these resources will help in data mining that will facilitate the exploration of questions that, at present, cannot readily be answered. Apart from these, specific online resources for the model plants and a sequenced crop can be used as reference in medicinal plants study (Table 5.2). Besides some of the model plants, there is lack of

Table 5.1: Online bioinformatics resources based on medicinal plants

S.No.	Resources	Description	URL
1.	BIAdb	Database for Benzyl Isoquin oline Alkaloids	http://crdd.osdd.net/raghava/ biadb/.
2.	CITES Plants	Convention on International Trade in Endangered Species of Wild Fauna and Flora	http://www.cites.org/
3.	CMKb	Customary Medicinal Knowledgebase stores information related to taxonomy, phytochemistry, biogeography, biological activities of customary medicinal plant species as well as images of individual species.	http://biolinfo.org/cmkb/
4.	Database on Antidiabetic Plants	Database for anti-diabetic plants with clinical/experimental trials	http://www.biotechpark.org.in/antidia/index.html
5.	DiaMedBase	Diabetes literature database of medicinal plants with link to other diseases for each medicinal plant	http://www.progenebio.in/DMP/DMP.htm
6.	DNP	Dictionary of Natural Products; contains chemical, physical, bibliographic and structural data of natural products	http://dnp.chemnetbase.com/dictionary-search.do?method =view&id=241895&si=
7.	Encyclopedia on Indian Medicinal plant	Comprehensive database on Indian medicinal plants	http://envis.frlht.org.in/indian-medicinal-plants-database.php
8.	Garden Info	Garden Information Manage/home.asp	http://tbgri.in/gardeninfo/Pinfo/
9.	IMPPDS	Indian Medicinal Plants Protein Dataset containing 181 protein models from 18 different Indian medicinal plants	http://mmppdb.googlepages.com/index.htm
10.	MAPA	Medicinal & Aromatic Plant Abstract; an abstracting journal on medicinal andaromatic plantsmapaintro.asp	http://www.niscair.res.in/sciencecommunication/AbstractingJournals/
11.	MAPPA	Medicinal and Aromatic Plants Program in Asia	http://www.mappa-asia.org/index.php?id=1

S.No.	Resources	Description	URL
12	Medicinal Plants Database	Database on medicinal plants developed by SRISTI (society for research and initiatives for sustainable technologies and institutions)	http://www.sristi.org/wsa/plantdb/index.php
13	Phyto-mellitus	Phyto-chemical database of medicinal plants for diabetes	http://www.bicmlacw.org/bt/1.htm
14.	Plants Databases	Developed by BGCI; an international organisation that ensure the world-wide conservation of threatened plants	http://www.bgci.org/resources/database_links/
15.	Plant Info	Centralised database on plant diversity of India particularly of Kerala state.	http://tbgri.in/plantinfo/plant01042003_Local/index.asp
16.	PRELUDE database	Prelude Medicinal Plants Database concerns the use of plants in different traditional veterinarian and human medicines in Africa.	http://www.metafro.be/prelude
17.	Science Reference Section Webliography	Selected internet resources on medicinal plants	http://www.loc.gov/rr/scitech/selected-internet/herbalmedicine.html
18.	TCMGeneDIT	Database of various association information about Traditional Chinese Medicines (TCMs), genes, diseases, TCM effects and TCM ingredients obtained from vast amount of biomedical literature	http://tcm.lifescience.ntu.edu.tw/

Table 5.2: Various bioinformatics resources for the model plants and a sequenced crop

| S.No. | Omics data | Rice | | |
		Resources		
		Arabidopsis	*Rice*	*Soybean*
1.	Genome sequence, gene annotaion	TAIR	RAP-Db, TIGR/MSU rice	SoyBase, Phytozome
2.	Molecular markers, variation data	TAIR, Nordborg lab	Gramene, Oryza SNP	Soymap
3.	Transcription factor data	RARTF, AGRIS, DATF	DRTF, GRASSIUS	SoybeanTFDB, LegumeTFDB
4.	Non-coding RNA	Arabidopsis MPSS	Rice MPSS	NA
5.	Microarray	AiGenExpress, Genevestigator	RICEATLAS, Genevestigator	Genevestigator, SGMD, SoyExpress
6.	Full-length cDNA clones, ESTs	RAFL clones, RARGE	KOME	Soybean full-length cDNA database
7.	Proteome profile	PPDB, PhoshAt	Rice proteome database	Soybean proteome database
8.	Subcellular localization	SUBA, NASC proteome database	Rice proteome database	Soybean proteome database
9.	Metabolic map	AraCyc	RiceCyc	PlantCyc, MedicCyc
10.	Natural variation	NASC, ABRC	Oryzabase, IRRI	Legume base
11.	Mutant lines	FOX line, Ac/Ds tag line, T-DNA line, TILLING	Ac/Ds tag line, TILLING, Tos 17 mutant panel	Ac/Ds tag line, TILLING
12.	Integrated database	TAIR	Gramene	SoyBase

information on proteins encoded by other plant genomes, but a huge amount of information is available on ESTs. An integrated sequence repository known as Sequence Platform for Phylogenetic analysis of Plant Genes (SPPG; http://bioinformatics.psb.ugent.be/cgi-bin/SPPG/index.htpl) has been created that combines EST data with protein information.[4] To have an efficient database system, various tools and softwares are in use which include LAMP (Linux-Apache-MySQL-PHP), an open source software, used to create many databases. PHP and HTML technologies have been used to build the dynamic web interface. MySQL is a relational database management system (RDBMS) that works at the backend. PHP is used for server-side scripting. The whole software system runs on Linux environment using Apache http server. PHP and MySQL combination is quite competent and powerful for database management. In some cases, MySQL, Perl and PHP programming languages in the Linux environment has also been implemented. Some databases have been constructed using HTML and JavaScript (for the development of database front end) on MS-Windows Operating System.

Phylogenetic Analysis

Phylogenetic analysis is the means of inferring the evolutionary relationships. The evolutionary history inferred, is depicted as tree-like diagram that represents an estimated pedigree of the inherited relationships among organisms. The evolutionary events (substitutions, insertions, deletions and re-arrangements) that are important to the history of a gene can also be used to resolve the problems about the evolutionary history and relationship between entire species. Different approaches are available for inferring the most likely phylogenetic relationship between genes and species using nucleotide and protein sequence information. Phylogenetic analysis helps to investigate the diversity of plant material within the species and between the different species. For example, all *Zingiber officinale* samples from different geographical origins were genetically indistinguishable, but other *Zingiber* species were significantly divergent, allowing all species to be clearly distinguished using this analysis.[5] A phylogenetic analysis of Huperziaceae family was also conducted as members of this family have reported to produce alkaloids, particularly huperzine A which play important role in neurodegenerative diseases. Sequence data from the chloroplast genes *rbc*L and *psb*A-*trn*H intergenic spacer was taken to construct phylogenetic framework for this family. In case of *rbc*L sequence analysis, 25 species were sampled and they were resolved into two major clades; one corresponded with sections *Huperzia* and *Serratae* of the genus *Huperzia*, while the other contained *Phlegmariurus* and the tropical species of *Huperzia*. Therefore, *Phlegmariurus* is derived from *Huperzia*. In the *psb*A-*trn*H sequence analysis, 17 species were analyzed and they were resolved into two major clades. Both Chinese species of *Huperzia* and

Phlegmariurus formed sister groups, and within Chinese *Huperzia*, sect. *Huperzia* and sect. *Serratae* formed sister groups. This study provided the insights into the taxonomic relationships of species in this family. [6] Also, a phylogenetic profile provides a list of species in which a particular protein is expressed. Proteins with the same profile are likely to be in a pathway that occurs only in certain organisms. Various packages for phylogenetic analysis are available at ExPASy Proteomics Server (http://www.expasy.ch/tools/#proteome) such as MEGA, PHYLIP, PAUP, EMBOSS etc.

Structural Genomics

Sequence Information: Genome is the elementary entity in bioinformatics studies; and its sequencing and analysis encompass the use of various bioinformatics methods to determine exact order of the bases in a strand of DNA, the proteins they code for and their biological role. Numerous software packages are available for managing, processing and analyzing the sequences (Table 5.3). Tracing the evolutionary history of uncharacterized genes, characterized genes and gene linkages among medicinal plants, are the most interesting and challenging aspects of genome analysis. The sequence information provides a great flexibility in terms of modulations in medicinal plants for genetic improvement. Also, the availability of genome sequences provides a unique opportunity to explore genetic variability both between different species and within the individual. Further, sequence comparison allows inference of function, structure and evolution of genes and genomes. Various methods and tools are employed in sequence deposition, retrieval and comparison (Table 5.4).

Next Generation Sequencing

First commercial launch of pyrosequencing platform in 2005 had a substantial impact on genomics analysis in terms of scale and feasibility, now referred to as Next-Generation Sequencing. Next Generation Sequencing (NGS)-based technologies are transforming biology by enabling individual researchers to sequence the genome of individual organisms or cells on a massive scale. Strategies such as Roche 454 FLX, Illumina GAIIx, Applied Biosystems SOLiD, Helicos HeliScope etc. are well suited to provide data required for genomics, epigenetics, transcription factor binding and transcriptomics. The diversity and rapid evolution of these technologies are posing challenges for bioinformatics in areas including sequence quality, scoring, alignment, assembly and data release. Because of introduction of NGS, a huge amount of data is generated at a very rapid pace. There should be a commensurate development in bioinformatics tools for analyzing these data.[7] Some of the tools which are available in public domain are presented in Table 5.5.

Table 5.3: Online tools for plant genome assembly

S.No.	Tools	Description	URL
1.	AMOS	Open-source whole genome assembler	http://www.jcvi.org/cms/research/software/
2.	Arachne	Tool for assembling genome sequences from whole genome shotgun reads	http://www.broadinstitute.org/science/software
3.	CAP3	DNA sequence assembly program	http://pbil.univ-lyon1.fr/cap3.php
4.	EGassembler	A web server that provides an automated as well as user-customized analysis tools for cleaning, repeat masking, vector trimming, organelle masking, clustering and assembling of ESTs and genomic fragments.	http://egassembler.hgc.jp/
5.	GAP4	Genome Assembly Program.	http://www.molgen.mpg.de/~service/scisoft/staden/gap4_unix_2.html
6.	Lucy	Tool for cleaning data produced by automated Sanger DNA sequencers prior to sequence assembly	http://www.jcvi.org/cms/research/software/
7.	Phred/ Phrap / Consed	(i) Phred reads DNA sequencing trace files, calls bases, and assigns a quality value to each called base.(ii) Phrap assembles shotgun DNA sequence data. (iii) Consed is used for viewing, editing, and finishing sequence assemblies created with phrap	http://www.phrap.org/phredphrapconsed.html
8.	SeqClean	For automated trimming and validation of ESTs or other DNA sequences by screening for various contaminants, low quality and low-complexity sequences	http://compbio.dfci.harvard.edu/tgi/software/

Table 5.4: Various bioinformatics sequence deposition, retrieval and comparison tools

S. No.	Tools	Description	URL
	Sequence deposition		
1.	BankIt	Web submission tool for GenBank developed by NCBI	http://www.ncbi.nlm.nih.gov/WebSub/?tool=genbank
2.	SAKURA	Nucleotide sequence data submission system at DDBJ	http://sakura.ddbj.nig.ac.jp/top-e.html
3.	Sequin	Stand-alone software tool for submitting and updating entries to GenBank, DDBJ and EMBL	http://www.ncbi.nlm.nih.gov/Sequin/index.html,http://www.ebi.ac.uk/Sequin/gettingstarted.html,http://www.ddbj.nig.ac.jp/sub/sequin-e.html
4.	Webin	Web-based submission system developed by EMBL for nucleotide sequences and biological annotation information	http://www.ebi.ac.uk/embl/Submission/index.html
	Sequence retrieval		
1.	Entrez	Text-based search and retrieval system used at NCBI	http://www.ncbi.nlm.nih.gov/sites/gquery
2.	SRS	Sequence Retrieval System at EBI, is a powerful searching tool to retrieve sequences and other types of data.	http://srs.ebi.ac.uk/
	Sequence comparison		
1.	FASTA	Pair-wise sequence comparison	http://www.ebi.ac.uk/Tools/fasta/index.html
	BLAST		http://blast.ncbi.nlm.nih.gov/Blast.cgi
2.	ClustalW	Multiple sequence alignment and phylogenetic analysis	http://www.ebi.ac.uk/Tools/clustalw2/index.html
	PAUP		http://paup.csit.fsu.edu/
	PHYLIP		http://evolution.genetics.washington.edu/phylip.html
3.	PSI-BLAST	Sequence-profile comparison	http://blast.ncbi.nlm.nih.gov/Blast.cgi
	HMMER		http://hmmer.janelia.org/
	SAM		http://compbio.soe.ucsc.edu/sam.html
	META-MEME		http://metameme.sdsc.edu/
4.	FORTE	Profile-profile comparison	http://www.cbrc.jp/forte/

Table 5.5: Bioinformatics tools for analyzing next-generation sequencing data

S.No.	Categories	Program	URL
1.	Alignment	Cross_match	http://www.phrap.org/phredphrapconsed.html
		ELAND	http://www.illumina.com/
		Exonerate	http://www.ebi.ac.uk/~guy/exonerate
		Mosaik	http://bioinformatics.bc.edu/marthlab/Mosaik
		RMAP	http://rulai.cshl.edu/rmap
		SHRiMP	http://compbio.cs.toronto.edu/shrimp
		SOAP	http://soap.genomics.org.cn
		SSAHA2	http://www.sanger.ac.uk/Software/analysis/SSAHA2
		SXOligoSearch	http://synasite.mgrc.com.my:8080/sxog/NewSXOligoSearch.php
2.	Alignment and variant detection	MAQ	http://maq.sourceforge.net
3.	Assembly	Edena	http://www.genomic.ch/edena
		SHARCGS	http://sharcgs.molgen.mpg.de
		SSAKE	http://www.bcgsc.ca/platform/bioinfo/software/ssake
		VCAKE	http://sourceforge.net/projects/vcake
		Velvet	http://www.ebi.ac.uk/%7Ezerbino/velvet
4.	Base caller	PyroBayes	http://bioinformatics.bc.edu/marthlab/PyroBayes
5.	Variant detection	PbShort	http://bioinformatics.bc.edu/marthlab/PbShort
		ssahaSNP	http://www.sanger.ac.uk/Software/analysis/ssahaSNP

DNA Barcoding

It is a technique for rapid species identification based on DNA sequences. Several molecular tools have been used to detect and identify the biological samples, but they cannot be used for routine analysis. The Barcode of Life Data Systems (BOLD; http://www.boldsystems.org/) is an integrated bioinformatics platform that supports all phases of the analytical pathway starting from specimen collection to tightly validated barcode library[8]. It consists of 3 components:

- BOLD-MAS (Management And Analysis) is a repository for barcode records coupled with analytical tools.

- BOLD-IDS (Identification Engine) provides a species identification tool that accepts DNA sequences from the barcode region and returns a taxonomic assignment to the species level.

- BOLD-ECS (External Connectivity) helps the web developers and bioinformaticians to build tools and workflows that can be integrated with the BOLD framework

Genome Mapping

It refers to mapping of a specific gene to particular region of a chromosome; and determining the location and relative distances between genes on the chromosome.

Types of maps

There are two types of maps

Genetic linkage map: It shows the arrangement of genes and genetic markers along the chromosomes as calculated by the frequency with which they are inherited together using linkage analysis. Various DNA markers used for genetic mapping are restriction fragment length polymorphisms (RFLPs), simple sequence length polymorphisms (SSLPs) and single nucleotide polymorphisms (SNPs). JoinMap 4 is the software for the calculation of genetic linkage maps in experimental populations (http://www.kyazma.nl/index.php/mc.JoinMap/).

Physical map: It represents the chromosomes, providing the physical distance between landmarks on the chromosome, ideally measured in nucleotide bases. Physical mapping techniques include Restriction mapping, Fluorescent *in situ* hybridization (FISH) and Sequence Tagged Sites (ESTs, SSLPs and random genomic sequences) mapping. The physical maps can be divided into three general types: chromosomal or cytogenetic maps, radiation hybrid (RH) maps and sequence maps. The ultimate physical map is the complete sequence itself.

For example, herb epimedii is prepared from the medicinal *Epimedium* species and is widely used to treat diseases such as coronary heart disease, chronic bronchitis, impotence and neurasthenia. A large set of Expressed Sequence Tags (ESTs) and Simple Sequence Repeats (SSRs) identified in these ESTs, were generated for *Epimedium sagittatum* using various bioinformatics platforms. EST processing, assembly and annotation were done using Lucy, SeqClean and CAP3 softwares, Pfam database and GOSlim program. EST-SSR detection was performed using Perl program MISA; and genetic diversity and average allele number were calculated with Arlequin software package. MICROSAT program was used to calculate genetic distance of SSRs genotype while polymorphic information content (PIC) was computed with PIC_CALC and GenAlex6. Phylogenetic analysis among different *Epimedium* species was done with PHYLIP package. Generated *Epimedium* EST dataset is a valuable resource of sequence information for deciphering secondary metabolism, especially for flavonoid pathway in *Epimedium* species.[9]

Quantitative Trait Loci (QTL)

Plants respond to both biotic and abiotic stress conditions through a common signaling system to provide defense against many adverse environments. Several genes/QTLs governing resistance/tolerance to abiotic and biotic stresses have been studied and mapped in various plants. QTL mapping is used to determine the loci responsible for variation in complex, quantitative traits. If the genes underlying the QTL are known, then transgenic approaches can also be used to directly introduce beneficial alleles across various species. Most QTL mapping studies in plants have used designed mapping populations, such as F_2 or backcross populations between two inbreds, molecular markers and the tediousness of their genotyping. The most obvious genomic resource for QTL mapping is a complete genome sequence; but mapped expressed sequence tags (ESTs) can be used when no complete sequence is available. *In silico* mapping via a mixed-model approach or mixed-model QTL mapping aims to exploit existing phenotypic, genotypic and pedigree data in phenotypic and genomic databases to discover genes in hybrid crops. *In silico* mapping has following advantages over designed mapping experiments:

- exploits larger populations than designed mapping experiments.

- experimental hybrids or inbreds are evaluated in multiple, diverse environments.

- hybrids and inbreds tested typically comprise a wide sample of the germplasm and genetic backgrounds.

- detect associations repeatable across different populations.

- data used are available at no extra cost.

The mixed-model approach leads to best linear unbiased predictions (BLUP) of random genetic effects and the best linear unbiased estimates (BLUE) of fixed environmental effects.[10,11] SOL Genomics Network (SGN; http://sgn.cornell.edu) is a bioinformatics platform for Solanaceae family and related families in the Asterid clade. It is mainly focused on map and marker data for members of Solanaceae such as tuber-bearing potato, a number of fruit-bearing vegetables, ornamental plants, plants with edible leaves and medicinal plants (*Datura, Capsicum*), their EST collection with computationally derived unigene sets, an extensive database of phenotypic information for a mutagenized tomato population and associated tools such as real-time quantitative trait loci.[12] The Generation Challenge Programme (GCP; www.generationcp.org), a global crop research consortium has developed an online resource (Dayhoff; http://dayhoff.generationcp.org/) which describes relationship between genes that may be involved in response to environmental stresses across various species. RiceGeneThresher is an online tool for mining of genes involved in controlling traits underlying quantitative trait loci (QTL) in rice genome (http://rice.kps.ku.ac.th). A list of softwares for genetic linkage analysis, QTL analysis for plant breeding data, genetic marker ordering and genetic association analysis is available at http://linkage.rockefeller.edu/soft/.

Proteomics

Proteomics is another important tool for plant biotechnology, considering the fact that cellular behaviour is more directly influenced by proteins rather than by mRNAs. This technology is based on high-throughput techniques for the separation and identification of proteins, allowing an integral study of many proteins at the same time. Two-dimensional polyacrylamide gel electrophoresis is the most powerful technique available for the separation of protein mixtures; after separation, proteins can be identified by mass spectrometry (MS), and the related bioinformatics data analysis tools allow us to interpret the data and get meaningful information. Medicinal plants proteomics is involved in the mapping of complete proteomes and comparison of proteomes which lead to the identification of differentially expressed proteins as well as elucidation of biosynthetic pathways leading to secondary metabolites. Various proteins contributing to cascades of reactions leading to these metabolites are enzymes, transport and regulatory proteins. Also, multi-enzyme complexes in plant secondary metabolism are of particular interest, e.g. in flavonoid biosynthesis. Various web-based plant proteome-related databases are available (MASCP; http://www.masc-proteomics.org/mascp/index.php/Main_Page and ExPASy server; http://www.expasy.ch/tools/). Further, it also deals with databases of protein sequences, predicted protein structures and protein expression analysis (Table 5.6). One of the most extensively studied medicinal plants, *C. roseus* (*Madagascar periwinkle*) produces two dimeric terpenoid indole alkaloids (TIAs), vinblastine and vincristine, which were the first natural drugs used in cancer therapy and are still among the most valuable agents used in the treatment of cancer. The dimerization reaction leading to α-3',4'-anhydrovinblastine in the biosynthesis of vinblastine is a key regulatory step for the production of anticancer alkaloids, involving an enzyme with anhydroblastine synthase activity, known as CrPrx1. It is a polypeptide of 363 amino acids with an N-terminal signal peptide and has been characterized using *in silico* approach (Fig. 5.1)[13].

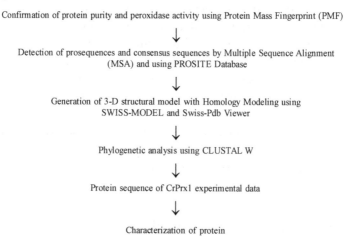

Fig. 5.1: Outline of *in silico* approach for characterization of vacuolar class III peroxidase isoenzyme (CrPrx1) having anhydrovinblastine synthase activity from *Catharanthus roseus*.

Various approaches such as isotope-coded affinity tags (ICAT) and isobaric tags for relative and absolute quantitation (iTRAQ) are also available to analyze the proteome of subcellular compartments of plant cells such as amyloplasts, chloroplasts, etioplasts, chromoplasts, mitochondria, vacuoles, plasma membrane, nucleus, peroxisomes and cell wall. Databases for subcellular proteome information include sub-cellular location database for Arabidopsis (SUBA; http://suba.plantenergy.uwa.edu.au/), Rice proteome database (http://gene64.dna.affrc.go.jp/RPD/) and Soybean proteome database (http://proteome.dc.affrc.go.jp/Soybean/). In addition to above, various types of post-translational modifications (phosphorylation, ubiquitination), modified proteins and their related functions are also studied. The Plant Protein Phosphorylation Database (P3DB; http://digbio.missouri.edu/p3db/) is a resource for plant phosphoproteomes from multiple plants and the Arabidopsis Protein

Table 5.6: Online tools and databases for proteomics study

Type of study		Tools	Description	URL
Protein identification	Electrophoresis analysis	SWISS-2DPAGE	Tool for locating the protein on 2D PAGE maps	http://expasy.org/ch2d/
		Melanie	Tool for analyzing, annotating and querying complex 2D gel samples	http://www.expasy.ch/melanie/2DImageAnalysisViewer.html
		Flicker	Stand-alone program for visually comparing 2D gel images	http://open2dprot.sourceforge.net/Flicker/
Protein identification through mass spectrometry		MS-Fit	Tool for Peptide Mass Fingerprinting (PMF)	http://prospector.ucsf.edu/prospector/cgi-bin/msform.cgi?form=msfitstandard
		SEQUEST	Tool for MS/MS-based peptide/protein identification	http://fields.scripps.edu/sequest/
		Mascot	Peptide Mass Fingerprinting (PMF) and MS/MS-based peptide/protein identification	http://www.matrixscience.com
		Lutefisk	Software for peptide de novo sequencing based on MS/MS spectra	http://www.hairyfatguy.com/lutefisk/
		PEAKS	Software for peptide de novo sequencing based on MS/MS spectra	http://www. bioinformaticssolutions.com/products/peaks/
Protein databases	Sequence-based	Pfam	Collection of protein families, represented by multiple sequence alignments and HMMs	http://pfam.janelia.org/
		ProDom	Comprehensive set of protein domain families generated from the global comparisonof all available protein sequences	*http://prodom.prabi.fr/*

(Contd.)

Type of study	Tools	Description	URL
	COG	Comparison of protein sequences encoded in complete genomes, representing major phylogenetic lineages	*www.ncbi.nlm.nih.gov/COG/*
	PROSITE	Protein domains, families and functional sites	http://www.expasy.ch/prosite/
	PRINTS	Protein fingerprints	http://www.bioinf. manchester.ac.uk/dbbrowser/ PRINTS/index.php
	BLOCKS	Contains multiple alignments of conserved regions in protein families	http://blocks.fhcrc.org/
	InterPro	Integrative database of predictive protein signatures used for the classification and automatic annotation of proteins and genomes	http://www.ebi.ac.uk/interpro/
	PANTHER	Database of protein families, subfamilies, functions and pathways	http://www.pantherdb.org/
Structure -based	SCOP	Manual classification of protein structural domains	http://scop.mrc-lmb.cam.ac.uk/ scop/

Phosphoylation site Database (PhosPhAt; http://phosphat.mpimp-golm.mpg.de) for Arabidopsis. Visualization of protein structure is important for evolutionary analysis as well as for possible role in an organism. Various computational methods such as homology modeling, *ab initio* prediction and fold recognition (threading) are used for protein structure prediction. A range of softwares are available online for visualization of these structures (Table 5.6).

Functional Genomics

It involves the development of global experimental approaches for analyzing gene function by utilizing information provided by structural genomics. It uses high-throuput techniques such as gene discovery approaches, regulatory sequence analysis, mutation analysis and microarrays. Because of the large quantity of data and the desire to be able to predict gene functions and interactions, bioinformatics is crucial to this type of analysis.

Gene Discovery

Gene discovery is one of the first and most important steps in understanding the genome of a species once it has been sequenced. It refers to identifying stretches of sequences (genes) in genomic DNA that are biologically functional like protein-coding genes, other functional elements such as RNA genes and regulatory regions. Mainly three computational approaches are being used to find genes:

• Homology-based Approaches

These are based on the similarity of sequences. Given a library of sequences of other organisms, target sequence is searched in this library and identify library sequences (known genes) that resemble the target sequence. Also, the target sequence is compared with expressed sequence tags (ESTs) of the same organism to identify regions corresponding to processed mRNA. If the identified sequences are genes, the target sequence is probably (putatively) a gene. Various homology-based gene prediction tools are available online (Table 5.7). However, some of them are unable to identify genes that code for proteins not already in the library. BLAST (Basic Local Alignment Search Tool, http:// blast.ncbi.nlm.nih.gov/Blast.cgi) is a well-known search tool in this category.

Table 5.7: Homology-based gene prediction tools

S.No.	Tools	Description	Plants	URL
1.	GeneSeqer@ PlantGDB	Gene identification tool based on spliced alignment or "spliced threading" of ESTs with a genomic query sequence	*Medicago, Arabidopsis,* Maize, Rice	http://www. plantgdb. org/cgi-bin/GeneSeqer /index.cgi
2.	Spidey	mRNA-to-genomic alignment program.	All plants	http://www. ncbi. nlm. nih.gov/IEB/Research/ Ostell/Spidey/ index.html

• *Ab Initio* Approaches

They search for certain signals of protein coding genes. These are more difficult for eukaryotes, as genes are separated by large intergenic regions and the signals (e.g., promoters, start codons, stop codons, splice sites, CpG islands and binding sites for a Poly-A tail) are more difficult to identify, since these signals are more complex and unspecified. Various bioinformatics approaches such as Hidden Markov Model (HMM), Neural Networks (NN), Dynamic Programming (DP) and Maximal Dependence Decomposition (MDD) are widely used for finding genes in various plant species (Table 5.8). A novel approach has been developed for Coding Potential Prediction (CPP) in short sequences (<200nt) for plants employing complementary sequence features and hybrid Markov-SVM (Support Vector Machine) model. [14]

• Comparative Genomics Approaches

They are involved in the analysis and comparison of genomes from different species. Whole-genome comparisons identifying chromosomal duplication and conserved synteny among related species provide evidence for evolutionary relationships. Various bioinformatics resources for comparative plant genomics are available online (Table 5.9). However, an automated bioinformatics pipeline for generation of legume Comparative Anchor Tagged Sequence (CATS) markers that combines multi-species ESTs and genome sequence data has been established. This pipeline has been applied on grass family as well as tested on peanut (*Arachis hypogea*). As it allows the large-scale generation of marker candidates useful for construction and comparison in legumes and grasses; and dependent on EST data, it can be extended to medicinal plants study. A web server GeMprospector (http://cgi-www.daimi.au.dk/cgi-chili/ GeneticMarkers/main) has been designed that allows users to submit other EST collections for comparison purposes. [15]

Table 5.8: *Ab initio* gene prediction tools

S.No.	Tools	Description	Gene model	URL
1.	EuGène	Gene finding software	DP	http://eugene.toulouse.inra.fr/
2.	FGENESH	Gene structure prediction program	HMM	http://mendel.cs.rhul.ac.uk/mendel.php?topic=fgenfile
3.	GeneId3	Gene identification and gene structure prediction program	DP	http://genome.crg.es/geneid.html
4.	GeneSplicer	Web-based detection of splice sites	HMM, MDD	http://www.cbcb.umd.edu/software/GeneSplicer/gene_spl.shtml
5.	GeneMark.hmm	Gene prediction program	HMM, DP	http://opal.biology.gatech.edu/GeneMark/eukhmm.cgi
6.	GenScan	Program for prediction of locations and exon-intron structures of genes in genomic sequences	HMM, DP	http://genes.mit.edu/GENSCAN.html
7.	GlimmerHMM	Eukaryotic gene finder software	HMM	http://www.cbcb.umd.edu/software/GlimmerHMM/
8.	NetPlantGene	Server for prediction of splice sites	NN	http://www.cbs.dtu.dk/services/NetPGene/
9.	NetGene2	Server for prediction of splice sites	NN, HMM	http://www.cbs.dtu.dk/services/NetGene2/

Table 5.9: Bioinformatics resources for integrative and comparative plant study

S.No.	Databases	Species/Family	Description	URL
1.	AgBase	Various plant species	Web-accessible resource for functional analysis of agricultural plants	http://agbase.msstate.edu/cgi-bin/team.pl
2.	Brassica Genome Gateway	Brassica	Information on Brassica species	http://brassica.bbsrc.ac.uk/
3.	ChloroplastDB	Various plant species	Chloroplast genome data	http://chloroplast.cbio.psu.edu/index.html
4.	CuGenDB	Cucurbitaceae	Cucurbit Genomics Database	http://www.icugi.org/
5.	Ensembl Plants	Various plant species	Whole genome data	http://plants.ensembl.org/index.html
6.	GabiPD	Various plant species	Genome Analysis of the Plant Biological System (GABI) Primary Database is a repository and analysis platform for data generated from high-throughput experiments in several plant species.	http://www.gabipd.org/
7.	GDR	Rosaceae	Genome Database for Rosaceae providing access to genomics and genetics data; and analysis tools	http://www.rosaceae.org/
8.	GPI	Various plant species	GénoPlante-Info is a collection of databases and bioinformatics programs for plant genomics.	http://genoplante-info.infobiogen.fr/.
9.	GrainGenes	Triticeae and Avena	Database providing information about genetic maps, mapping probes and primers, genes, alleles and QTLs	http://wheat.pw.usda.gov/GG2/index.shtml
10.	Gramene	Gramineae	A resource for comparative grass genomics	http://www.gramene.org/
11.	GRIN	Various plant species	Germplasm Resources Information Network is a web server provides germplasm information about plants, animals, microbes and invertebrates.	http://www.ars-grin.gov/index.html

(Contd.)

S.No.	Databases	Species/Family	Description	URL
12.	KEGG PLANT	Various plant species	Whole genome data and EST data	http://www.genome.jp/kegg/plant/
13.	MaizeGDB	Maize	Maize Genetics and Genomics Database	http://www.maizegdb.org/
14.	MBGP	Brassica	Multinational *Brassica* Genome Project	http://www.brassica.info/index.php
15.	Phytozome	Various plant species	Whole genome data	http://www.phytozome.net/
16.	PlantGDB	More than 24,000 plant species	Plant Genome Database of plant molecular sequences such as EST, GSS, promoter regions; and analysis tools	http://www.plantgdb.org/
17.	Plant Genomes-NCBI	Various plant species	Collection of plant-specific genomic data and resources	http://www.ncbi.nlm.nih.gov/genomes/PLANTS/PlantList.html
18.	PlantMarkers	Various plant species	Genetic marker database that contains a comprehensive pool of predicted molecular markers from plants such as SNP, SSR, conserved orthologs set markers	http://markers.btk.fi/
19.	PlantSat	Various plant species	Database for plant satellite repeats	http://w3lamc.umbr.cas.cz/PlantSat/
20.	PMRD	Various plant species	Plant micro RNA Database	http://bioinformatics.cau.edu.cn/PMRD/
21.	RAP-DB	Rice	Rice Annotation Project Database	http://rapdb.dna.affrc.go.jp/

Regulatory Sequence Analysis

Regulatory Sequence is a segment of DNA where regulatory proteins such as transcription factors bind. These regulatory proteins bind to short stretches of DNA called regulatory regions, which are positioned in the genome, usually a short distance 'upstream' of the gene being regulated. Hence, these regulatory proteins can recruit another protein complex, called the RNA polymerase. In this way, they control gene expression and thus protein expression. The main examples of regulatory sequence are promoters and transcription Factors (TFs). In *Catharanthus roseus,* a part of terpenoid indole alkaloid biosynthesis is under the control of ORCA3 (a jasmonate-responsive APETALA2 (AP2)-domain transcription factor), whose constitutive overproduction in cell cultures has resulted in enhanced expression of several biosynthetic genes as well as enhanced accumulation of terpenoid indole alkaloids [16]. LegumeTFDB is a database that provides access to transcription factor (TF) repertoires of three major legume species: *Glycine max, Lotus japonicus* and *Medicago truncatula* (http://legumetfdb.psc.riken.jp/)[17]. Various tools and databases are available at http://bip.weizmann.ac.il/toolbox/seq_analysis/promoters.html for promoter prediction and analysis. PlantProm DB (a plant promoter database) is a collection of proximal promoter sequences for RNA polymerase II with experimentally determined transcription start sites (TSSs) from various plant species (http://linux1.softberry.com/berry.phtml? topic=plantprom & group=data&subgroup=plantprom). Another plant promoter database is ppdb, which contains information on promoter structures, transcription start sites that have been identified from full-length cDNA clones and also a large amount of TSS tag data. This database helps to identify the core promoter structure, the presence of regulatory elements and the distribution of TSS clusters. It is based on species-specific sets of promoter elements, rather than on general motifs for multiple species (http://ppdb.gene.nagoya-u.ac.jp/cgi-bin/index.cgi#REG). Eukaryotic Promoter Database (EPD) is an annotated non-redundant collection of RNA POL II system of higher eukaryotes (multicellular plants and animals), that are experimentally defined by a transcription start site (http://epd.vital-it.ch/). PlantCARE is a database of plant cis-acting regulatory elements, enhancers and repressors; and also a source of tools for *in silico* analysis of promoter sequences (http://bioinformatics.psb.ugent.be/webtools/plantcare/html/). The web resource Regulatory Sequence Analysis Tools (RSAT; http://rsat.ulb.ac.be/rsat/) offers a collection of software tools dedicated to the prediction of regulatory sites in non-coding DNA sequences and performs various tasks such as sequence retrieval, pattern discovery, pattern matching, genome-scale pattern matching, feature-map drawing, random sequence generation and other utilities. RSA-tools-genome-scale dna-pattern - search a pattern within all upstream or downstream regions of a genome as well as whole chromosomes. AlignACE (Aligns Nucleic Acid Conserved Elements; http://atlas.med.harvard.edu/) is a program which finds sequence elements conserved in a set of DNA sequences. It works on an algorithm that scans non-coding nucleic acid sequences at high resolution for motifs which occur with non-random frequency.

Forward and Reverse Genetics

Two main genetic approaches are gene discovery; forward genetics (starts with the phenotype and identify gene sequence) and reverse genetics (starts with the gene sequence rather than a mutant phenotype).

TILLING Array

Targeting Induced Local Lesions in Genomes (TILLING) was developed as a reverse-genetic strategy which provides an allelic series of induced point mutations in genes of interest. It permits rapid and low-cost discovery of induced point mutations in populations of chemically mutagenized individuals. It is an efficient technique to discover novel genes and elucidate their structure. This technology can also be used to explore allelic variations that appears in natural variation, which is known as EcoTILLING. TILLING projects in rice, tomato and Arabidopsis have been performed (http://tilling.ucdavis.edu/index.php/Main_Page). RevGenUK at the John Innes Center provides TILLING service for *Medicago truncatula*, *Lotus japonicas* and *Brassica rapa* (http://revgenuk.jic.ac.uk/about.htm). Similar study has been done in *Arabidopsis* which lead to the *Arabidopsis* Tiling Array Express (At-TAX) resource. [18]

RNA Interference

Various methods to interrupt gene expression have been developed and applied to the functional analysis of plant genes. RNA interference (RNAi) is a method for RNA-mediated gene silencing by sequence-specific degradation of homologous mRNA triggered by double-stranded RNA (dsRNA). This gene silencing system has facilitated the improvement of specific medicinal plants for their greater exploitation to produce commercially valuable, plant-derived drugs and flavoring agents. This has also provided an alternative to block the activity of those enzymes that are not only encoded each by a multigene family but are also expressed across a number of tissues and developmental stages. For example, the activity of berberine bridge enzyme (BBE) in the California poppy (*E. californica*) has been blocked that resulted in the accumulation of (S)-reticuline, an important intermediate of metabolic pathways of isoquinoline alkaloid biosynthesis. *Papaver somniferum* has been transformed with an RNAi construct designed to reduce transcript levels of the gene encoding the morphine

114   Agriculture Bioinformatics

biosynthetic enzyme, salutaridinol 7-Oacetyltransferase (SalAT) that led to accumulation of the intermediate compounds, salutarydine and salutaridinol. [19] *M. truncatula* RNAi Database is also available online as resource for RNAi-based gene silencing (https://mtrnai.msi.umn.edu/). RNAi Web is a resource center for RNA interference technology that offers scientific and technical information on RNAi technology, guidelines and tips on siRNA design; and also contains RNAi related web resources (http://www.rnaiweb.com/). Also, the information on various databases and tools for RNAi-based technique are available at http://biophilessurf.info/rnai.html. *Arabidopsis* Genomic RNAi Knock-out Line Analysis (AGRIKOLA) project has also been started with the objective of analyzing Arabidopsis genes by RNAi-based technology (http://www.agrikola.org/index.php?o=/agrikola/ html/index). FiorcDB is a database that stores phenotypic information created by Chimeric REpresspr gene Silencing Technology (CRES-T); a novel gene silencing technology targeting transcription factor, in different plants (http://www.cres-t.org/fiore/public_db/f_help.shtml). [20]

Microarray Analysis

The field of transcriptomics provides information about both the presence and the relative abundance of RNA transcripts within the cell or organism. The application of sequence-based methods, microarrays, Serial Analysis of Gene Expression (SAGE) and Diversity Array Technology (DArT) are main approaches to current genomic data for expression profiling. Hybridization-based microarrays are dominant transcriptomic tools. It is possible to interpret microarray experiments for a single gene at a time, but most studies generate long lists of differentially regulated genes that can provide insight into the biological phenomena under investigation. Their interpretation requires the integration of prior biological knowledge, which is stored in online databases and covers various aspects of gene function and biological information. Main applications of microarray studies are: to determine the biological meaning of groups of related genes, identify common expression patterns between genes in order to deduce their biological function, capture the relationship of cellular components in order to explain biological phenomena, construct models for gene regulatory networks, visualization of large amount of biological components and their interactions, mapping of biological pathways information into inferred network and analysis of transcriptional regulation. Variety of online tools have been developed for the specific task of drawing biological meaning from microarray data as mentioned in Table 5.10. For example, microarray analysis using Dchip software, hierarchical clustering with CLUSTER and TREEVIEW software and Average Linkage Clustering suggested that *Hypericum* extract can be effective in the treatment of major depressive disorder

Integrating Knowledge of Bioinformatics in Medicinal Plant Research

Table 5.10: Online tools and databases for analysis of microarray data

Tools and Databases		Description	URL
Tools	BASE	BioArray Software Environment	http://base.thep.lu.se/
	Bioconductor	Open source and open development software project	http://www.bioconductor.org/
	CaArray	Open source microarray data management system	https://cabig.nci.nih.gov/tools/caArray
	Cluster	Program that performs different types of cluster analysis on large microarray data.	http://rana.lbl.gov/EisenSoftware.htm
	Dchip	Software for analysis and visualization of gene expression and SNP microarrays	http://biosun1.harvard.edu/complab/dchip/
	EP:NG	Expression Profiler: Next Generation is a web-based platform for microarray data analysis	http://www.ebi.ac.uk/expressionprofiler/
	GCOS	GeneChip Operating Software	http://genome.hku.hk/portal/index.php/software-for-microarray-analysis/gcos
	GeneSpring	Platform for expression analysis	http://www.chem.agilent.com/en-US/products/software/lifesciencesinformatics/pages/gp35082.aspx
	Gene Traffic	Web-based microarray data analysis and management software	http://www.flyarrays.com/index.php?option=com_content&task=view&id=18&Itemid=35
	RankProd	A bioconductor package for detecting differentially expressed genes with application in meta-analysis	http://www.bioconductor.org/packages/2.2/bioc/html/RankProd
Databases	ArrayExpress	Database of functional genomics experiments	http://www.ebi.ac.uk/microarray-as/ae/
	ExpressDB	Collection of RNA expression datasets	http://arep.med.harvard.edu/cgi-bin/ExpressDByeast/EXDStart
	GeneX	Open source Gene Expression Database	http://genex.sourceforge.net/
	GEO	Gene Expression Omnibus	http://www.ncbi.nlm.nih.gov/geo/
	MGED	Microarray Gene Expression Database	http://www.mged.org/
	SMD	Standford Microarray Database	http://smd.stanford.edu/resources/restech.shtml

associated with hypothalamus. Human orthologs for these genes have been identified using dbSNP database, Online Mendelian Inheritance in Man (OMIM) database and BLAST. These genes may be potential novel candidates for future pharmacogenetic studies. [21] Similarly, bioactivities of herbal extract of *Anoectochilus formosanus* and plumbagin, single-compound drug isolated from *Plumbago rosea*, was characterized with metabolite profiling and DNA microarray analysis. The extract is a crude phytocompound mixture used in the treatment of cancer and liver disease while plumbagin is a plant naphthoquinone with antitumor and antibacterial properties.[22] Also, co-expression analysis is one of the useful approach for system-level understanding of cellular processes. It involves calculation of similarity of gene expression profiles using correlation coefficient or any other distance parameters. If the correlation between two genes is above a threshold, the genes can be connected together to construct a network. There are two practical approaches for this type of analysis, guide-gene approach and non-targeted approach as described in Fig. 5.2.

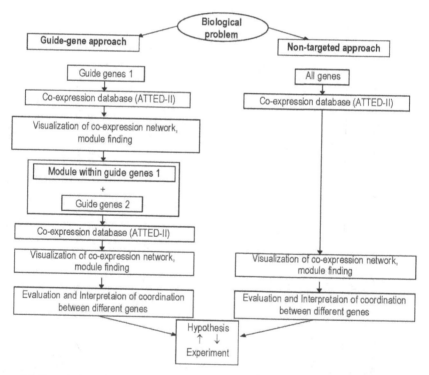

Figure 5.2: Practical approach of co-expression analysis of genes associated with biosynthetic pathway in plants. Left: guide gene approach; Right: non-targeted approach. A module represents a set of genes having discrete function that arises from interactions among them. ATTED-II- a database of co-expressed genes and cis elements for identifying co-regulated gene groups in Arabidopsis.

In guide-gene approach, a set of genes relating to the biological problem is selected based on experimental knowledge and literature information, known as guide genes and then correlation coefficients between these genes are retrieved from a correlation coefficient data set calculated from gene expression data. For example, analysis on genes associated with the phenylpropanoid biosynthesis pathway was done. Flavonoid biosynthesis genes and cinnamate-monolignol pathway/sinapoyl ester biosynthesis genes from pathway viewer KaPPA-View (http://kpv.kazusa.or.jp/kpv4/), were selected as guide genes1. Co-expression patterns within guide genes 1 is investigated by searching co-expression database ATTED-II (http://atted.jp/) for pair-wise Pearson correlation coefficients (PCCs) and the resulted co-expression network was visualized using a network viewer Pajek (http://vlado.fmf.uni-lj.si/pub/networks/pajek/default.htm). Futher analysis revealed that the guide genes 1 were classified into four distinct co-expression modules, most of the genes in modules 1, 2 and 3 belonged to the flavonoid biosynthesis pathway while genes in module 4 belonged to the cinnamate–monolignol pathway. This result suggested that expression of genes in both pathways (flavonoid biosynthesis and cinnamate–monolignol pathway) is differentially coordinated to produce pathway-specific metabolites. Next, to find the relationship between the phenylpropanoid pathway and metabolically upstream pathways, the genes in the four modules were combined with guide genes 2 containing 'aromatic amino acid biosynthesis' genes, 'Calvin cycle' genes and 'pentose phosphate pathway' genes taken from KaPPA-View. Again, ATTED-II was searched for PCCs and this analysis revealed that two genes in the aromatic amino acid biosynthesis pathway, 5-enolpyruvylshikimate-3-phosphate synthase gene (EPSP synthase) and the 3-deoxy-D-arabino-heptulosonate 7-phosphate synthase gene (DAHP synthase), were co-expressed with cinnamate–monolignol pathway genes in module 4. But, these genes did not have co-expression links with modules containing flavonoid biosynthesis genes which proved the hypothesis that aromatic amino acid biosynthesis is more tightly coordinated with the cinnamate– monolignol pathway than with flavonoid biosynthesis. Further, it suggested that EPSP synthase and DAHP synthase could be regulatory points to control metabolic flow from sugar phosphate to monolignol. In another approach, known as non-targeted approach, a knowledge-independent module search of the entire network is performed.[23]

Metabolomics

Plant metabolomics is of particular importance because of enormous chemical diversity due to complex set of metabolites produced in each plant species. The number of plant metabolites is estimated to be more than 1,00,000. However, secondary metabolites from medicinal plants have been identified as rich source of useful compounds such as pharmaceuticals, fragrances, dyes, flavor compounds and agrochemicals. Metabolome refers to the comprehensive pool of low molecular weight metabolites present in a cell under a given set of physiological conditions and metabolomics is the identification and characterization of these metabolites. Several approaches for plant metabolomics studies include metabolite profiling, metabolite fingerprinting, metabolite footprinting, targeted analysis and flux analysis.[24,25] But, none of these alone can completely fulfills the need for an accurate, simple and rapid method with a broad dynamic range because of divergent phytochemical properties of different plant metabolites unlike nucleic acids and proteins. Also, these techniques has resulted in the accumulation of metabolites data on a large scale for some species. In this context, it is imperative to use bioinformatics tools for further analysis and interpretation of the data. Bioinformatics play an important role in following areas:

Pattern Recognition and Metabolic Data Classification

Metabolite data generated by various techniques such as mass spectrometry (MS), Nuclear Magnetic Resonance (NMR), Liquid Chromatography (LC) and Gas Chromatography (GC) demand the use of bioinformatics tools. Various statistical techniques such as pattern recognition are used for comparing the resulting analytical spectra with the known samples as described in Tables 5.11.[26] Different bioinformatics metabolomics tools are also available online (Table 5.12).

Pathway Studies

Biochemical Pathway is a series of chemical reactions which occur within the organism (Table 5.13). Pathway studies help us:

- To understand the mechanism of an organism

  - Offer the guidance for biologists to design experiments.

  - Helps pharmacist to design drugs easier.

  - Makes us know the difference of mechanism between species

- Enable us to do simulation

  - Saves time and money for biologists to do experiments

  - Predicts the behaviors by a special treatment for an organism.

  - Warns the dangers to try a new drug on patients.

Table 5.11: Various approaches in metabolomics with applications in functional genomics

S.No.	Approaches	Data Analysis	Applications
1.	Metabolite profiling	Co-response analysis (FANCY)	Gene functional classification
		Peak deconvolution, calibration and database searches	Catalogue of metabolome
		Multivariate statistics, chemical structure analysis, co-response analysis	Pathway identification
		Temporal analysis, non-linear optimisation	Construction of predictive models
2.	Metabolite fingerprinting	Multivariate statistics	Gene functional classification
		Multivariate statistics and machine learning	Characterization of phenotypes
		Multivariate statistics and machine learning	Characterization of the responses to environment

Table 5.12: Online bioinformatics metabolomics tools

S.No.	Tools	Description	URL
1.	AMDIS	Automated Mass Spectral Deconvolution and Identification System software	http://www.amdis.net/
2.	Metabolomics society	Various metabolomics softwares and servers are available	http://www. metabol omicssociety.org/software.html
3.	MetaFIND	Metabolomics Feature INterrogation and Discovery software, designed for feature analysis within NMR metabolomics data	http://mlg.ucd.ie/metabol.html
4.	MetAlign	Software for the analysis, alignment and comparison of mass spectrometry datasets	http://www.pri.wur.nl/UK/products/MetAlign/
5.	MSFACTS	Metabolomics Spectral Formatting, Alignment and Conversion Tools Sumner/msfacts/index.html#AboutMSFACTs	http://www.noble.org/PlantBio/
6.	MZmine 2	Open-source project for metabolomics data processing, with focus on LC-MS data	http://mzmine.sourceforge.net/
7.	PRIMe	Platform for RIKEN Metabolomics, a web-based service for metabolomics and transcriptomics. It measures standard metabolites by means of multi-dimensional NMR spectroscopy, GC/MS, LC/MS, and CE/MS	http://prime.psc.riken.jp/
8.	SpecAlign	Software for the alignment of multiple mass spectra.	http://physchem.ox.ac.uk/~jwong/specalign/

Table 5.13: Types of Biochemical Pathways on the basis of function and route

S.No.	Types of Biochemical Pathways	Classification	Description
1.	On the basis of function	Metabolic Pathway	It is the pathway that shows how to generate energy and remove metabolites from the body.
		Signal Transduction Pathway	It is the pathway that is involved in controlling the behavior of a cell in terms of time and location with the signals like hormones.
		Regulatory Pathway (Gene Regulatory Network)	It is the pathway that shows how to adapt the change of environment based on the information in hereditary materials.
2.	On the basis of route	Linear Pathway	It consists of a series of chemical reactions going from a starting material to end-product, much like a traditional assembly line in a factory.
		Non-Linear Pathway	It involves a series of chemical pathways leading to the recycling of some critical starting material.
		Electron Transport System	It consists of a series of proteins which along with a series of electron acceptors carry electrons and as the electrons are passed along the electron transport system, energy from the electrons is used to do work, often to make ATP.

Data collection: It involves data mining from the existing bioinformatics resources and available literature for pathway data and pathway diagrams (visual representation of the biochemical network of an organism; Fig. 5.3) drawn using various graphical modelers are needed to draw pathways such as NetBuilder, CellDesigner, Cellware etc.

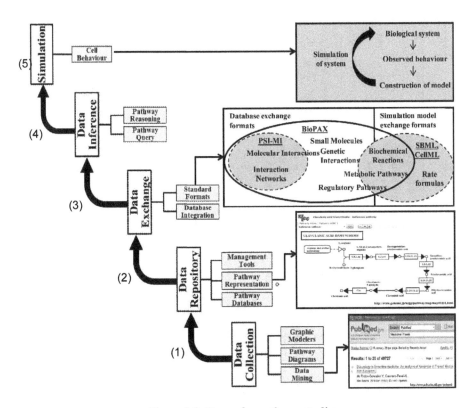

Figure 5.3: Stages for pathway studies

(1) Data Collection using data mining from the existing web resources
(2) Data Repository and its associated components
(3) Data Exchange using standard formats
(4) Data Inference using appropriate query and reasoning
(5) Simulation of cell behaviour using simulators

Data repository: Pathway Databases reflect the information about reactions and pathways, data related to organism-specific information about genes, their related gene products, protein functions, expression data, data about enzymatic activities, kinetic data etc. Various situations of pathways in the form of pathway diagrams using various tools and softwares are represented in these databases (Table 5.14). [27]

Table 5.14: Pathway databases and other databases for data retrieval.

Classification	Databases	Description	URL
Pathway Databases	BioCyc	Collection of 506 Pathway Genome Databases (PGDB), each describes the genome and metabolic pathways of a single organism.	http://biocyc.org/
	EMP	Enzymes and Metabolic Pathways Database of enzymatic and metabolic data from different organisms	http://www.empproject.com
	ExPASy-Biochemical Pathways	Contains ENZYME database and Boehringer Mannheim Biochemical Pathways Chart	http://expasy.org/cgi-bin/search-biochem-index
	KEGG	Kyoto Encyclopedia of Genes and Genomes is an integrated database resource consisting of 16 main databases, that are broadly categorized into systems information, genomic information, and chemical information	http://www.genome.jp/kegg/
	SPAD	Signaling Pathway Database is an integrated database for genetic information and signal transduction systems	http://www.grt.kyushu-u.ac.jp/spad/
Other databases used for data retrieval			
Genome Databases	GenBank	Annotated collection of all publicly available DNA sequences in NCBI	http://www.ncbi.nlm.nih.gov/genbank/
	EMBL-Bank	Europe's primary nucleotide sequence resource	http://www.ebi.ac.uk/embl/
	DDBJ	DNA Data Bank of Japan	http://www.ddbj.nig.ac.jp/
Protein and Protein-related Databases	Swiss- Prot/	Protein knowledgebase with computer-annotated supplement	http://www.expasy.ch/sprot/
	TrEMBL DDBJ	Europe's primary nucleotide sequence resource	http://www.ebi.ac.uk/embl/
DNA Data Bank of Japan		http://www.ddbj.nig.ac.jp/	
Protein and Protein-related Databases			
	Swiss- Prot/ TrEMBL	Protein knowledgebase with computer-annotated supplement	http://www.expasy.ch/sprot/
	PROSITE	Database of protein domains, families and functional sites	http://www.expasy.ch/prosite/

(Contd.)

Classification	Databases	Description	URL
	PDB	Protein Data Bank is a repository for the 3-D structural data of proteins and nucleic acids	http://www.pdb.org/pdb/home/home.do
	SCOP/ CATH	Protein structure classification databases	http://scop.berkeley.edu/ http://www.cathdb.info/
	PIR	Protein Information Resource is an integrated public bioinformatics resource that supports genomics, proteomics and systems biology research	http://pir.georgetown.edu/
Enzyme Databases	ExPASy-ENZYME	Repository of information about nomenclature of enzymes	http://expasy.org/enzyme/
	EC-PDB	Database of known enzyme structures deposited in PDB	http://www.ebi.ac.uk/thornton-srv/databases/enzymes/
	BRENDA	Information on properties of all classified enzymes, their occurrence, catalyzed reaction, kinetics, substrates, products, inhibitors, cofactors, activators, structure and stability	http://www.brenda- enzymes.org/
Chemical Reaction Databases	Beilstein	Provides chemical data on organic substances and reactions, including structures, properties, bioactivity records, preparation details and specific reaction pathways	http://info.crossfiredatabases.com/
	CAS	CAS databases offer many scientific disciplines, including biomedical sciences, chemistry, engineering, materials science, agricultural science etc.	http://www.cas.org/ expertise/ cascontent/ index.html
	CIRX	The ChemInform Reaction Database is an excellent tool for planning new synthesis and optimizing synthetic processes	http://www.fiz-chemie.de/en/home/ products- services/ chemical-data/ chemische-daten/ cheminform-rx.html
Literature Databases	PubMed/ Medline	PubMed is a database accessing the MEDLINE database of citations, abstracts and some full text articles on life sciences and biomedical topics	http://www.ncbi.nlm.nih.gov/sites/ entrez?db=pubmed

Data exchange: Moving towards data integration, a number of common file formats and standard languages to store biological information are available including Biological Pathway Exchange (BioPax); pathway language for the description of protein-protein interactions, genetic interactions, gene regulatory, metabolic and signaling pathways, Systems Biology Markup Language (SBML); format for describing qualitative and quantitative models of biochemical networks; CellML; XML-like language for exchange of computer-based mathematical models, formulas and equations and Proteomics Standards Initiative Molecular Interaction (PSI-MI); format for exchange, comparison and verification of proteomics data.[28]

Data inference: It involves querying of various pathway databases and resources using bioinformatics tools and softwares (Table 5.15). For example, given a set of genes, find all the pathways that will influence these genes, or find all crosstalk pathways in a specific cell type.

Simulation: Simulation and modeling have become a standard approach to understand complex biochemical processes. Standard approaches used in the field comprises the deterministic and stochastic simulation of reaction networks, the computation of steady states and their stability, stoichiometric network analysis (e.g. computing elementary modes), sensitivity analysis (e.g. metabolic control analysis), optimization and parameter estimation. To make experimental predictions, a model in combination with a simulator is used. This is a kind of deductive inference. If the deductive predictions of a model are not in agreement with observed behaviour, then the model is falsified. The process of forming a model to explain a given set of experimental results is called model identification. This is a kind of inductive inference. The application of automatic model identification is generally recognized to be essential for large-scale *in silico* modeling. Various bioinformatics simulators are available online that uses formal models to represent cellular activities as temporal trajectories (Table 5.15). Different statistical methods have been applied to these trajectories in order to infer knowledge.[29, 30]

Discovery of New Pathways, Modeling of Metabolic Networks and Predictive Metabolic Engineering

For elucidation of plant cellular systems, metabolomics allow us to engineer the pathways to improve their productivity and functionality. There are three main types of cellular processes modeled in bioinformatics: biochemical pathways (the metabolome), gene-networks (the transcriptome) and protein signal-transduction (the proteome). *In vivo*, these processes are interrelated and entangled, but they are still generally modeled separately (Fig. 5.4). Modeling all three types of processes is similar, with models constraining the change over time of levels of metabolite, mRNA or protein.

Table 5.15: Bioinformatics tools and softwares available online for performing pathway analysis

Tools and Softwares	Description	URL
	Pathway representation, management and querying	
Ariadne Genomics Pathway Studio	Software for visualization and analysis of biological pathways	http://www.ariadnegenomics.com/products/pathway-studio/
BIANA	Biologic Interactions and Network Analysis is a Python framework designed for biological information integration and network management	http://sbi.imim.es/web/BIANA.php
BioLayout Express 3D	2D/3D visualization and cluster analysis of networks derived from biological systems	http://www.biolayout.org/
Biological Networks	Integrated Research Environment that allows easy retrieval, construction and visualization of complex biological networks, including genome-scaleintegrated networks of protein–protein, protein–DNA and genetic interactions	http://biologicalnetworks.org/
BisoGenet	Tool for gene network building, visualization and analysis http://bio.cigb.edu.cu/bisogenet-cytoscape/	
BisoGenet	Tool for gene network building, visualization and analysis	http://bio.cigb.edu.cu/bisogenet-cytoscape/
CellDesigner	Modeling tool for pathways	http://www.celldesigner.org/
ComPath	Comparative Pathway Workbench is a web-based pathway analysis workbench, integrating KEGG and many other resources	http://compath.informatics.indiana.edu/cgi- bin/platcom/ComPath/ Programs/compath_home.cgi
Cyclone	Popular and intuitive software tool dedicated to biological networks visualization	http://sourceforge.net/projects/nemo-cyclone
Cytoscape	Network visualization and analysis software, extensive list of plug-ins for advanced visualization	http://www.cytoscape.org/
EGAN	Exploratory Gene Association Networks is a Java desktop application that provides a point-and-click	http://akt.ucsf.edu/EGAN/

(Contd.)

Tools and Softwares	Description	URL
	environment for contextual graph visualization of high-throughput assay results. By loading the entire network of genes, pathways, interactions, annotation terms and literature references directly into memory, it allows the user to repeatedly query and interpret multiple experimental results	
Eu.Gene Analyzer	Stand-alone application that allows microarray data analysis in the context of biological pathways.	http://www.duccioknights.org/eu-gene/
GeNGe	GEne Network Generator is a web application for the generation and analysis of gene regulatory networks (GRNs) providing different simulation steps with interactive user interfaces	http://genge.molgen.mpg.de
GenMAPP	Gene Map Annotator and Pathway Profiler, a computer application designed to visualize gene expression data on maps representing biological pathways and groupings of genes	http://www.genmapp.org/about.html
Ingenuity Pathways Analysis	Software for building, designing and analysis of pathways	http://www.ingenuity.com/products/pathways_analysis.html
JCell	Java-based application to reconstruct genetic interactions from microarray data	http://www.ra.cs.uni-tuebingen.de/software/JCell/
JNets	Network visualization and analysis tool, available as a stand-alone application and a web deployable applet, and is applicable to any type of biological or non-biological network data	http://www.bioinf.manchester.ac.uk/jnets/
KEGG Atlas	Visualization of data on KEGG pathways	http://www.genome.jp/kegg/
MetaRoute	User-friendly tool for exploring genome-scale metabolic networks	http://www-bs.informatik.uni-tuebingen.de/Publications/MetaRoute/

(Contd.)

Integrating Knowledge of Bioinformatics in Medicinal Plant Research

Tools and Softwares	Description	URL
MetNetAligner	Web service tool which for metabolic network alignments, predicting unknown pathways, comparing and finding conserved patterns, and resolving ambiguity caused by unidentified enzymes	http://alla.cs.gsu.edu:8080/MinePW/pages/gmapping/GlMMain.htm
Osprey	Tool for visualization of networks	http://tinyurl.com/osprey1/
PATIKA	Pathway Analysis Tool for Integration and Knowledge Acquisition is a server containing pathway editing and anlaysis softwares	http://www.patika.org/software/
PHT	Pathway Hunter Tool for analysing the shortest paths in metabolic pathways	http://pht.tu-bs.de/PHT/
ProteoLens	JAVA-based visual analytic software tool for creating, annotating and exploring multi-scale biological networks.	http://bio.informatics.iupui.edu/proteolens/
Proviz	Protein-protein interaction graphs visualisation tool	http://cbi.labri.fr/eng/proviz.htm
Rahnuma	Tool for prediction and analysis of metabolic pathways, and comparison of metabolic networks	http://portal.stats.ox.ac.uk:8080/rahnuma/
RMBNToolbox	Software to generate Random Models for Biochemical Networks	http://sourceforge.net/projects/rmbntoolbox/
VisANT	Integrative Visual ANalysis Tool for biological networks and pathways	http://visant.bu.edu/
VitaPad	Software for editing pathway diagrams, integration of microarray data	http://tinyurl.com/vitapad/
WGCNA	WeiGhted Correlation Network Analysis is a R software package i.e. collection of R functions for performing various aspects of weighted correlation network analysis	http://www.genetics.ucla.edu/labs/horvath/CoexpressionNetwork/Rpackages/WGCNA/

(Contd.)

Tools and Softwares	Description	URL
	Simulation	
COPASI	Complex Pathway Simulator Software for simulation and analysis of biochemical networks and their dynamics	http://www.copasi.org/tiki-view_articles.php
E-Cell	Software for whole-cell simulation	http://www.e-cell.org/ecell/
GEPASI	Software package for modeling biochemical systems	http://www.gepasi.org/
KINSIM	Tool to simulate the kinetics of biochemistry	http://www.biochem.wustl.edu/cflab/message.html
MCell	Monte Carlo Cell is a software for realistic simulation of cellular signaling in the complex 3-D subcellular microenvironment in and around living cells-and uses specialized Monte Carlo algorithms to simulate the movements and reactions of molecules within and between cells	http://www.mcell.cnl.salk.edu/
MetaModel	A Metabolic Modelling Program for MS-DOS to model the behaviour of metabolic systems	http://bip.cnrs- bio.org/mrs.fr/bip10/modeling.htm
PySCeS	Python Simulator for Cellular Systems	http://pysces.sourceforge.net/
SCAMP	General purpose simulator of metabolic and chemical networks	http://www.sys- sbwWiki/sysbio/winscamp
Snazer	Simulations and Networks Analyzer is a software for visualizing and manipulating reactive models	https://www.cosbi.eu/index.php/research/prototypes/snazer
VCell	The Virtual Cell is a software for modeling and simulation of cell biology	http://www.nrcam.uchc.edu/

Qualitative Modeling

Construction of network structure or pathway map including inhibitors, activators, reversibility and feedback regulation from available biological resources

The above information is converted into a machine-readable modeling language such as SBML ready for simulation software, with specifications for simulation such as type of integrators, integration step size and simulation procedures

Incorporation of quantitative data such as metabolite and enzyme concentrations, reaction equations and kinetic parameters to formulate a mathematical system model

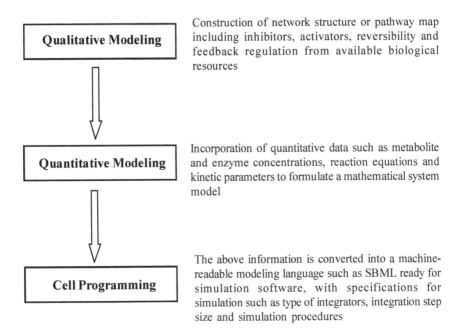

Figure 5.4: General approach in pathway modeling involves three steps: qualitative modeling, quantitative modeling and cell programming

Genome Mining

It is one of the powerful approach to access the structural diversity of various natural products.[31] A Genome-based Modeling (GEM) System has been developed for automatic prototyping simulation models of cell-wide metabolic pathways from genome sequences and other biological information available in public domain. It is based on a generic bioinformatics workbench i.e.

G-language Genome Analysis Environment, the system can directly access genome sequences and perform computational genome data mining.[32] The generated models have reduced the labor-intensive tasks required for systems biology research (Fig. 5.5).

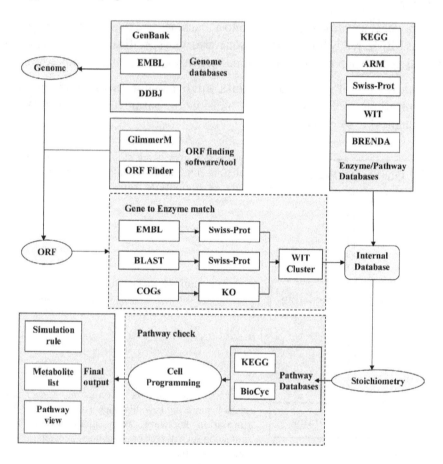

Figure 5.5: Schematic representation of Genome-based Modeling (GEM) System. The Workflow diagram involves Genome Analysis Environment for construction of stoichiometric simulation model from genome sequences and other biological information available in public domain.

Steps in Genome-based modeling (GEM) system

- The complete genome, both annotated and unannotated is taken as input to produce a simulation model of the organism.
- The second step matches the genes to the protein product using a combined method of annotation reference, homology search and orthology search.

- The system also contains information on enzymes extracted from various enzyme and pathway databases and curates it for consistency of nomenclature in the form of an internal database

- Enzyme Commission (EC) numbers obtained from search are matched to corresponding stoichiometry enzyme reaction equation from these databases.

- Each gene is assumed to have one-to-one enzyme relationship, so it is difficult to distinguish between isozymes and heteropolymeric enzymes. This problem is resolved by imposing a pathway check on the stoichiometric reaction list, that compares the extracted list with the general reference pathway of KEGG and BioCyc databases.

- The system generates a stoichiometric simulation model, which is applicable to metabolic-flux analysis on a number of simulation platforms.

Limitations of GEM system

- Only for reactions that have complete EC number assigned.

- It cannot identify reactions that are experimentally observed but with no corresponding gene found.

A major milestone in natural product research involves the discovery of novel metabolites by *in silico* genome mining of Polyketide Synthases (PKSs). These enzymes form a large family of multifunctional proteins and synthesize polyketide products with enormous diversity in their chemical structures. Bioinformatics analysis of a large number of PKS clusters with known metabolic products was carried out to correlate the chemical structures of these metabolites to the sequence and structural features of the PKS proteins. The remarkable conservation observed in the PKS sequences across different organisms, combined with unique structural features in their active sites and contact surfaces, is instrumental in deriving predictive rules for deciphering metabolic products of uncharacterized PKS clusters[33].

Flux Balance Analysis

Flux or metabolic flux is the rate of turnover of molecules through a metabolic pathway and is regulated by the enzymes involved in a pathway. Within cells regulation of flux is vital for all metabolic pathways to regulate the metabolic pathway's activity under different conditions. The quantitative knowledge of intracellular fluxes is required for a complete characterization of metabolic networks and their functional operation. Flux is therefore of great interest in metabolic network modeling, where it is analyzed via flux balance analysis

(FBA). It is a constraint-based modeling approach where constraints of mass balance describe the potential behavior of an organism. The reconstruction of a biochemical network is the first step in FBA. Once, all reaction and transport mechanisms are identified, a feasible set of steady-state fluxes is derived for all the metabolites. It gives rise to a set of differential equations that can be represented in the form of a matrix. Biochemical network reconstruction is presented in a matrix 'S', the $m \times n$ stoichiometric matrix, where m is the number of metabolites and n is the number of reactions in the network. The vector v represents all fluxes in the metabolic network, including the internal fluxes, transport fluxes and the growth flux. The equation for mass balance is given as

$$S . v = 0$$

Once this set of steady-state fluxes is identified, optimization techniques may be employed to evaluate the performance of the biological system at different environments. The resultant sets of fluxes may be compared with each other and with experimental data. The final collection of possible fluxes may yield predictive models of large-scale biochemical networks exposed to different conditions. Thus, FBA approach allows the quantitative interpretation of metabolic physiology and provides a guide to metabolic engineering[34,35].

Machine Learning (ML) Methods for Metabolic Pathway Prediction

A process of applying ML methods to pathway prediction has been developed. Data from various curated pathway/genome databases (PGDBs) are gathered into a "gold standard" collection that contains known information about which pathways are present and absent in a variety of organisms. Features are defined using biological knowledge and their values are calculated for all the pathways in the gold standard. The final dataset is split into training and tests sets. Feature selection and parameter estimation for various predictor types are performed using training data set, and predictors are evaluated using test sets. The predictor that performs best on test set will be applied to data from newly sequenced and annotated genomes to perform metabolic network reconstruction (Fig. 5.6)[36].

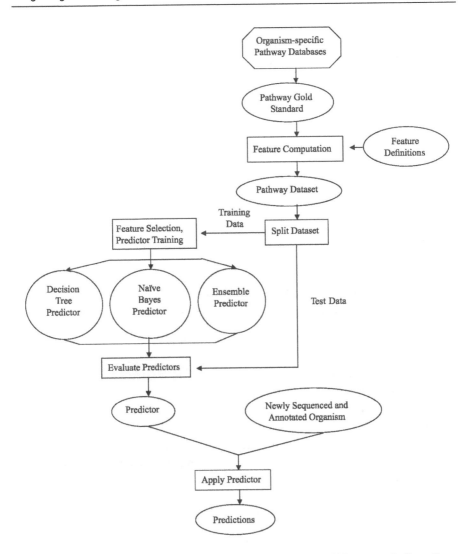

Figure 5.6: Process for applying Machine Learning (ML) methods to metabolic pathway prediction using data from various curated pathway/genome databases (PGDBs) that are gathered into a "gold standard" collection; and used for further testing and predictions.

Molecular Networks of Anti-tumor Components of *Withania somnifera*

Key gene targets for *Withania* crude extract and its two anti-tumor components, withanone and withaferin were identified using a combined approach of siRNA and ribozyme library screening. Ingenuity Pathway Analysis (IPA) software (http://www.ingenuity.com/products/pathways_analysis.html) was used for network construction to explore the interaction between these gene targets as well as difference in their bioactivities[37].

Gene-to-metabolite Networks for Biosynthesis of Terpenoid Indole Alkaloids (TIAs) in *C. roseus*

C. roseus contains more than 120 TIAs synthesized in seven different sub cellular compartments and exhibit strong pharmacological activities. To gain insights into the regulatory system governing TIA metabolism, transcript profiling method such as cDNA-AFLP and targeted and non-targeted liquid chromatography-mass spectrometry for metabolic profiling was used. PLS toolbox package was used for performing principal component analysis to explore the variability structure of data (expression profiles of 417 transcripts and 178 metabolite peaks). Correlation network analysis was done to set up gene-to-gene and gene-to-metabolite corregulation patterns using MATLAB (http://www.mathworks.com/products/matlab/) and Tom Sawyer Visualization software (http://www.tomsawyer.com/home/index.php[38].

Investigation of Biosynthetic Pathway of Camptothecin from *in silico* Tracer Experiment in *Ophiorrhiza pumila*

This is one of the best examples, of integration of computational metabolomics and experimental biochemistry. Camptothecin is a monoterpenoid indole alkaloid with antitumour activity and produced in large quantity in hairy root culture of *Ophiorrhiza pumila*. The incorporation of [1-13C] glucose (Glc) into central sugar metabolism, its labeling patterns in successive metabolites and metabolic flux to camptothecin was studied using Atomic Reconstruction of Metabolism (ARM) software (http://www.metabolome.jp/software). The tracing is performed by a shortest path algorithm that can output all possibilities through the network[39].

From Metabolite to Metabolite (FMM)

FMM is a web server that has many applications in synthetic biology and metabolic engineering. It can reconstruct metabolic pathways from one metabolite to another among different species. It is freely available at http://fmm.mbc.nctu.edu.tw/. Many plant flavonoids biosynthetic enzymes that are individually expressed in *E.coli* as functional enzymes, have been reported. For example, the genes that can be cloned into *E.coli* to produce naringenin from 4-coumarate, or the species from which they can be cloned, are unknown, FMM identifies them most easily. 4-coumarate (C00811) and naringenin (C00509) were taken as input into FMM; *E.coli* was selected from 'major species' and '*A. thaliana*' and '*O. sativa*' from 'comparative species' in comparative analysis web page. It results that the metabolic pathway from 4-coumarate to naringenin passes through three enzymes, none of which are present in *E.coli*. FMM results also reveal that these genes can be cloned from other plant species[40].

Ionomics

Ionome refers to mineral nutrient and trace element composition of an organism and so represents the inorganic component of cellular and organismal systems. Ionomics is the study of the ionome that involves the quantitative and simultaneous measurement of the elemental composition of living organisms and changes in this composition in response to physiological stimuli, developmental state and genetic modifications. This study requires the application of high-throughput elemental analysis technologies and their integration with both bioinformatics and genetic tools. The Purdue Ionomics Information Management System (PiiMS; http://www.ionomicshub.org/home/PiiMS) is a working example in which the ionomics workflow has been divided into four stages: planting, harvesting, drying and analysis. Plant ionome is the summation of many biological processes, so a high-throughput ionomics platform allows to analyze the multiple physiological and biochemical activities that affect the ionome. Ionomics, in combination with other omics technologies such as transcriptomics, proteomics, and metabolomics can facilitate to close the growing gap between our knowledge of genotype and the phenotypes it controls[41].

Transport Systems Study

Plants experience multiple and complex combinations of nutrient supply in their natural environment. To understand and potentially manipulate the roles of these nutrients, a detailed knowledge is required about how they move in, out and throughout plants. These processes are mediated by membrane transporters such as pumps, carriers and channels which are involved in various functions like osmoregulation, cell homeostasis, storage, cell signaling and stress responses. A combination of molecular, genetic, physiological and bioinformatics approaches is required for identification and characterization of transporter genes. An overview of various approaches developed for identification of transporter genes is given in Fig. 5.7.

Usually, the first step after identifying a new protein sequence is the search for its homologs in bioinformatics databases. Also, protein sequence patterns, profiles, motifs and signatures can be used for determining the functions of the unknown protein. Many of these methods are based on principles of probabilistic modelling such as HMMs. Online functional libraries for plant transporter systems include PlantsT for information on uptake and translocation of mineral nutrients and toxic metals in plants, with the goal of identifying gene networks that control these processes (http://plantst.genomics.purdue.edu/), P-type ATPase database (http://traplabs.dk/patbase/) and ABCISSE database http://www1.pasteur.fr/recherche/unites/pmtg/abc/database.iphtml) for ABC transporters [42].

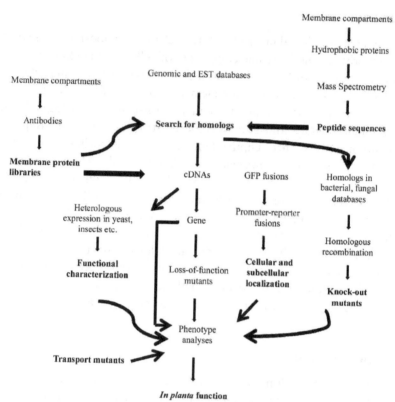

Fig. 5.7: Overview of various approaches developed for identification of plant transporter genes.

Computer-Aided Drug Discovery and Development (CADD)

Drug discovery and development are very time and resources consuming processes. Both *in silico* and experimental approaches have important roles in drug discovery and development; and represent complementary approaches. There is an ever growing effort to apply *in silico* techniques to the combined chemical and biological space in order to rationalize drug discovery, design, development and optimization. Computer-aided drug design methods contribute to the early stage of the drug discovery process to identify new bioactive molecules. Commonly used computational approaches include ligand-based design (e.g. pharmacophore), structure (target)-based design (e.g. docking), and quantitative structure-activity/property relationships (QSAR/QSPR)[43]. Various tools and resources are available for these approaches (Table 5.16). However, e-LEA3D (http://bioinfo.ipmc.cnrs.fr/lea.html) is a web server that integrates three complementary tools to perform computer-aided drug design based on molecular fragments: virtual screening, de novo drug design tool and combinatorial library design.

Table 5.16: Various bioinformatics tools/resources for drug designing

Tools/Resources		Description	URL
Tools	ACD/ChemSketch	Drawing package that allows you to draw chemical structures	http://www.acdlabs.com/resources/freeware/chemsketch/
	DOCK	Autodock Suite of automated docking tools	http://autodock.scripps.edu/
		Docking studies	http://dock.compbio.ucsf.edu/
	e-LEA3D	Computational-aided drugdesign web server	http://bioinfo.ipmc.cnrs.fr/lea.html
	GLIDE	Ligand-receptor docking studies	https://www.schrodinger.com/products/14/5/
	GOLD	Protein-Ligand Docking	http://www.ccdc.cam.ac.uk/products/life_sciences/gold/
	INVDOCK	Method and software for computer automated prediction of protein targets of small molecules	http://bidd.nus.edu.sg/group/softwares/invdock.htm
	MOE	Molecular Operating Environment; drug discovery software package	http://www.chemcomp.com/software.htm
	Molinspiration	Editor that allows drawing, editing of the chemicals as well as calculate their properties and predict bioactivity	http://www.molinspiration.com/cgi-bin/properties
	Phase	Program for ligand-based drug design	https://www.schrodinger.com/products/14/13/php?family=Modules,SimplePage,discovery_info
	SYBL	Complete computational chemistry and molecular modeling environment	http://www.tripos.com/index.
Resources	DrugBank	Database that contains detailed drug (i.e. chemical, pharmacological and pharmaceutical) data with comprehensive drug target (i.e. sequence, structure, and pathway) information	http://www.drugbank.ca/
	Organic Chemistry Portal	Overview of recent topics, interesting reactions, and information on important chemicals	http://www.organic-chemistry.org/
	PubChem	Database on chemical structures of small organic molecules and information on their biological activities	http://pubchem.ncbi.nlm.nih.gov/
	ZINC	Database of commercially-available compounds for virtual screening	http://zinc.docking.org/

General approach for characterization of biological activity of natural products from plant extracts exhibiting relevant biomedical effects, is given in Fig. 5.8.

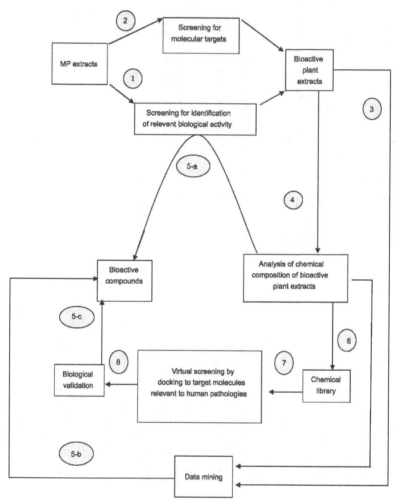

Fig. 5.8: General approach for characterization of biological activity of natural products from plant extracts exhibiting relevant biomedical effects. Numbering (1-8) denotes the presentation of different activities leading to final identification of bioactive compounds starting from MP extracts, through intermediate steps

Some examples of *in silico* drug-target search for medicinal plants are described below:

- Drug target identification in various medicinal plants using INVDOCK (Table 5.17). This software is based on a ligand-protein inverse docking strategy i.e. a compound of interest is docked to known ligand-binding pockets of each of the proteins in a protein 3D structural database.[44]

Table 5.17: Use of ayurvedic herbs in computer aided drug designing

Medicinal Plants	Compound of interest	Tools	Therapeutic implications
Acronychia pedunculata	Acronycine	INVDOCK	Cancer, diabetes, viral infection
Scutellaria baicalensis or Oroxylum indicum	Baicalin		Cardiovascular disease, erectile dysfunction Cancer, diabetes, viral infection
Rheum palmatum, Rumex dentatus. Cassia tora	Emodin		Asthma, neurodegenerative disorder, atherosclerosis, cancer, diabetes
Allium sativum	Allicin		Cancer, diabetes
Camptotheca acuninata	Camptothecin		Induction of apoptosis in tumor
Withania somnifera		Modeller, GLIDE SYBYL software package standard library, Glide (Grid-Based Ligand Docking With Energetics) software	Anti-inflammatory
Cupressus pyramidalis, Aegle marmelos	27 molecules		Anti-tumor, anti-inflammatory andpro-apoptotic activity
Canscora decussate, Nardostachys jatamansi, Mucuna pruriens	Active compounds having properties of memory enhancement	Structure-based drug designing (Modeller, Swiss-Pdb Viewer, ACD/ChemSketch software, RAMPAGE server, ArgusLab, HEX softwares, PATCHDOCK server)	Alzheimer's disease
Andrographis paniculata	Andrographolide	Ebola hemorrhagic fever (Ebola HF)	

- Identification of bioactive compounds from *Withania somnifera*: *Withania* is an Ayurvedic Indian medicinal plant which is widely used for several ailments. Its main bioactive compounds are steroidal lactone compounds known as withanolides and withanolide glycosides and alkaloids. Fifty seven compounds with anti-inflammatory activity from *Withania* were docked against homology modeled human 5-LOX (lipoxygenase) / COX (cyclooygenase)-2 enzymes using GLIDE software. The alkaloids: withasomine, cuseohygrine and anahygrine were identified as novel inhibitors of the target proteins after docking studies.[45]

Conclusion

Advancement of knowledge and advent of modern tools have made plant medicinal biotechnology a challenging and exciting area of research. The increasing demand for plant based pharmaceuticals has promoted scientist to greatly rely on precision and simulation-based tools for manipulation of plants. Understanding and integration of bioinformatics approaches in this field has revolutionized our knowledge on various natural compounds and their applications. Web-based resources, information pathways, enzymes, genes and regulatory sequences involved in secondary metabolites has provided a great impetus in our research. With increasing need for novel drugs for newly identified molecular targets, this field has shown great promise. Identification of right tools, integration of knowledge from various disciples and coordinated effort is the key enhancement of the commercial value of medicinal plants. Large infusion of interest in the field of bioinformatics in recent times and its applications in medicinal plants research has opened a new horizon in medicinal plants improvement.

References

1. Kamboj, V.P. (2000) Herbal medicine. Curr. Sci., 78(1): 35-51.
2. Sheth, P.P. (2006) Global opportunities and challenges for medicinal uses of ayurveda, herbal products, Neutraceuticals and Alternatives. Health Administrator, XIX(1): 74-75.
3. Bhattarai, S., Chaudhary, R.P., Taylor, R.S.L. and Ghimire, S.K. (2009) Biological activities of some nepalese medicinal plants used in treating bacterial infections in human beings. Nepal. J. Sci. Tech., 10: 83-90.
4. Vandepoele, K. and Peer, Y.V. (2005) Exploring the plant transcriptome through phylogenetic profiling. Plant Physiol., 137: 31-42.
5. Jiang, H., Xie, Z., Koo, H.J., McLaughlin, S.P., Timmermann, B.N. and Gang, D.R. (2006) Metabolic profiling and phylogenetic analysis of medicinal *Zingiber* species: Tools for authentication of ginger (*Zingiber officinale* Rosc.). Phytochemistry, 67(15):1673-1685.
6. JI, S., Huo, K., Wang, J. and Pan, S. (2008) A molecular phylogenetic study of Huperziaceae based on chloroplast *rbc*L and *psb*A-*trn*H sequences. Journal of Systematics and Evolution, 46 (2): 213–219.

7. Voelkerding, K.V., Dames, S.A. and Durtschi, J.D. (2009) Next-Generation Sequencing: From basic research to diagnostics. Clin. Chem., 55(4): 641–658.

8. Ratnasingham, S. and Hebert, P.D.N. (2007) BOLD: The barcode of life data system (www.barcodinglife.org). Mol. Ecol. Notes, 7(3): 355–364.

9. Zeng, S., Xiao, G., Guo, J., Fei, Z., Xu, Y., Roe, B.A. and Wang, Y. (2010) Development of a EST dataset and characterization of EST-SSRs in a traditional Chinese medicinal plant, *Epimedium sagittatum* (Sieb. Et Zucc.) Maxim. BMC Genomics, 11:94 doi:10.1186/1471-2164-11-94.

10. Yu, J., Arbelbide, M. and Bernardo, R. (2005) Power of in silico QTL mapping from phenotypic, pedigree, and marker data in a hybrid breeding program. Theor. Appl. Genet., 110(6): 1061-1067.

11. Arbelbide, M., Yu, J. and Bernardo, R. (2006) Power of mixed-model QTL mapping from phenotypic, pedigree and marker data in self-pollinated crops. Theor. Appl. Genet., 112(5): 876-884.

12. Mueller, L.A., Solow, T.H., Taylor, N., Skwarecki, B., Buels, R., Binns, J., Lin, C., Wright M.H., Ahrens, R. and Wang, Y. *et al.* (2005) The SOL genomics network. A comparative resource for solanaceae biology and beyond. Plant Physiol., 138: 1310–1317.

13. Costa, M.M.R., Hilliou, F., Duarte, P., Pereira, L.G., Almeida, I., Leech, M., Memelink, J., Barcelü, R. and Sottomayor, M. (2008) Molecular cloning and characterization of a vacuolar class III peroxidase involved in the metabolism of anticancer alkaloids in catharanthus roseus. Plant Physiol., 146: 403–417.

14. Saeys, Y., Rouzé, P. and Peer, Y. (2007) In search of the small ones: improved prediction of short exons in vertebrates, plants, fungi and protists. Bioinformatics, 23(4): 414–420.

15. Fredslund, J., Madsen, L.H., Hougaard, B.K., Nielsen, A.M., Bertioli, D., Sandal, N., Stougaard, J. and Schauser, L. (2006) A general pipeline for the development of anchor markers for comparative genomics in plants. BMC Genomics, 7: 207 doi:10.1186/1471-2164-7-207.

16. Oksman-Caldentey, K. and Inze′, D. (2004) Plant cell factories in the post-genomic era: new ways to produce designer secondary metabolites. Trends Plant Sci., 9(9): 433-440.

17. Mochida, M., Yoshida, T., Sakurai, T., Yamaguchi-Shinozaki, K., Shinozaki, K. and Tran, L. (2010) LegumeTFDB: an integrative database of *Glycine max*, *Lotus japonicus* and *Medicago truncatula* transcription factors. Bioinformatics, 26(2): 290–291.

18. Rhee, S.Y., Dickerson, J. and Xu, D. (2006) Bioinformatics and its applications in plant biology. Annu. Rev. Plant Biol., 57: 335–360.

19. Borgio, J.F. (2009) RNA interference (RNAi) technology: a promising tool for medicinal plant research. JMPR, 3(13): 1176-1183.

20. Mochida, K. and Shinozaki, K. (2010) Genomics and bioinformatics resources for crop improvement. Plant Cell Physiol., 51(4): 497–523.

21. Wong, M.L., O'Kirwan, F., Hannestad, J.P., Irizarry, K.J.L., Elashoff, D. and Licinio, J. (2004) St John's wort and imipramine-induced gene expression profiles identify cellular functions relevant to antidepressant action and novel pharmacogenetic candidates for the phenotype of antidepressant treatment response. Mol. Psychiatry, 9: 237–251.

22. Yang, N., Shyur, L., Chen, C., Wang, S. and Tzeng, C. (2003) Medicinal herb extract and a single-compound drug confer similar pharmacogenomic activities in MCF-7 Cells. J. Biomed Sci., 11: 418-422.

23. Aoki, K., Ogata, Y. and Shibata, D. (2007) Approaches for extracting practical information from gene Co-expression networks in plant biology. Plant Cell Physiol., 48(3):381-390.

24. Weckwerth, W. (2003) Metabolomics in systems biology. Annu. Rev. Plant Biol., 54:669-689.

25. Kell, D.B., Brown, M., Davey, H.M., Dunn, W.B., Spasic, I. and Oliver, S.G. (2005) Metabolomics footprinting and systems biology: The Medium Is The Message. Nat. Rev. Microbiol., 3: 557-565.

26. Mendes, P. (2002) Emerging bioinformatics for the metabolome. Brief Bioinform., 3(2): 134–145.
27. Witting, U. and Beuckelaer, A.D. (2001) Analysis and comparison of metabolic pathway databases. Brief Bioinform., 2(2): 126-142.
28. Pavlopoulos, G.A., Wegener, A. and Schneider, R. (2008) A survey of visualization tools for biological network analysis. BioData Mining, 1: 12 doi:10.1186/1756-0381-1-12.
29. Hoops, S., Sahle, S., Gauges, R., Lee, C., Pahle, J., Simus, N., Singhal, M., Xu, L., Mendes, P. and Kummer, U. (2006) COPASI—a COmplex PAthway Simulator. Bioinformatics, 22(24): 3067–3074.
30. King, R.D., Garrett, S.M. and Coghill, G.M. (2005) On the use of qualitative reasoning to simulate and identify metabolic pathways. Bioinformatics, 21(9): 2017–2026.
31. Wilkinson, B. and Micklefield, J. (2007) Mining and engineering natural-product biosynthetic Pathways. Nat. Chem. Biol., 3(7): 379-386.
32. Arakawa, K., Yamada, Y., Shinoda, K., Nakayama, Y. and Tomita, M. (2006) GEM System: automatic prototyping of cell-wide metabolic pathway models from genomes BMC Bioinform, 7:168 doi:10.1186/1471-2105-7-168.
33. Yadav, G., Gokhale, R.S. and Mohanty, D. (2009) Towards prediction of metabolic products of polyketide synthases: An *In silico* Analysis. PLoS Comput. Biol., 5(4): e1000351. doi:10.1371/journal.pcbi.1000351.
34. Kauffman, K.J., Prakash, P. and Edwards, J.S. (2003) Advances in flux balance analysis. Curr. Opin. Biotechnol, 14: 491–496.
35. Lee, J.M., Gianchandani, E.P. and Papin, J.A. (2006) Flux balance analysis in the era of Metabolomics. Brief Bioinform, 7(2): 140-150.
36. Dale, J.M., Popescu, L. and Karp, P.D. (2010) Machine learning methods for metabolic pathway prediction. BMC Bioinform, 11: 15 doi:10.1186/1471-2105-11-15.
37. Deocaris, C.C, Widodo, N., Wadhwa, R. and Kaul, S.C. (2008) Merger of Ayurveda and tissue culture-based functional genomics: Inspirations from systems biology. J. Transl. Med., 6: 14 doi:10.1186/1479-5876-6-14.
38. Rischer, H., Orešiė, M., Seppänen-Laakso, T., Katajamaa, M., Lammertyn, F., Ardiles-Diaz, W., Montagu, M.C.E.V., Inzé, D., Oksman-Caldentey, K. and Goossens, A. (2006) Gene-to-metabolite networks for terpenoid indole alkaloid biosynthesis in *Catharanthus roseus* cells. PNAS 103(14): 5614-5619.
39. Yamazaki, Y., Kitajima, M., Arita, M., Takayama, H., Sudo, H., Yamazaki, M., Aimi, N. and Saito, K. (2004) Biosynthesis of camptothecin. *In Silico* and *In Vivo* Tracer Study from [1-13C]Glucose. Plant Physiol., 134: 161–170.
40. Chou, C., Chang, W., Chiu, C., Huang, C. and Huang, H. (2009) FMM: a web server for metabolic pathway reconstruction and comparative analysis. NAR 37(2): W129-W134.
41. Salt, D.E., Baxter, I. and Lahner, B. (2001) Ionomics and the Study of the Plant Ionome. Annu. Rev. Plant Biol., 59: 709–733.
42. Barbier-Brygoo, H., Gaymard, F., Rolland, N. and Joyard, J. (2001) Strategies to identify transport systems in plants. Trends Plant Sci., 6(12): 577-585.
43. Kapetanovic, I.M. (2008) Computer-aided drug discovery and development (CADDD): *in silico*-chemico-biological approach. Chem. Biol. Interact., 171(2): 165–176.
44. Chen, X., Ung, C.Y. and Chen, Y. (2003) Can an *In Silico* drug-target search method be used to probe potential mechanisms of medicinal plant ingredients? Nat. Prod. Rep., 20:432–444.
45. Santhi, N., Rajeshwari, M., Kalaiselvi Senthil and Parvatham, R. (2010) Identification of potential anti-inflammatory bioactive compounds from *Withania somni fera* (L.) Dunal: Molecular docking studies using Glide *(http://cs.nyu.edu/parida/APBC2010/ poster_abstracts/85.pdf)*

Chapter – 6

Bioinformatics Applications in Plant Biology

Sharmila Anishetty

Abstract

The completion of large number of genome sequencing projects has generated large volumes of raw sequence data. One of the fundamental tasks in bioinformatics is the creation of databases for data storage and development of tools for analysis. The first plant genome to be completed is that of Arabidopsis thaliana in 2000 [49]. This was followed by the rice genome and others. The availability of complete genomes has provided us with an opportunity to analyze them and understand plant biology in a holistic way. It has enhanced our understanding of factors affecting plant growth and development, abiotic and biotic stress, disease resistance, host pathogen interactions amongst others.

Genomes, Gene finding and Annotation

With advances in sequencing technologies, a number of genome sequencing projects are being completed. Once a genome sequence is complete, the next task is to identify the genes. Gene finding is more complex in eukaryotes compared to prokaryotes because of the additional problem of finding intron-exon boundaries in eukaryotes. Gene prediction is done through homology searches or through *abinitio* methods [13,36]. Homology searches are performed using BLAST family of programs which compare protein and genomic sequences with existing protein, gene, EST and cDNA sequences present in databases. BLASTn queries the nucleotide database with a nucleotide sequence whereas BLASTx translates the nucleotide sequence in all six reading frames and performs a search against the protein database. BLASTp, PSIBLAST and PHIBLAST facilitate searching of protein databases. PSIBLAST is particularly sensitive in detecting remote homologs.

Abinitio methods rely on identification of gene specific features like promoter sites, splice sites, polyadenylation sites etc. Some of the gene finding programs are GENSCAN [12], HMMGene[28], GENIE [29], GlimmerM [39] etc. Gene finding programs like the GeneSeqer at PlantGDB which are specific to plant genomes have also been developed [41]. Nucleotide sequence data is deposited in nucleotide databases GenBank [6], DDBJ [46] and EMBL [45]. UniprotKB [50] is a very well curated protein resource. Protein Data Bank [7] provides information on proteins whose structures are resolved. KEGG [23] is a metabolic pathway resource. For every genome that is completely sequenced metabolic pathway reconstruction is done at KEGG.

Functional Annotation

Assigning functions to genes is called functional annotation. Homology based searches using BLAST programs is one method of annotating proteins. Genomic inference methods use genomic or protein context information like gene order conservation or synteny, rosetta stone information, phylogenetic profiling to infer functional linkages in proteins [12]. Information about sequence motifs can be obtained at the Prosite database [43] while domain information is housed in domain databases like Pfam [5] and InterPro [20].

Plant Genomes and Analysis

Bioinformatics analyses of plant genome sequences have provided invaluable insights into various aspects of plant biology. The genome of the model flowering plant *Arabidopsis thaliana* was completed in 2000 [49]. Gene finding algorithms were optimized to *Arabidopsis* gene structure. Out of the 25,498 genes that were predicted, 69% were assigned a functional category based on sequence similarity to proteins with known function. A comprehensive bioinformatics analysis of the Arabidopsis genome revealed interesting features. For example, only 8-23% of Arabidopsis proteins involved in transcription have related genes in other eukaryotic counterparts suggesting an independent evolution of plant transcription factors. A stringent BLASTp analysis showed

Bioinformatics Applications in Plant Biology

segmental duplications and tandem arrays. Conserved protein domain analysis and comparison with other multicellular eukaryotes like *C.elegans*, *S.cerevisiae* and *Drosophila* revealed an abundance of PPR repeat containing proteins which are involved in RNA stabilization and RNA processing. In addition, protein kinases and associated domains, disease resistance signature containing proteins and Toll/IL-1R (TIR) domain which is a component of pathogen recognition molecules were also relatively abundant in *Arabidopsis*. The draft sequence of the rice genome *japonica* and *indica* was published in 2002 [18,53]. A finished quality genome sequence of *japonica* was completed and published by the International Rice Genome Sequencing Project in 2005 [22].

A list of genome sequencing projects which are complete or ongoing can be obtained from NCBI Plant Central http://www.ncbi.nlm.nih.gov/genomes/PLANTS/PlantList.html, or at TAIR http://www.arabidopsis.org/portals/genAnnotation/other_genomes/index.jsp or at the GOLD database [8].

Plant Specific Databases and Resources

There are currently a number of plant specific resources available in the public domain. While many of them provide information about plant genomes and their annotation, some databases are specific to plant transcription factors, disease resistance, protein protein interactions, plant pathogen interactions, mutant phenotypes, Quantitative Trait Loci (QTL) and others. Tools for analysis and for comparative genomics are also available at some sites. Table 6.1 lists some of the plant specific resources.

Table 6.1: Plant Specific Resources

Databases	Comments	URL
TAIR	The Arabidopsis Information Resource	http://www.arabidopsis.org/
MIPS PlantDB	MIPS plants databases	http://mips.helmholtz-muenchen.de/plant/genomes.jsp
MSU rice genome project	Rice genome annotation project database	http://rice.plantbiology.msu.edu/
Graingenes	Database on *Triticeae* and *Avena*	http://wheat.pw.usda.gov/GG2/index.shtml
Gramene	Model and crop plants data and a Comparative Genomics platform	http://www.gramene.org/
MOsDB	The MIPS *Oryza sativa* database	http://mips.helmholtz-muenchen.de/plant/rice/
RAP-DB	The Rice Annotation Project	http://rapdb.dna.affrc.go.jp/
Oryza tag line	An integrated database for the functional analysis of the rice genome	http://urgi.versailles.inra.fr/OryzaTagLine/
GRIN	Germplasm Resources Information Network	http://www.ars-grin.gov/

(Contd.)

Databases	Comments	URL
PlantGDB	Resources for comparative plant Genomics	http://www.plantgdb.org/
Entrez Genome	Plant Genome Central	http://www.ncbi.nlm.nih.gov/ genomes/PLANTS/ PlantList.html
JCVI Plant Genomics	Castor Bean Database, *Medicago Truncatula* database, Rice database	http://www.jcvi.org/cms/ research/groups/plant-genomics/
TIGR	TIGR Plant Transcript Assemblies	http://plantta.jcvi.org/
Invasive plant Atlas	Invasive plant Atlas of the United States	http://www. invasivep lan tatlas.org/index.html
PlantTFDB	Plant Transcription Factor Database	http://planttfdb.cbi.pku.edu.cn/
DRTF	Database of Rice Transcription Factors	http://drtf.cbi.pku.edu.cn/
Phytozome	Comparative Genomics	http://www.phytozome.net/
JGI	Joint Genome Institute Plant Genomics and Other Genomics Programs.	http://genome.jgi-psf.org/
PPAP	The UniProtKB/Swiss-Prot Plant Proteome Annotation Program	http://expasy.org/sprot/ppap/
GreenPhylDB	A phylogenomic database for plant comparative genomics	http://greenphyl.cirad.fr/v2/cgi-bin/index.cgi
Plant Specific DB	Plant specific database	http://genomics.msu.edu/ plant_specific/
PlantsP	Plant Protein Motif Databases	http://plantsp. genomics. purdue. edu/html/
APIRS	Aquatic Plant Information Retrieval System)	http://plants.ifas.ufl.edu/
PLEXdb	Plant Expression database	http://www.plexdb.org/
FRLHT	Encyclopedia on Indian Medicinal Plants	http://envis.frlht.org.in/ bot_search.php
PlaPID	Plant Protein Interaction Database	http://www.plapid.org/
AtPID	*Arabidopsis thaliana* Protein Interactome Database	http://atpid.biosino.org/
SIGnAL	Salk Plant Interactome database	http://signal.salk.edu/ interactome.html
CCSB	CCSB plant interactome database	http://interactome. dfci. harvard. edu/A_thaliana/
FPPI	Protein protein interactions for *Fussarium graminearum*	http://csb.shu.edu.cn/fppi/
IntAct	Protein protein interaction database	http://www.phi-base.org/
Metacrop gatersleben.de	Crop plant Metabolism	http://metacrop.ipk-
NIAS Genbank	Databases of Plant Diseases in Japan	http://www.gene.affrc.go.jp/ databases-micro_pl_ diseases _en.php
PathoPlant	Plant pathogen interactions	http://www.pathoplant.de/
PRGdb	Plant Resistance Genes db	http://prgdb.cbm.fvg.it/
PHI-base	Pathogen Host Interactions	http://www.phi-base.org/

A brief account of some of the plant specific resources are given below:

The Arabidopsis Information Resource (TAIR) maintains a database for genetic and genomic data for *Arabidopsis thaliana* [47]. It has information on *Arabidopsis* genes, clones, ecotypes, gene ontologies, genome maps, genetic and physical markers, metabolic information, variant information etc. TAIR is also responsible for the updation of genome annotation of *Arabidopsis thaliana*. The initial prediction of 25,498 genes has now changed due to improved gene models developed from information obtained through experiments. According to the TAIR website, the latest version TAIR10 Genome release has 27,416 protein coding genes.

Munich Information Center for Protein Sequence (MIPS) hosts PlantsDB [44]. The plant genomics group at MIPS focuses on analysis of plant genomes and comparative genomics of plants using bioinformatics techniques. Currently, PlantsDB has databases on the Triticeae genome project, maize, rice, sorghum and *brachypodium* genomes amongst monocots and MIPS *Arabidopsis* genome database, *Medicago truncatula* genome database, *Lotus japonicas* genome database and *Solanum lycopersicum*, the tomato genome database amongst the dicots.

Oryza databases: MosDB at MIPS is *an Oryza sativa* database and has information on two subspecies of *Oryza sativa, japonica* and *indica* [24]. The MSU rice genome project database [33] provides genome sequences of *Nipponbare* subspecies of rice and annotation of 12 rice chromosomes. RAP-DB the Rice Annotation Project Database provides annotation of rice gene sequences [32]. For functional analysis of the rice genome under the genomics initiative, Genoplante a library of 30,000 T-DNA insertion lines were produced in the reference *japonica* cultivar *Nipponbare* [38]. OryzaTagLine database was developed to organize data originating from the phenotypic characterization of this library [30].

Gramene is a major resource for model and crop plants including Arabidopsis, Brachypodium, maize, sorghum, poplar, grape and several species of rice [52]. It provides information on genes, proteins, metabolic pathways, QTL, ontologies and others. It is a good platform for comparative plant genomics.

Graingenes is a database for Triticeae species: It holds genetic and genomic information for wheat (*Triticum aestivum*), barley (*Hordeum vulgare*), rye (*Secale cereal*), and their wild relatives and oat *Avena sativa* and its wild relatives [14].

PlantGDB is a genomics database for green plants and has annotated transcript assemblies for more than 100 plant species [16].

Entrez Genome: The Plant Genomes Central at NCBI provides information about completed and ongoing plant genome sequencing projects, EST sequencing, genetic maps, PlantESTBlast and others.

ESTs: dbEST [9] at NCBI houses EST information while TIGR Plant Transcript Assemblies [35] construct transcript assemblies for all plant species for which more than 1000 ESTs or cDNA sequences are available.

PlantTFDB [56] is an integrated transcription factor database for plants. The current version has information on 53,319 transcription factors from 49 species classified into 58 families. It covers 9 species of green algae, one species each of moss and ferns, 3 species of gymnosperms and 35 species of angiosperms. DRTF [17] is exclusively for rice transcription factors while wDBTF is a resource for systemic study of transcription factor families and their expression in wheat [37].

Plant Pathogens and Disease Resistance Genes

Plants are sessile and are exposed to pathogens, pests and parasites. They have developed defense mechanisms to counter these attacks. From a molecular perspective, plant defense systems depend upon a set of dominant disease resistance genes called R genes and a corresponding set of pathogen specific genes called Avr (avirulent) genes expressed in the pathogen [25]. They adopt a gene-for-gene resistance mechanism where for every pathogen Avr gene expressed there is a corresponding plant R gene expressed. The product of the R gene triggers a cascade of events which help in plant defense against the pathogen. Most of the proteins encoded by the R genes possess at their C terminii, a Nucleotide Binding Site followed by many Leucine Rich Repeats [57]. The N terminus of these proteins harbors the TIR (Toll and Interleukin 1 Receptor protein) domain or a coiled coil domain. Understanding of disease resistance genes has contributed to the development of many transgenic crops where over expression of relevant R genes have resulted in better resistance towards disease. For example, over expression of the R gene Pto in transgenic tomato plant has been shown to confer resistance towards *Pseudomonas syringae Pv. tomato* [31].

PRGdb, a Plant Resistance Genes database [40] is a bioinformatics resource providing a comprehensive overview of disease resistance genes (R-genes) in plants. It currently holds a collection of 16864 known and putative R genes from 192 plant species challenged by 119 pathogens. As more information about R genes and Avr genes accumulate in databases we will be able to understand plant pathogen interactions in greater detail.

Bioinformatics Applications in Plant Biology

Protein Protein Interactions

It is believed that R gene products directly interact with the corresponding Avr gene products of the pathogen. Evidence for this was first shown through yeast two hybrid analysis where tomato bacterial speck R gene product Pto was shown to interact with AvrPto of *Pseudomonas syringae Pv. tomato* [42,48]. PathoPlant [11] is a database on plant pathogen interactions and components of signal transduction pathways in plant pathogenesis. It also provides gene expression datasets from microarray experiments of *Arabidopsis thaliana* to enable users to query on genes regulated upon pathogen infection or elicitor treatment. While IntAct [3] at EBI and others provides protein to protein interactions data, some plant specific protein protein interaction databases like AtPID [15], and others are listed in Table 6.1.

Gene Expression

Gene expression omnibus (GEO) at NCBI and ArrayExpress at EBI provide a platform for submission, storage and retrieval of heterogenous datasets from high throughput gene expression and genomic hybridization experiments [4,14]. BarleyBase and PLEXdb [55] is a public resource for large scale analysis of gene expression in plants and pathogens. TAIR also provides gene expression information for *Arabidopsis*. Analysis of datasets related to gene expression studies in plants can provide insights into plant abiotic and biotic stress responses amongst others. In one large scale bioinformatics analysis of large and heterogenous microarray datasets from the public domain, principles that govern regulation of gene expression in plants were evaluated [1]. Analysis of the quantitative effect of various experimental factors like mutations, stress and organ identity on the plant transcriptome revealed the key role of developmental processes for establishing mRNA levels throughout the plant. Their findings also revealed an important role for epigenetic mechanisms in the regulation of gene expression in plants. In yet another study, environmental abiotic stress induced patterns and a model bioinformatics analysis of gene expression in response to UV-B light, drought and cold was presented [26].

Systems Biology

Finally as large volumes of heterogenous data accumulate in the public domain databases, it is now possible to integrate data, find co-relations and develop models. Systems Biology involves data collection and integration, development of systems models, global level experimentation and generation of new hypothesis [21,27]. Instead of looking at individual components, systems biology provides a holistic view. Plant systems biology has increased our knowledge about circadian rhythms, multigenic traits, stress responses and plant defenses[54].

Systems approach is also expected to make predictions about the effect of a perturbation, for example, the effect of a gene mutation on traits which might affect seed yield, plant growth may now be predicted. Such predictive models may be used for improving traits associated with agriculture. The Virtual Plant 1.0 at http://virtualplant.bio.nyu.edu/cgi-bin/vpweb2/ is a platform that integrates genomic data and provides tools to aid researchers in generating biological hypothesis. AtPID [15] contains data relevant to protein protein interactions, sub cellular localization, ortholog maps, domain attributes and gene regulation of Arabidopsis. It is an integrative platform for systems biology in *Arabidopsis thaliana*. Systems Biology approaches can also be used to aid metabolic engineering of plants. Metacrop is a manually curated database with information on the metabolism of six major crop plants of agronomical importance[19]. It allows model creation and export of data thereby facilitating systems biology studies.

Conclusions

Starting from the *Arabidopsis* genome, plant genome sequencing has come a long way. The number of completed genomes is steadily increasing. Sophisticated computational techniques for analysis are keeping pace with advances in experimental omics technologies. Knowledge gained through an interdisciplinary approach will definitely enhance our understanding of plant biology and aid in the development of crops with desirable traits.

References

1. Aceituno, F.F., Moseyko, N., Rhee, S.Y. and Gutiérrez, R.A. (2008) The rules of gene expression in plants: organ identity and gene body methylation are key factors for regulation of gene expression in *Arabidopsis thaliana*. BMC Genomics, 9:438.
2. Altschul, S.F., Madden, T.L., Schäffer, A.A., Zhang, J., Zhang, Z., Miller, W. and Lipman, D.J. (1997) Gapped BLAST and PSI-BLAST: a new generation of protein database search programs. Nucleic Acids Res., 25(17):3389-402.
3. Aranda, B., Achuthan, P., Alam-Faruque, Y., Armean, I., Bridge, A., Derow, C., Feuermann, M., Ghanbarian, A.T., Kerrien, S., Khadake, J., Kerssemakers, J., Leroy, C., Menden, M., Michaut, M., Montecchi-Palazzi, L., Neuhauser, S.N., Orchard, S., Perreau, V., Roechert, B., van Eijk, K. and Hermjakob, H. (2010) The IntAct molecular interaction database in 2010. Nucleic Acids Res., 38(Database issue): D525-31.
4. Barrett, T. and Edgar R. (2006) Gene expression omnibus: microarray data storage, submission, retrieval, and analysis. Methods Enzymol, 411:352-69.
5. Bateman, A., Coin, L., Durbin, R., Finn, R.D., Hollich, V., Griffiths-Jones, S., Khanna, A., Marshall, M., Moxon, S., Sonnhammer, E.L., Studholme, D.J., Yeats, C. and Eddy, S.R. (2004) The Pfam protein families database. Nucleic Acids Res., 32 (Database issue): D138-41.
6. Benson, D.A., Karsch-Mizrachi, I., Lipman D.J., Ostell, J. and Wheeler D.L. (2008) GenBank, Nucleic Acids Res., 36: D25–D30.

Bioinformatics Applications in Plant Biology

151

7. Berman, H.M., Westbrook, J., Feng, Z., Gilliland, G., Bhat, T.N., Weissig, H., Shindyalov, I.N. and Bourne, P.E. (2000) The Protein Data Bank. Nucleic Acids Res., 28(1):235-42.

8. Bernal A, Ear, U. and Kyrpides N. (2001) Genomes Online database (GOLD): a monitor of genome projects world-wide. Nucleic Acids Res., 29(1):126-7.

9. Boguski, M.S., Lowe, T.M. and Tolstoshev, C.M. (1993) dbEST—database for "expressed sequence tags". Nat Genet, (4):332-3.

10. Bowers, P.M., Pellegrini, M., Thompson, M.J., Fierro, J., Yeates, T.O., Eisenberg and D. Prolinks (2004) a database of protein functional linkages derived from coevolution. Genome Biol., 5(5):R35.

11. Bülow, L., Schindler, M., Choi, C. and Hehl, R. (2004) PathoPlant: a database on plant-pathogen interactions., *In Silico* Biol., 4(4):529-36.

12. Burge, C. and Karlin, S. (1997) Prediction of complete gene structures in human genomic DNA. J. Mol. Biol., 268(1):78-94.

13. Burset, M. and Guigó, R. (1996) Evaluation of gene structure prediction programs. Genomics, 34(3):353-67.

14. Carollo, V., Matthews, D.E., Lazo, G.R., Blake, T.K., Hummel, D.D., Lui, N., Hane, D.L. and Anderson, O.D. (2005) GrainGenes 2.0. an improved resource for the small-grains community. Plant Physiol., 139(2):643-51.

15. Cui, J., Li, P., Li, G., Xu, F., Zhao, C., Li, Y., Yang, Z., Wang, G., Yu, Q., Li, Y. and Shi, T. (2008) AtPID: Arabidopsis thaliana protein interactome database—an integrative platform for plant systems biology, 36(Database issue):D999-1008.

16. Duvick, J., Fu, A., Muppirala, U., Sabharwal, M., Wilkerson, M.D., Lawrence, C.J., Lushbough, C. and Brendel V. (2008) PlantGDB: a resource for comparative plant genomics. Nucl. Acids Res., 36: D959-D965.

17. Gao, G., Zhong, Y., Guo, A., Zhu, Q., Tang, W., Zheng, W., Gu, X., Wei, L. and Luo, J. (2006) DRTF: a database of rice transcription factors. Bioinformatics, 22(10):1286-7.

18. Goff, S.A., Ricke, D., Lan, T.H., Presting, G., Wang, R., Dunn, M. and Glazebrook, J. (2002) Sessions A, Oeller P, Varma H, *et al*. A draft sequence of the rice genome (*Oryza sativa* L. ssp. *japonica*), Science, 296:92–100.

19. Grafahrend-Belau, E., Weise, S., Koschützki, D., Scholz, U., Junker, B.H. and Schreiber, F. (2008) MetaCrop: a detailed database of crop plant metabolism. Nucleic Acids Res., 36(Database issue): D954-8.

20. Hunter, S., Apweiler, R., Attwood, T.K., Bairoch, A., Bateman, A., Binns, D., Bork, P., Das, U., Daugherty, L., Duquenne, L., Finn, R.D., Gough, J., Haft, D., Hulo, N., Kahn, D., Kelly, E., Laugraud, A., Letunic, I., Lonsdale, D., Lopez, R., Madera, M., Maslen, J., McAnulla, C., McDowall, J., Mistry, J., Mitchell, A., Mulder, N., Natale, D., Orengo, C., Quinn, A.F., Selengut, J.D., Sigrist, C.J., Thimma, M., Thomas, P.D., Valentin, F., Wilson D., Wu, C.H. and Yeats, C. (2009) Inter Pro: the integrative protein signature database. Nucleic Acids Res., 37 (Database issue):D211-5.

21. Ideker, T., Galitski, T. and Hood, L. (2001) A new approach to decoding life: systems biology, Annu Rev Genomics Hum Genet., 2:343-72.

22. International Rice Genome Sequencing Project. (2005) The map-based sequence of the rice genome. Nature, 436: 793 – 800.

23. Kanehisa, M. and Goto, S. (2000) KEGG: Kyto encyclopedia of genes and genomes, Nucleic Acids Res., 28(1): 27-30.

24. Karlowski, W.M., Schoof, H., Janakiraman, V., Stuempflen, V. and Mayer, K.F. (2003) MOsDB: an integrated information resource for rice genomics.Nucleic Acids Res., 31(1):190-2.

25. Keen, N.T. (1990) Gene-for-gene complementarity in plant-pathogen interactions. (1990) Annu. Rev. Genet., 24: 447–463.

26. Kilian, J., Whitehead, D., Horak, J., Wanke, D., Weinl, S., Batistic, O., D'Angelo, C., Bornberg-Bauer, E., Kudla, J. and Harter, K. (2007) The AtGenExpress global stress expression data set: protocols, evaluation and model data analysis of UV-B light, drought and cold stress responses. The Plant Journal, 2:347-63.

27. Kitano H. (2002) Systems biology: a brief overview. Science, 295(5560):1662-4.

28. Krogh, A. (1997) Two methods for improving performance of an HMM and their application for gene finding. Proc. Int. Conf. Intell. Syst. Mol. Biol, 5:179–186.

29. Kulp, D., Haussler, D., Reese, M.G. and Eeckmann, F.H. (1996) A generalized hidden Markov model for the recognition of human genes in DNA. Intell. Systems Mol. Biol., 4:134–142.

30. Larmande, P., Gay, C., Lorieux, M., Périn, C., Bouniol, M., Droc, G., Sallaud, C., Perez, P., Barnola, I., Biderre-Petit, C., Martin, J., Morel, J.B., Johnson, A.A., Bourgis, F., Ghesquière, A., Ruiz, M. and Courtois, B. (2008) Guiderdoni, E. Oryza Tag Line, a phenotypic mutant database for the Genoplante rice insertion line library. Nucleic Acids Res., 36 (Database issue):D1022-7.

31. Martin, G.B., Brommonschenkel, S.H., Chunwongse, J., Frary, A., Ganal, M.W., Spivey, R., Wu, T., Earle, E.D. and Tanksley, S.D. (1993) Map-based cloning of a protein kinase gene conferring disease resistance in tomato. Science, 262:1432-1436.

32. Ohyanagi, H., Tanaka, T., Sakai, H., Shigemoto, Y., Yamaguchi, K., Habara, T., Fujii, Y., Antonio, B.A., Nagamura, Y., Imanishi, T., Ikeo, K., Itoh, T., Gojobori, T. and Sasaki, T. (2006) The Rice Annotation Project Database (RAP-DB): hub for *Oryza sativa* ssp. *japonica* genome information. Nucleic Acids Res., 34 (Database issue): D741-4.

33. Ouyang, S., Zhu, W., Hamilton, J., Lin, H., Campbell, M., Childs, K., Thibaud-Nissen, F., Malek, R.L., Lee, Y., Zheng, L., Orvis, J., Haas, B., Wortman, J. and Buell, C.R. (2007) The TIGR Rice Genome Annotation Resource: improvements and new features. Nucleic Acids Res., 35 (Database issue): D883-7.

34. Parkinson, H., Kapushesky, M., Shojatalab, M., Abeygunawardena, N., Coulson, R., Farne, A, Holloway, E., Kolesnykov, N., Lilja, P., Lukk, M., Mani, R., Rayner, T., Sharma, A., William, E., Sarkans, U. and Brazma, A. (2007) ArrayExpress—a public database of microarray experiments and gene expression profiles. Nucleic Acids Res., 35 (Database issue): D747-50.

35. Pertea, G., Huang, X., Liang, F., Antonescu, V., Sultana, R., Karamycheva, S., Lee, Y., White, J., Cheung, F., Parvizi, B., Tsai, J. and Quackenbush, J. (2003) TIGR Gene Indices clustering tools (TGICL): a software system for fast clusteringof large EST datasets. Bioinformatics, 19(5):651-2.

36. Rogic, S., Mackworth, A.K. and Ouellette, FB. (2001) Evaluation of gene-finding programs on mammalian sequences. Genome Res., 11(5):817-32.

37. Romeuf, I., Tessier, D., Dardevet, M., Branlard, G., Charmet, G. and Ravel, C. (2010) wDBTF: an integrated database resource for studying wheat transcription factor families. BMC Genomics, 11:185.

38. Sallaud, C., Gay, C., Larmande, P., Bès, M., Piffanelli, P., Piégu, B., Droc, G., Regad, F., Bourgeois, E., Meynard, D., Périn, C., Sabau, X., Ghesquière, A., Glaszmann, J.C., Delseny, M. and Guiderdoni, E. (2004) High throughput T-DNA insertion mutagenesis in rice: a first step towards in silico reverse genetics. Plant Journal, 39(3):450-64.

39. Salzberg, S.L., Delcher, A.L., Kasif, S. and White, O. (1998) Microbial gene identification using interpolated Markov models. Nucleic Acids Res., 26(2):544-8.

40. Sanseverino, W., Roma, G., De Simone, M., Faino, L., Melito, S., Stupka, E., Frusciante L. and Ercolano, M.R. (2010) PRGdb: a bioinformatics platform for plant resistance gene analysis. Nucleic Acids Res., 38(Database issue):D814-21.

Bioinformatics Applications in Plant Biology

41. Schlueter, S.D., Dong, Q. and Brendel, V. (2003) GeneSeqer@PlantGDB: gene structure prediction in plant genomes. Nucleic Acids Res., 31:3597–600.
42. Scofield, S.R., Tobias, C.M., Rathjen, J.P., Chang, J.H., Lavelle, D.T., Michelmore, R.W. and Staskawics, B.J. (1996) Molecular basis of gene-for-gene specificity in bacterial speck disease of tomato. Science, 274:2063-2065.
43. Sigrist, C.J., Cerutti, L., de Castro, E., Langendijk-Genevaux, P.S., Bulliard, V., Bairoch A. and Hulo, N. (2010) PROSITE, a protein domain database for functional characterization and annotation. Nucleic Acids Res., 38(Database issue):D161-6.
44. Spannagl, M., Noubibou, O., Haase, D., Yang, L., Gundlach, H., Hindemitt, T., Klee, K., Haberer, G., Schoof, H. and Mayer, K.F. (2007) MIPSPlantsDB—plant database resource for integrative and comparative plant genome research. Nucleic Acids Res. Jan., 35(Database issue): D834-40.
45. Stoesser, G., Tuli, M.A, Lopez, R. and Sterk, P. (1999) The EMBL nucleotide sequence database. Nucleic Acids Res., 27(1):18-24.
46. Sugawara, H., Ogasawara, O., Okubo, K., Gojobori, T. and Tateno, Y. (2008) DDBJ with new system and face. Nucleic Acids Res., 36(Database issue):D22-4.
47. Swarbreck, D., Wilks, C., Lamesch, P., Berardini, T.Z., Garcia-Hernandez, M., Foerster H, Li, D., Meyer, T., Muller, R., Ploetz, L., Radenbaugh, A., Singh, S., Swing, V., Tissier C., Zhang, P. and Huala, E. (2008) The Arabidopsis Information Resource (TAIR): gene structure and function annotation. Nucleic Acids Res., 36(Database issue):D1009-14.
48. Tang, X., Frederick, R.D., Zhou, J., Halterman, D.A., Jia, Y. and Martin, G.B. (1996) Initiation of plant disease resistance by physical interaction of AvrPto and Pto kinase. Science, 274:2060-2063.
49. The Arabidopsis Genome Initiative, Analysis of the genome sequence of the flowering plant *Arabidopsis thaliana*. (2000) Nature, 408:796–815.
50. The Universal Protein Resource (UniProt) consortium, (2010), Nucleic Acids Res., 38: D142–D148.
51. Xiao, S., Ellwood, S., Calis, O., Patrick, E., Li, T., Coleman, M. and Turmer, J.G. (2001) Broad spectrum powdery mildew resistance in *Arabidopsis thaliana* mediated by RPW8. Science, 291:118-120.
52. Youens-Clark, K., Buckler, E., Casstevens, T., Chen, C., Declerck, G., Derwent, P., Dharmawardhana, P., Jaiswal, P., Kersey, P., Karthikeyan, A.S., Lu, J., McCouch, S.R., Ren, L., Spooner, W., Stein, J.C., Thomason, J., Wei, S. and Ware, D. (2010) Gramene database in 2010: updates and extensions. Nucleic Acids Res., Nov 13. [Epub ahead of print].
53. Yu, J., Hu, S., Wang, J., Wong, G.K., Li, S, Liu, B., Deng, Y., Dai, L., Zhou, Y., Zhang, X., *et al.* (2002) A draft sequence of the rice genome (*Oryza sativa L.* ssp. *indica*). Science, 296:79–92.
54. Yuan, J.S., Galbraith, D.W., Dai, S.Y., Griffin, P. and Stewart, C.N. (2008) Jr. Plant systems biology comes of age. Trends Plant Sci., 13(4):165-71.
55. Wise, R.P., Caldo, R.A., Hong, L., Shen, L., Cannon, E. and Dickerson, J.A. (2007) BarleyBase/PLEXdb. Methods Mol Biol., 406:347-63.
56. Zhang, H., Jin, J., Tang, L., Zhao, Y., Gu, X., Gao, G. and Luo, J. (2010) PlantTFDB 2.0: update and improvement of the comprehensive plant transcription factor database. Nucleic Acids Res., doi:10.1093/nar/gkq1141.

Chapter – 7

Genome Mapping in Plants

K. Nirmal Babu, S. Asha, V. Jayakumar, D. Minoo and K.V. Peter

Abstract

Manipulation of a large number of genes is often required for the improvement of even the simplest character. Molecular marker trace valuable alleles in a segregating population and mapping them. Genome mapping has emerged as a powerful new approach for research in botany, agriculture and other related fields. Genome mapping methods help to locate specific DNA markers to delineate when one has reached particular gene of interest and forming a molecular map. DNA mapping encompasses a wide range of techniques useful for studying DNA at different levels of magnification. Genetic Linkage Mapping Using Molecular Markers, Development of Genetic Maps, Mapping populations and comparative genomics are elucidated.

Introduction

Understanding biology and genetics at molecular level has become very important for better understanding and manipulation of genome architecture. Molecular markers have contributed significantly in this respect and have been widely used in plant science in a number of ways including genetic fingerprinting, identification of duplicates and selecting core collections, determination of genetic distances, genome analysis, identification of markers associated with desirable breeding traits which are useful in marker assisted breeding, genomics and development of transgenics. Use of these markers also significantly reduce breeding time and cycles required for crop improvement. Molecular level understanding of the inheritance of agriculturally important traits creates new opportunities to streamline plant breeding.

Plant improvement, either by natural selection or through the efforts of breeders, has relied upon creating, evaluating and selecting the right combination of alleles. The manipulation of a large number of genes is often required for the improvement of even the simplest of characteristics. With the use of molecular markers it is now a routine to trace valuable alleles in a segregating population and mapping them. These markers once mapped enable dissection of the complex traits into component genetic units more precisely, thus providing breeders with new tools to manage these complex units more efficiently in a breeding programme.

In the past decade, genome mapping has emerged as a powerful new approach for research in botany, agriculture as well as in many other fields. Many of the concepts associated with genome mapping are more than a century old. However recent advances in Molecular Biology have resulted in description of the structure and function of plant genome in unprecedented detail. Molecular level understanding of the inheritance of agriculturally important traits creates new opportunities to streamline plant breeding. Further, this understanding will enable us to identify specific DNA elements responsible for a particular plant characteristics. This is reflected in the fact that complete molecular and genetic maps have now been constructed for at least 10 major crop species and many more have been characterized at less defined level and the genomes of a few model plant species were completely sequenced.

Genome Mapping Methods

A "Genome map" can be like a road map, reflecting the relative proximity of different landmarks to one another. It is made possible by the fact that the nuclear genome in higher organisms are organized and transmitted as linear units called chromosomes. Genetic information supported by molecular tools enable the geneticists to establish specific DNA markers at defined places along each chromosome so as to tag agronomically important genes with molecular markers. Thus DNA markers can then be used to delineate when one has reached particular gene of interest and this forms a molecular map.

Genome Mapping in Plants

DNA mapping encompasses a wide range of techniques useful for studying DNA at different levels of magnification. The beginning of genome mapping lies in genetic linkage. Because the chromosomes and not the genes are the unit of transmission during meiosis, linked genes is not free to undergo independent assortment, *i.e.* all alleles at all loci of one chromosome should be transmitted as a unit during gamete formation. And crossing over during meiosis will result in the reshuffling or recombination of the alleles between the homologs. The degree of crossing over between any two loci on single chromosomes is proportional to the distance between them known as inter-locus distance. Thus the percentage of recombination gametes varies, depending on which loci are considered. This correlation serves the basis for the construction of chromosome map [12,1].

Conventional genetic mapping involves the crosses between individuals heterozygous for two or more genes. This allows us to estimate the relative likelihood that a cross over will occur between them. The principle underlying the mapping is, if two genes are very close together on the same chromosomes a cross over is unlikely to begin in the region between them. If two genes are far apart a cross over is more likely to initiate in this region and thereby recombine the alleles of the two genes. The presence of recombinant offsprings is correlated with the distance between the two genes. The map distance of two genes under considerations can be calculated by:

$$\frac{\text{Number of recombinant offspring}}{\text{Total number of progenies}} \times 100$$

The distance is expressed in map unit or centimorgan.

When the distance between the two genes is large the chance of multiple cross over in the region between the two genes is large. So multiple crossover sets a quantitative limit on the relationship between map distance and the percentage of recombinant offsprings. Tri hybrid crosses can be used to determine the gene order and the correct distance between them.

For example, in a three point test cross:

Parents	XYZ/xyz	X xyz/xyz
Gametes	XYZ	xyz/xyz
	XyZ	xYZ
	Xyz	xYz
	XyZ	Xyz

Progenies will be:

Sl. No.	Genotypes	Number of observations	Recombination between X and Y	Recombination between X and Z	Recombination between Y and Z
1.	XYZ/xyz	853	-	-	-
2.	xyz/xyz	926	-	-	-
3.	Xyz/xyz	51	Yes	Yes	Yes
4.	xYZ/xyz	42	Yes	Yes	-
5.	xYz/xyz	52	Yes	-	Yes
6.	XyZ/xyz	60	Yes	-	Yes
7.	XyZ/xyz	7	-	Yes	Yes
8.	Xyz/xyz	9	-	Yes	Yes

Recombination between X and Y

$$\frac{51 + 42 + 52 + 60}{2000} \times 100 = 10.25\%$$

Recombination between X and Z

$$\frac{51 + 2 + 7 + 9}{2000} \times 100 = 5.45\%$$

Recombination between Y and Z

$$\frac{52 + 60 + 7 + 9}{2000} \times 100 = 6.4\%$$

Then the Gene order will be:

$$X \underset{5.45 \text{ cM}}{\rule{2cm}{0.4pt}} Z \underset{6.4 \text{ cM}}{\rule{2cm}{0.4pt}} Y$$

Genetic Linkage Mapping Using Molecular Markers

A molecular marker is a segment of DNA, found in a specific site in the genome and has properties which enable it to be uniquely recognized using molecular tools. As with the alleles, the characteristics of molecular markers vary from one individual to another. DNA markers arise from different classes of mutations such as single base pair change altering the restriction site for an enzyme or a primer binding sites in a PCR, rearrangements in the DNA intervening between two restriction sites or primer targets or due to change in the number of tandem repeats viz. minisatellites and microsatellites. The term polymorphism refers to the idea that the individuals with in a population differ with regard to a particular fragment. The DNA segment identical among all the members of a population is said to be monomorphic. The polymorphic markers allow the experimenter to follow a character as in conventional mapping.

\# lod score is calculated as:

$$\text{lod} = \log_{10} \frac{\text{Probability of a certain degree of linkage}}{\text{Probability of independent assortment}}$$

For e .g. if the lod score $+3$, $\log_{10}$ of 3 $=1000$ and so there is a 1000 fold greater probability that the two markers are linked and then they are assorting independently. Geneticists expect that two markers are linked if the lod score

Genome Mapping in Plants 159

is +3 or higher. Mapping of genome by using molecular marker is done by making crosses and analyzing the offspring, but importance is given to the DNA fragments visualized in an agarose gel rather than phenotypical characters as with the conventional mapping[8].

The major steps involved in preparation of molecular maps are

- Prepare a suitable mapping population
- Develop molecular profiles (RFLP, RAPD, SSR, AFLP,EST and Isoenzyme)
- Score the data of the progenies as parentals and recombinants for each loci.
- Analyse the data to prepare a molecular map. Usually readymade softwares like Mapmaker, Joinmap etc are used.
- Prepare primary linkage groups – 'preliminary map'.
- Add data on more markers to saturate the map so that the markers are distributed uniformly in the genome.
- Collect segregation and recombination data on an important single gene controlled agronomic characters. Add to the map as before.
- Locate the DNA markers associated with important agronomic characters

For QTLs, a slightly different way of scoring is done to assign the different levels of expression to different multiple allells. QTL softwares are used for analysis and thus it adds characters to the molecular map. These markers can be used as probes for isolating the genes from DNA libraries and in marker assisted selection.

Genetic distances versus Physical distance

Linear arrangements of the data's obtained from the meiotic recombinations cannot pinpoint the physical whereabouts of the gene. Cytogenetic mapping provides a way to determine the locations of the genes by microscopically examining the chromosomes. Among the several techniques available, Fluorescence *In Situ* Hybridization is the most commonly used method using the cloned and labeled gene as a probe.

Development of Genetic Maps

Genetic map construction requires that the researchers select an appropriate mapping population, establish linkage groups, estimate map distances and determine map order based on pairwise recombination frequencies using these population. Selection of suitable marker system and the software for analysis are the key requirements for a molecular mapping.

Since large mapping populations are often characterized by different marker systems, map construction has been computerized. Computer software packages, such as Linkage1, Mapmaker, Mapmanager, Joinmap etc have been developed to aid in the analysis of genetic data for map construction. These programs use data obtained from the segregating populations to estimate recombination frequency that are then used to determine the linear arrangement of genetic markers.

Mapping Populations

Mapping populations are usually obtained from controlled crosses. However where such populations are difficult to obtain for example in perennials natural populations can also be used. The parents of mapping population must have sufficient variation for the traits of interest at both the DNA sequence and the phenotype level to trace the recombination events.

Since a map's economic significance will depend upon marker-trait association, as many qualitatively inherited morphological traits as possible should be included in the genetic stocks chosen as parents for generating mapping population. Consideration must be given to the source of parents (adapted vs exotic) used in developing mapping population. Sometimes even wide crosses can be used for generating mapping population. Wide crosses, generally yield greatly reduced linkage distances since chromosome pairing and recombination rates can be disturbed. To have significant value in crop improvement programme, a map made from a wide cross must have similar order of loci with map constructed using adapted parents.

Types of mapping populations

The various types of mapping populations used in linkage mapping are: (i) F_2 population; (ii) F_2 derived F_3 populations; (iii) Backcrosses; (iv) Doubled haploids (DHs); (v) Recombinant Inbred Lines (RILs); and (vi) Near-isogenic Lines (NILs).

F_2 population

F_2 population produced by selfing or sib mating the individuals in segregating populations generated by crossing the selected parents are better suited for preliminary mapping and requires less time for development. The ratio expected for dominant marker is 3:1 and for co dominant marker is 1:2:1. But they are

Genome Mapping in Plants 161

of limited use for fine mapping as linkage established using F_2 population is based on only one cycle of meiosis. Quantitative traits cannot be precisely mapped using F_2 population as each individual is genetically different and cannot be evaluated for G x E interaction in replicated trials

F_2 derived F_3 population

F_2: 3 population is obtained by selfing the F_2 individuals for a single generation and is suitable for mapping quantitative traits and recessive genes but like F_2 population, this population is not 'immortal'

Backcross mapping population

Backcross populations are generated by crossing the F_1 with either of the parents. Usually in genetic analysis, backcross with recessive parent (testcross) is used. In test cross, the progeny would segregate in a ratio of 1:1 irrespective of the nature of marker. Test cross population like an F_2 population requires less time and is not 'immortal' but has the advantage of its utility in marker-assisted backcross breeding also.

Doubled haploids (DHs)

Chromosome doubling of haploid plants from F_1 generates DHs. DHs are comparable to F_2 in terms of recombination information with an expected ratio for the marker as 1:1, irrespective of genetic nature of marker. DHs mapping populations are permanent and can be replicated and hence useful for mapping both qualitative and quantitative characters. Here recombination from the male parent alone is accounted. Often non availability of suitable culturing and haploid production methods are the limiting factors.

Recombinant inbred lines (RILs)

RILs are produced by continuous selfing or mating the progeny of individual members of an F_2 population until complete homozygosity is achieved. In RILs the genetic segregation ratio for both dominant and co dominant marker would be 1:1. Once homozygosity is achieved, RILs can be propagated indefinitely without further segregation and hence can be called immortal population. They can be replicated over locations and years and therefore are of immense value in mapping QTLs. Since RILs are obtained after several cycles of meiosis, they are very useful in identifying tightly linked makers. Developing RILs is long and relatively difficult in crops with high inbreeding depression.

Near-Isogenic lines (NILs)

NILs are generated by repeated selfing or backcrossing the F_1 plants to the recurrent parents. NILs are similar to recurrent parent but for the gene of interest. The expected segregation ratio of the markers is 1:1 irrespective of the nature of marker. NILs though take long time to develop, are 'immortal mapping population' suitable for tagging the trait of interest. They are useful in functional genomics

Characterization of mapping populations

Precise molecular and phenotypic characterization of mapping population is important for any mapping project. Since the molecular genotype of any individual is independent of environment, it is not influenced by G x E interaction. However, trait phenotype could be influenced by the environment, particularly in case of quantitative characters. Therefore, it becomes important to precisely estimate the trait value by evaluating the genotypes in multi-location testing over years using immortal mapping populations to have a valid marker-trait association.

Segregation distortion of markers

Significant deviation from expected segregation ratio in a given marker-population combination is referred to as segregation distortion. There are several reasons for segregation distortion, including: gamete/zygote lethality, meiotic drive/preferential segregation, sampling and selection during population development. Segregation distortion can also be specific with respect to some markers. It is important that the marker showing high degree of segregation distortion be eliminated from the analysis.

Mapping polygenes (Quantitative Trait Loci)

Characters whose phenotypic variation are continuous and are determined by the segregation of multiple loci have often been referred to quantitative traits and the inheritance as polygenic. The individual loci controlling a quantitative trait are referred to as polygenes or quantitative trait loci (QTLs). Phenotypic variation, which is due to segregation at a single genetic locus, is very easy to map. Quantitative traits show a continuum of the phenotypic variations with in a group of individual, for such traits it is impossible to place organisms into a discrete phenotypic class [6,4,2].

With polymorphic molecular markers and linkage maps as tools, mapping QTL is simply a matter of growing and evaluating large populations of plants, and of applying the appropriate statistical tools. Many softwares are designed to assist map making. The main stages of QTL mapping are Detection; Detection of a QTL depends on a statistical test. The power of this statistical test, defined as the probability of detecting a QTL, is dependent on the number of progeny in the population, the strength of the QTL, the type of cross, and the dominance of the QTL. For additive QTLs, an F_2 is somewhat more powerful than a backcross, but for dominant QTLs a backcross can be twice as powerful as an F_2. The number of progeny required for detecting a QTL is roughly speaking, proportional to the variance of the non-genetic (environmental) contributions and inversely proportional to the square of the strength of the QTL. The likelihood ratio statistics needed for significance is about 20 for an F_2 and about 15 for a backcross. Weak QTLs, which account for a few percent of the trait variance, may require several hundred progeny for a 50% chance of detection.

QTL localization

When several QTLs appear to contribute to a trait, evaluate them one at a time, starting with the most significant. Choose a marker locus near the most significant and copy it to a new chromosome called "Background". Use this chromosome as the background chromosome for composite interval mapping to localize the next QTL. If the next QTL is also significant, add a copy of a nearby locus to the Background chromosome and try to localize a third QTL. The process stops when the next QTL is not statistically significant.

QTL fine mapping

Fine mapping of QTLs requires the construction of special populations with large numbers of recombination in the region identified by the genome scan. The underlying assumption of using marker loci to detect polygenes is that linkage disequilibrium exists between alleles at the marker locus and alleles of the linked polygenes. Linkage disequilibrium can be defined as the nonrandom association of alleles at different loci in a population and can be caused by a number of factors including selection and genetic drift.

To discover a marker/QTL linkage, one must have a segregating (i.e. experimental) population. In plants, experimental populations such as F_2, backcross (BC), recombinant inbred (RI), and doubled haploid (DH) are easy to produce. Arguably, the most useful type of population is the DH. A simple ANOVA or t-test would be used to establish whether or not there is a significant difference in plant size between the two marker classes. It is possible to perform the same test using simple linear regression [22,24,25,18].

Softwares used in QTL mapping

Computer software packages, such as Linkage1[20] GMendel[10] Mapmanager[10] and Joinmap, have been developed to aid in the analysis of genetic data for map construction.

MapMaker/QTL: It will analyze F2 or backcross data using standard interval mapping.

QTL Cartographer: It permits analysis from F2 or backcross populations. It displays map positions of QTLs using the GNUPLOT software.

QGene: Eleven different population types (all derived from inbreeding) can be analyzed. It has a number of other features for displaying and analyzing molecular marker and phenotypic data.

MapQTL: It can analyze a variety of pedigree types including out-bred pedigrees (cross pollinators).

Map Manager QT: It will conduct single-marker regression analysis, regression-based simple interval mapping, composite interval mapping and permutation tests.

MultiCrossQTL: It can deal with a wide variety of simple mapping populations for inbred and outbred species.

Bulk Segregant Analysis

Bulk Segregant Analysis (BSA) approach uses most of the above-mentioned populations in gene tagging. In BSA, two parents (resistant and susceptible lines), showing high degree of molecular polymorphism and contrast for the target trait are crossed and F_1 is selfed to generate F_2 population. In F_2, individual plants are phenotyped for resistance and susceptibility. The DNA isolated from 10 plants in each group is pooled to constitute resistant and susceptible bulks. The resistant parent, susceptible parent, resistant bulk and susceptible bulk are surveyed for polymorphism using polymorphic markers. A marker showing polymorphism between parents as well as bulks is considered putatively linked to the target trait and is further used for mapping using individual F_2 plants. Conceptually, the genetic constitution of the two bulks is similar, but for the genomic region associated with the target trait. In absence of isogenic lines, the BSA approach provides a very useful alternative for gene tagging [11].

Genome Mapping in Plants

Mapping Strategies in Perennials

The major horticultural and plantation crops of interest like mango, apple, citrus, coconut, oil palm, cashew, cocoa, eucalyptus etc are perennials. Most of them, except a few, are perennial polyploids with long pre bearing period. They are essentially open pollinated heterozygous populations. The seed progenies segregate and are not uniform hence are propagated vegetatively for commercial purposes. The polyploid nature and long pre bearing period make breeding of these crops cumbersome, time consuming and understanding of their genetics difficult. The perennial and heterozygous nature with long pre bearing period makes conventional molecular approaches and mapping strategies not suitable for these group of crops.

Pseudo – test cross approach

A Pseudo – Test Cross approach as a Mapping Strategy was used to construct Genetic Linkage Maps in out breeding perennial forest trees like *Eucalyptus* as well as in any highly heterozygous sexually reproducing living organisms[5]. In this method, crosses are made between two heterozygous individuals and many single-dose RAPD markers will be heterozygous in one parent, null in the other and therefore segregate 1: 1 in their F_1 progeny following a testcross configuration. A large number of markers which segregate in test cross fashion in such 'pseudo test crosses' will be scored and used for making linkage groups and preparation of frame work maps with softwares like 'JOIN MAP' or 'MAPMAKER'. This strategy is efficient at the intra specific level and increasingly so with crosses of genetically divergent individuals. The ability to quickly construct single tree genetic linkage maps opens up a paradigm shift in species index map to the proposal of constructing several maps for individual trees of a population, thus mitigating the problem of LD between marker and the trait loci for marker assisted strategy for tree breeding.

Natural populations as a tool for gene mapping – association mapping

Genetic linkage maps developed on the basis of segregating families as described above have served plant geneticists and breeders well. However, there are three limitations to this method. It requires the growth of three generations before linkage analysis is possible. Very large segregating populations are needed to achieve a high resolution map. Where large mapping families do not exist, a complementary approach is to analyze Linkage Disequilibrium (LD) in natural populations. LD between two loci in natural populations is affected by all the recombination events which have happened since the two alleles appeared in some individuals of the population. Both linkage analysis and association studies rely on co-inheritance of neighboring

DNA variants. In both analyses, recombination is the main force eliminating linkage over generations. The difference is that in linkage analysis, there have been only a few opportunities for recombination to occur (one or two generations), resulting in comparatively low resolution of the map (i.e. markers may not be very closely linked with genes of interest). In association analysis with populations, recombination events over many generations eliminate the linkage between a mutated gene and molecular markers, except over very short distances. Only very close markers are in LD with the mutated gene, and form the basis for association mapping. LD measures the closeness of the genetic association between markers and a particular trait, and may be used to identify markers in close proximity to the gene(s) responsible for the trait. Thus Linkage Disequilibrium mapping or association mapping is a novel approach in plants, presenting opportunities to exploit the genetic variation in natural populations for high-resolution mapping of simple and complex traits in crop plants [14,16]. The most common soft wares used are 'TASSEL' and 'STRUCTURE'.

Comparative Genomics

In perennial polyploids where other conventional mapping strategies are found to be laborious and unsuccessful, genomics approach and using information from other sources with comparisons at heterologus genomes or genes can give us the necessary leads for tagging.

Information available from *Arabidopsis* genome for resistance to biotic and abiotic stresses can be used for tagging and isolating genes for drought, salinity and disease resistance in arid crops also. Since the pathogens are common,candidate genes responsible for pathogenesis can also be identified from the sequence information available on the similar pathogen in other crops also [9,17].

Similarly information available on mapping of heterologous loci for example, *Ph-2* locus controlling partial resistance to *Phytophthora infestance* in tomato [13], genetic and physical mapping of Molecular markers linked to the *Phytophtho*ra resistant gene *Rps 1-κ* in soybean [21] can also be used to tag *Phytophthora* resistance in citrus.

Information on similar heterologous genes and pathogens is currently available in various databases and on available BAC genomic libraries, EST libraries, short insert DNA libraries, repetitive DNA clone and molecular cytogenetic markers. The information generated will also help in better understanding of other related families.

Genome Mapping in Plants

The genomics information available in data bases and genome consortia, for example, INIBAP, *Solanum* genome initiative, Composite databases etc, can be used for crops of similar nature and family to tag genes of interest, to avoid delay. Genomics coupled with bioinformatics will help us through most of our understanding mentioned above especially when direct leads are not available.

The genomes of many crop species were sequenced or are in the final stages of sequencing completely. While the genome sequencing of *Arabidopsis thaliana*, Rice, Poplar, Potato, Grape, Papaya, *Lotus japonica, soybean etc* were completed, sequencing of Sorghum, Cassava, Barley, Wheat, Cotton, Tomato, Maize, Shepherd's purse, *Brachypodium distachyon, Medicago truncatul* and Peach are nearing completion. Sequence drafts of Cocoa and Oilpalm are getting ready.

These provide a wealth of information and its careful utilization will help in understanding the genomes of related species, genera and families.

References

1. Allard, R.W. (1999) Principles of plant breeding, 2nd edition John Wiley & Sons, Inc. New York, PP. 254
2. Collard B.C.Y, Jahufer M.Z.Z, Brouwer J.B. and Pang E.C.K. (2005) An introduction to markers, quantitative trait loci (QTL) mapping and marker-assisted selection for crop improvement: The basic concepts. Euphytica, 142:169-196.
3. Echt C.S., Erdahl L.A. and McCoy, T.J. (1992) Genetic segregation of random amplified polymorphic DNA in diploid cultivated alfalfa. Genome, 35: 84-87
4. George, A.W., Visscher, P.M. and Haley, C.S. (2000) Mapping quantitative trait loci in complex pedigrees: a two-step variance component approach. Genetics, 156: 2081-2092.
5. Grattapaglia, D. and Sederoff, R. (1994) Genetic Linkage Maps of *Eucalyptus grandis* and *Eucalyptus urophylla* using a Pseudo – Test Cross: Mapping Strategy and RAPD markers. Genetics, 137: 1121- 1137.
6. Kao, C.H., Zeng Z.B. and Teasdale, R.D. (1999) Multiple interval mapping for quantitative trait loci. Genetics, 152: 1203-1216.
7. Lander, E.S., Abrahamson, J. and Barlow, A. (1987) MAPMAKER: a computer package for constructing genetic linkage maps. Cytogenetic and Cell Genetics, 46: 642.
8. Lander, E.S. and Botstein D. (1989) Mapping Mendelian factors underlying quantitative traits using RFLP linkage maps. Genetics, 121: 185-199.
9. Long A.D. and Langley, C.H. (1999) The power of association studies to detect the contribution of candidate genetic loci to variation in complex traits Genome Res., 9:720-731.
10. Manly K.F. and Elliott, R.W. (1991) RI manager, a microcomputer program for analysis of data from recombinant inbred strains. Mammalian Genome, 1:123–126.
11. Michelmore, R.W., Paran, I. and Kesseli, R.V. (1991) Identification of markers linked to disease resistance genes by bulked segregant analysis: A rapid method to detect markers in specific genomic regions by using segregating populations. Proc. Natl Acad. Sci., USA 88:9829-9832.
12. Patterson, A.H. (1996) Genome mapping in plants. Academic Press. pp 40-90.

13. Philippe, M., Philippe, T., Jocelyne, O., Henr,i L. and Nigel, G. (1998) Genetic mapping of Ph-2, a single locus controlling partial resistance to *Phytophthora infestans*. MPMI, 11(4): 259-269.
14. Pritchard, J.K., Stephens, M. and Donnelly, P. (2000) Inference of population structure using multilocus genotype data. Genetics, 155: 945-959.
15. Pritchard, J.K., Stephens, M., Rosenberg, N.A. and Donnelly P. (2000) Association mapping in structured populations Am. J. Hum. Genet., 67: 172-181.
16. Remington, D.L., Thornsberry, J.M., Matsouka, Y., Wilson, L.M., Whitt, S.R., Doebley, J., Kresovich, S., Goodman, M.M. and Buckler, I.V.E.S. (2001) Structure of linkage disequilibrium and phenotypic associations in the maize genome. Proc. Natl. Acad. Sci. USA, 20:11479-11484.
17. Salanoubat, M., Genin, S., Artiguenave, F., Gouzy, J., Mangenot, S., Arlat, M., Billaultk, A., Brottier, P., Camus, J.C., Cattolico, L., Chandler, M., Choisen, N., Claudel-Renardl, C., Cunnac, S., Demange, N., Gaspin, C., Lavie, M., Moisan, A., Robert, C., Saurin, W., Schiex, T., Siguier, P., Thebault, P., Whalen, M., Wincker, P., Levy, M., Weissenbach, J. and Boucha, C.A. (2002) Genome sequence of the plant pathogen *Ralstonia solanacearum*. Nature, 415: 497-502.
18. Seaton,G., Haley, C.S., Knott, S.A, Kearsey, M. and Visscher, P.M. (2002) QTL express: mapping quantitative trait loci in simple and complex pedigrees. Bioinformatics, 18: 339-340.
19. Stam, P. (1993) Construction of integrated genetic linkage maps by means of a new computer package: Join Map The Plant Journal, 3(5): 739–744.
20. Suiter, K.A., Wendel, F.J. and Case. S.J. (1983) LINKAGE-1: a PASCAL computer program for the detection and analysis of genetic linkage. J. Hered., 74: 203-204.
21. Takao, K., Shanmukhaswami, S.S., Jinrui, S., Mark, G., Richard, I.B. and Madan, K.B. (1997) High resolution genetic and physical mapping of Molecular markers linked to the *Phytophtho*ra resistant gene *Rps 1*-k in soybean. MPMI, 10(9): 1035-1044.
22. Tanksley S.D. (1993) Mapping Polygenes, Ann. Rev. Genet., 27: 205-211.
23. Thornsberry, J.M., Goodman, M.M., Doebley, J., Kresovich, S., Nielsen, D. and Buckkler, E.S. (2001) *Dwarf 8* polymorphism associate dwith variation in flowering time. Nature Genetics, 28: 286-289.
24. Zeng, Z.B. (1993) Theoretical basis of separation of multiple linked gene effects on mapping quantitative trait loci. Proc. Natl. Acad. Sci. USA, 90: 10972-10976.
25. Zou, F, Yandell B.S. and Fine. J.P. (2001) Statistical issues in the analysis of quantitative traits in combined crosses. Genetics, 158:1339-1346.

Chapter – 8

Inter-Species Conservation of Splice Sites: An Analysis Using Support Vector Machine Based Pattern Recognition

Bhumika Arora and Pritish Kumar Varadwaj

Abstract

Development of reliable automated techniques for the identification and annotation of exon and intron regions in a gene has been an important area of research in the field of computational biology. Several classical approaches like probabilistic model, artificial intelligence and digital signal processing are already in use to cater to this challenging task. In this work, we propose a Support Vector Machine based kernel learning approach for detecting splice site patterns and their conservation across the species. Electron-Ion Interaction Potential (EIIP) values of nucleotides were used to map DNA character sequences to corresponding numeric sequences. Radial Basis Function (RBF) SVM kernel was trained with cross validation using these EIIP numeric sequences. This was then tested on test gene datasets of multiple species for detection of splice site pattern. Receiver Operating Characteristic (ROC) curves and various other statistical measures have been utilized for displaying the performance of the classifier. Results indicate an inter-species splice site conversion profile.

Introduction

Identification of genes from sequence data is an important area of research in the field of computational biology. The successful completion of several genome projects in the recent past yielded vast amount of sequence data. With this enormous amount of genome data available in the public domain, the development of reliable automated techniques for interpreting the long anonymous genome sequences became imperative. Rational analysis of the genome data to extract relevant information can have profound implications on automated annotation and functional motif identification. Consequently, computational approaches to gene finding have attracted a lot more attention among the genomics and molecular biology community. In eukaryotes, the complexity of gene finding approaches increases due to presence of coding regions called *exons* interrupted by non-coding regions called *introns* complemented by *intergenic* regions. The performance of the ab-inito gene identification approaches mostly depends on the effectiveness of detecting the splice sites which mark the boundaries between exons and introns. The *exon-intron* border is known as donor splice site whereas *intron-exon* border is known as acceptor splice site. Identification of *exons, introns* and splice site regions of a gene is not a new problem. There exist several probabilistic based, artificial intelligence (AI) based and Digital Signal Processing based approaches for addressing the above problem. Approaches such as Dynamic Programming, Hidden Markov Models (HMM) and Bayesian Networks fall in the first category while Artificial Neural Network (ANN), Support Vector Machine (SVM) based approaches came under the category of artificial intelligence. Furthermore, the discrete nature of DNA sequence representation has motivated many signal processing engineers to derive an equivalent numeric sequence for DNA strands and then apply various Digital Signal Processing (DSP) techniques to find some interpretable results. Proposed SVM based method for splice site detection is basically a computational intelligence based pattern recognition approach. This further explored the patterns of splice sites in model species and then the model kernel was tested using multiple species to mine the interspecies conservation profile. Herewith, we summarize a brief review of relevant previous work.

An Artificial Neural Network (ANN) based approach was adopted for gene recognition in [1-3]. Taking a sliding window of fixed size, a classifier has been developed to classify central nucleotide residue, either as coding or non-coding region. Several other variations of ANN or rule based system combined with ANN have also been used for the purpose of gene identification[4-7]. Various statistical coefficients and frequency indicators calculated from genomic sequence have been used as input features[1-7], but unfortunately, none of these methods have satisfactory levels of accuracy across the species studied. A survey of various computational approaches used in gene identification has been reported [8]. Support vector machine (SVM) is an alternative method used in machine learning which has much robust theoretical background and often it gives better results than ANN. If the input data belong to two classes with n common features then each input sample can be represented as a vector in an n-dimensional space and the classification problem reduces in finding a hyperplane in that space which should separate two classes. The SVM finds two parallel hyperplanes in the same space with any orientation such that the margin between them could be maximized. The input samples (called vectors) which fall on these parallel hyperplanes are called support vectors. Several kernel functions have been recommended for SVMs but radial basis function (RBF) often gives better performance. The performance of RBF is heavily dependent on two parameters C and γ. The optimum values of these parameters vary from problem to problem, this aspect has been ignored by most researchers while using SVM. The SVM has been used for several bioinformatics problems [9-13]; even the problem of splice site detection has been recently addressed elsewhere [14-22]. These methods involve feature and model selection (simple or hidden Markov Model) and SVM kernel engineering for splice site prediction using conditional positional probabilities. For DSP based approaches, various schemes of converting a DNA character sequence into numeric sequence and then applying various DSP techniques for gene finding and other such applications have been summarized [23-24]. The binary or Voss representation [25] one, two or three dimensional tetrahedron representation [26], EIIP method [27-28], paired numeric, paired and weighted spectral rotation (PWSR), Paired spectral content [29] etc are popular conversion techniques. All such techniques utilizes the statistical properties of *exon* and *intron* regions observed in many genes of different species e.g. period-3 property, frequency of particular nucleotides in *exon* or *intron* regions etc. Either time domain methods e.g. correlation structures [30], average magnitude difference function (AMDF) [29] etc. or frequency domain methods e.g. DFT [23, 31-32] wavelet [27-28], autoregressive [33] etc have been used for exploring above mentioned properties of genes. For overcoming spectral spreading problem of transform method, a few filtering approaches have also been suggested [34]. To summarize, although existing protocols through one of the above mentioned approaches are being adopted with satisfactory results, the datasets used for most of such studies are probably biased towards a single chromosome or single gene of a species, thus not giving a desired output even for the other genes of the same species. The novel method adopted in proposed scheme is giving high accuracy identification for a large variety of genes across the chromosomes belonging to different species with less preprocessing required. Being natural characteristic feature of nucleotides,

EIIP values appear more appealing for obtaining numeric sequences, thus this method has been used in present the study. Furthermore we found SVM to be a better classifier than artificial neural network (ANN) at least for aforesaid purpose. In this work, EIIP mapped numeric datasets of genomic sequences have been prepared for training the SVM and the optimum values of various SVM parameters have been explicitly determined. Use of these optimized values during testing phase has increased the accuracy of results by many folds.

Methodology

We have selected *Arabidopsis thaliana* as the model species for our work. On getting quite promising results for this species [35], we further extended our work to two other species viz. *Drosophila melanogaster* and *Caenorhabditis elegans* in order to perform a comparative analysis with the view of establishing an inter-species splice site pattern finding model. Genomic sequence data was collected across all the chromosomes of these three species from *The Exon-Intron Database* (EID) [36-37]. In case of genes with alternatively-spliced forms, only one isoform per gene was taken into consideration while creating the datasets for our study. Further the sequence entries were subjected to similarity screening to ensure the final set of data are with less than 23% inter similarities. The selected training and test datasets for this study consist of four types of sequences each of 12 residues window length: 1) Sequences bearing the donor splice sites, 2) Sequences bearing the acceptor splice sites, 3) Non- splice site sequences from exons and 4) Non-splice site sequences from introns. To get these 4 types of data sequences, we need to slide a window across the collected genomic sequences. We hypothesized that it should be the chemical environmental effect of the residues in the vicinity of the splice sites which makes the specific pattern detection possible. For this reason, we have considered sequences of 12 residues length (sequence window size = 12) after validating the result with different window lengths of 6, 8, 10, 12, 14 and 16 residues sequences. The ROC plot of this has been shown in Fig.8.1.

Inter-Species Conservation of Splice Sites

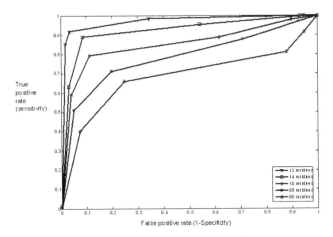

Fig. 8.1: ROC plot of perform ance vs different window size

The window size equal to 12 is selected which is moved from one end of a gene to other and the presence or absence of splice site is detected for each windows sliding with one residue ahead per move (Fig.8.2).

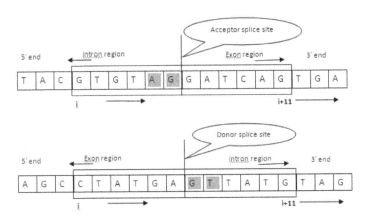

Fig. 8.2: Acceptor and splice sites

For this study three datasets were created, namely AT (from species *Arabidopsis thaliana*), DM (from species *Drosophila melanogaster*) and CE (from species *Caenorhabditis elegans*). Each of the datasets comprised of 5,000 sequences of 12 residues length each and was used for both intra and inter-species classifier validation. Each of these three datasets consist of 50% splice site bearing sequences (25% acceptor splice sites and 25% donor splice sites) and 50% non-splice site sequences (25% from *exon* and 25% from *intron* regions). Further, each of these datasets was divided into training and test sets in the ratio of 2:3. The data distribution of all the three datasets described above has been summarized in Table 8.1. Training and test data can be obtained from the

174   Agriculture Bioinformatics

supplementary files AT_train.txt, AT_test.txt, DM_train.txt, DM_test.txt, CE_train.txt and CE_test.txt respectively.

Table 8.1: Distribution of data into training and test sets.

S.N.	Species	Dataset	No. of Training set data sequences		No. of Test set data sequences		Total no. of sequences
			Splice site bearing	Non-splice site	Splice site bearing	Non-splice site	
1.	*Arabidopsis thaliana*	AT2	1000	1000	1500	1500	5000
2.	*Drosophila melanogaster*	DM	1000	1000	1500	1500	5000
3.	*Caenorhabditis elegans*	CE	1000	1000	1500	1500	5000

SVM requires that each data instance should be represented as a vector of real numbers. The genomic character sequences of training and test sets were, therefore, converted into numeric sequences using *Electron-Ion Interaction Potential* (EIIP) values of the corresponding nucleotides as given in Table.8.2.

Table 8.2:   EIIP values of nucleotide residues.

Nucleotide Letters	Name	EIIP value
A	Adenine	0.1260
G	Guanine	0.0806
C	Cytosine	0.1340
T	Thymine	0.1335

EIIP values have been used by earlier researchers [38] to represent the energy of the delocalized electrons in nucleotides and amino acids. Being natural characteristic feature of nucleotides, EIIP values appear more appealing for obtaining numeric sequences, thus this method has been used in present work. As described earlier, other potential character to numeric mapping schemes appears to be biased in attempt to exploit various statistical properties of gene. Also as compared to the use of 'binary indicator sequences' for mapping, the EIIP method reduces the computational overhead by 75% as in case of binary indicator method, 4 binary sequences are obtained per character sequence whereas in EIIP method we get only one numeric sequence per character sequence. The EIIP values, probably, carry the chemical information patterns required to be recognized by *spliceosome* [39]. Seeking a chemical environmental

Inter-Species Conservation of Splice Sites

effect is more logical than finding statistical features of the whole sequence. Therefore our 12 residue length character sequences were converted into EIIP mapped sequences of the same length which were then used for training the classifier.

Training set data were subjected to SVM classifier, which involved fixing several hyper-parameters and values of these hyper-parameters determining the function that SVM optimizes and therefore have a crucial effect on the performance of the trained classifier [40]. The classifier was trained using the training set data only whereas the testing examples were not exposed to the system during learning, kernel selection and hyper-parameter selection phases. We have used several kernels: linear, polynomials and radial basis function (RBF). We found RBF as the suitable classifier function (as the number of features was not very large), for which training errors on splice site data (false negatives) were outweighed by the errors on non-splice site data (false positives).

The classical Radial Basis Function (RBF) used in this work has similar structure as SVM with Gaussian kernel and is defined by the following equation.

$$K(x_i, x_j) = \exp\left(-\gamma \| x_i - x_j \|^2\right), \gamma > 0 \tag{1}$$

This kernel is basically suited best to deal with data that have a class-conditional probability distribution function approaching the Gaussian distribution. It maps such data into a higher dimensional space where the data becomes linearly separable. To actually visualize this, it is convenient to observe that the kernel (which is exponential in nature) can be expanded into an infinite series, thus giving rise to an infinite-dimension polynomial kernel: each of these polynomial kernels will be able to transform certain dimensions to make them linearly separable. However, this kernel is difficult to design, in the sense that it is difficult to arrive at an optimum '$\tilde{a}$' and choose the corresponding C that works best for a given problem.

While using RBF kernels, there are two parameters: C and γ. C is the cost factor i.e. the penalty parameter of the error term and $\tilde{a}$ is the kernel parameter. The best values of C and γ are not known before hand, so some kind of parameter search must be done. The goal is to identify optimum (C, γ) pair so that the classiûer can accurately predict unknown data. Achieving high training accuracy may not be that useful as the performance of a classifier is more precisely reflected by the prediction accuracy on unknown data. To address this issue, cross-validation approach was adopted. Cross-validation procedure also solves the problem of over fitting. The testing accuracy of the classifier which overfits the training data is not good.

Since searching the best hyperplane parameters is associated with the problem of over fitting, a grid parameter search exploring all combinations of C and γ with ten folds cross-validation routine, where $\tilde{a}$ ranged from 2^{-15} to 2^4 and C ranged from 2^{-5} to 2^{15} [41] has been implemented. In ten-fold cross-validation, the training dataset was divided into ten subsets of equal sizes, where one of such subsets was adopted as the test dataset while the other subsets were used for training the classifier. The process was repeated ten times, each time taking different subsets of corresponding test and training datasets. In short, iteratively each subset was tested using the classifier trained on the remaining 9 subsets. In this way, each instance in the whole training set is predicted once and so the cross-validation accuracy is the percentage of correctly classified data. Pairs of (C, γ) have been tried and the one with the best cross-validation accuracy was picked.

The best combinations of γ and C obtained from the grid based optimization process were then used for training the RBF kernel based SVM classifier using the entire training dataset. Model file of the training dataset thus obtained was used for testing. For intra-species classifier performance analysis the test was performed using model file of the training dataset from same species whereas, for inter-species classifier performance analysis test was performed taking model file of the training data from other species.

Results and Discussion

To optimize the SVM parameters γ and C, a grid search using 10-fold cross-validation routine has been applied on each of the training datasets bin, exploring various combinations of C (2^{-5} to 2^{15}) and γ (2^{-15} to 2^4). The contour plots of the grid search results for training sets AT, DM and CE are shown in Fig. 8.3, Fig. 8.4 and Fig. 8.5 respectively. The optimum values of the hyper-parameters and the corresponding ten-fold cross-validation accuracies obtained for these three training sets are given in Table 8.3.

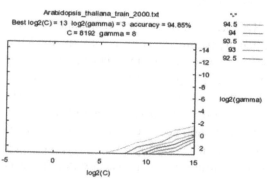

Fig. 8.3: The contour plots of the grid search results for training sets AT

Inter-Species Conservation of Splice Sites

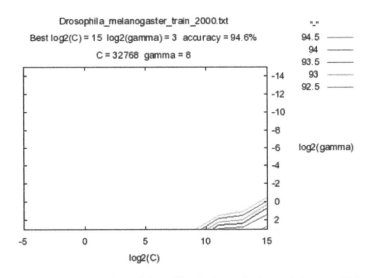

Fig. 8.4: The contour plots of the grid search results for training sets DM

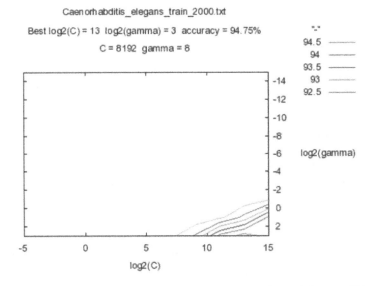

Fig. 8.5: The contour plots of the grid search results for training sets CE

Table 8.3: *Grid parameter* search results.

Training dataset	Best C	Best ã	Cross-validation accuracy
AT	8192.0	8.0	94.85%
DM	32768.0	8.0	94.6%
CE	8192.0	8.0	94.75%

The SVM classifier efficiency was evaluated by various quantitative variables: *a) TP*, true positives – the number of correctly classified splice sites, *b) FP*, false positives – the number of incorrectly classified non-splice sites, *c) TN*, true negatives – the number of correctly classified non-splice sites, *d) FN*, false negatives – the number of incorrectly classified splice sites. Using these variables, several statistical metrics were calculated to measure the effectiveness of the proposed RBF-SVM classifier. *Sensitivity (Sn)* and *Specificity (Sp)* metrics, which indicates the ability of a prediction system to classify the splice site and non-splice site information were calculated and receiver operating characteristic curves (ROC) for the same have been plotted. To indicate an overall performance of the classifier system; Accuracy *(Ac)*, for the percentage of correctly classified splice sites and non-splice sites, the Matthews Correlation Coefficient *(MCC)* and the Youden's Index were computed.

Values obtained for the four quantitative variables using different training and test datasets are shown in Table 8.4.

Table 8.4: Values of quantitative variables obtained using different training and test datasets.

S.No.	Test set / No. of data	Training set/ No. of data	TP	FP	TN	FN
1.	AT / 3000	AT / 2000	1454	119	1381	46
2.	DM / 3000	DM / 2000	1433	125	1375	67
3.	DM / 3000	AT / 2000	1470	172	1328	30
4.	DM / 3000	CE / 2000	1394	118	1382	106
5.	CE / 3000	CE / 2000	1425	123	1377	75
6.	CE / 3000	AT / 2000	1439	176	1324	61

Sensitivity, defined as the proportion of actual positives that are correctly identified by the test, was calculated by the following equation:

$$Sensitivity\,(\%)\frac{TP}{TP+FN}\times100 \tag{2}$$

Sensitivities obtained for our classification model using different training and test sets are shown in Table 8.5. The comparative values of these sensitivities indicate towards the generalization of our model for splice site detection over different species.

Sensitivity does not tell us as to how well the classifier predicts for negative cases. This issue was addressed by specificity which measures the proportion of negatives correctly identified and was calculated as follows:

$$Specificity\,(\%)=\frac{TN}{TN+FP}\times100=(\textit{false positive rate}) \tag{3}$$

Inter-Species Conservation of Splice Sites 179

Specificity values obtained for our classification model using different training and test sets are also shown in Table 8.5.

Table 8.5: Sensitivity and Specificity values for the classifier using different training and test datasets.

S.No.	Test set / No. of data	Training set/ No. of data	Sensitivity	Specificity
1.	AT/3000	AT/2000	96.93 %	92.07 %
2.	DM/3000	DM/2000	95.53 %	91.67 %
3.	DM/3000	AT/2000	98.00 %	88.53 %
4.	DM/3000	CE/2000	92.93 %	92.13 %
5.	CE/3000	CE/2000	95.00 %	91.80 %
6.	CE/3000	AT/2000	95.93 %	88.27 %

For analyzing the performance of the classifier and displaying the sensitivity rate, the ROC curves were plotted and area under curves calculated. ROC curve is two dimensional depiction of classifier performance. For comparing the classifiers, it may be required to reduce the ROC performance to a single scalar value representing expected performance. An easiest possible method for this is to calculate the area under the ROC curve. If AUC is close to 1, it indicates very good classifier performance. ROC plots for the respective datasets have been shown in Fig. 8.6-11. Cut-off point (encircled) shown in these plots represent the point of best sensitivity and specificity. The diagonal line is the *line of no discrimination*. The Area under curve (AUC) values, standard errors and 95% confidence intervals obtained for our classifier using different training and test datasets are shown in Table 8.6.

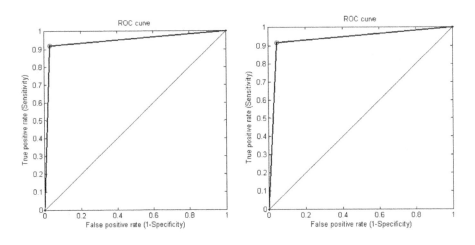

Fig. 8.6: ROC plot for set 1   Fig. 8.7: ROC plot for set 2

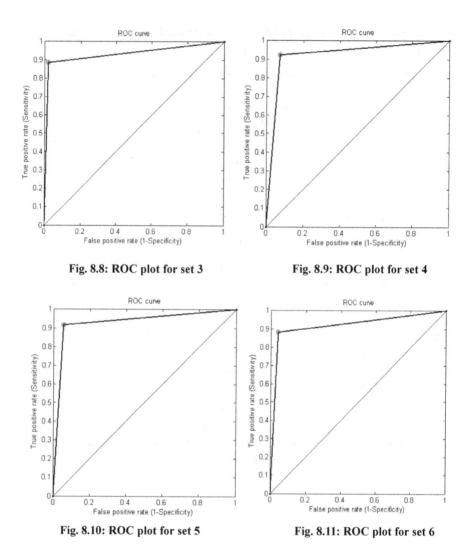

Fig. 8.8: ROC plot for set 3

Fig. 8.9: ROC plot for set 4

Fig. 8.10: ROC plot for set 5

Fig. 8.11: ROC plot for set 6

Table 8.6: AUC values and the related metrics for the classifier using different training and test datasets.

S.No	Test set / No. of data	Training set / No. of data	AUC	Standard error	95% Confidence Interval
1.	AT / 3000	AT / 2000	0.95912	0.00372	0.95183 - 0.96641
2.	DM / 3000	DM / 2000	0.95647	0.00384	0.94895 - 0.96400
3.	DM / 3000	AT / 2000	0.94152	0.00445	0.93279 - 0.95025
4.	DM / 3000	CE / 2000	0.95789	0.00378	0.95049 - 0.96529
5.	CE / 3000	CE / 2000	0.95695	0.00382	0.94947 - 0.96443
6.	CE / 3000	AT / 2000	0.93895	0.00455	0.93003 - 0.94787

Inter-Species Conservation of Splice Sites 181

Youden's Index (J), a function of sensitivity and specificity, which is a commonly used measure of overall effectiveness of a classifier, was calculated as:

$$J = \text{sensitivity} + \text{specificity} - 1 \qquad (4)$$

The values of the Youden's Index obtained for the classifier using different training and test datasets are shown in Table 8.7. As the values obtained for Youden's Index are close to 1, they indicate high effectiveness of our classifier.

Matthews correlation coefficient (MCC) which is regarded as a balanced measure, for the quality of binary classification, that can be used even when classes are of very different sizes was evaluated using the following equation:

$$MCC = \frac{TP + TN - FP + FN}{\sqrt{(TP + FP) + (TP + FN)(TN + FP)(TN + FN)}} \qquad (5)$$

Matthews's correlation coefficients obtained for our classification model using different combinations of training and test datasets are shown in Table.7. These values are found to be close to +1, therefore, the predictive ability of the classifier can be regarded as being reasonably good.

Table 8.7: Values of the Youden's Index and Matthews Correlation Coefficient values for the classifier using different training and test datasets.

S.No.	Test set / No. of data	Training set / No. of data	Youden's Index	MCC
1.	AT / 3000	AT / 2000	0.8900	0.8911
2.	DM / 3000	DM / 2000	0.8720	0.8727
3.	DM / 3000	AT / 2000	0.8653	0.8692
4.	DM / 3000	CE / 2000	0.8506	0.8507
5.	CE / 3000	CE / 2000	0.8680	0.8684
6.	CE / 3000	AT / 2000	0.8420	0.8445

For demonstrating the overall predictive performance of our classification model, the accuracy for the percentage of correctly classified splice sites and non splice sites was computed. It is defined as the percentage of correctly identified samples (both true positives and true negatives) and was evaluated using the following formula:

$$Ac(\%) = \frac{TP + TN}{TP + TN + FP + FN} \times 100 \qquad (6)$$

Accuracies obtained for the classifier using different combinations of training and test sets are shown in Table.8.8.

Table 8.8: Accuracies obtained for the classifier using different training and test datasets.

S.No.	Test set/No. of data	Training set/No. of data	Accuracy
1.	AT / 3000	AT / 2000	94.50 %
2.	DM / 3000	DM / 2000	93.60 %
3.	DM / 3000	AT / 2000	93.2667 %
4.	DM / 3000	CE / 2000	92.5333 %
5.	CE / 3000	CE / 2000	93.40 %
6.	CE / 3000	AT / 2000	92.10 %

The overall accuracies, area under ROC curves and other statistical metrics obtained as a result of present study are significantly higher than those for the existing methods. While the reported accuracy on the training datasets may indicate the effectiveness of a prediction method, it may not accurately portray how the method will perform on novel, hitherto undiscovered splice sites. Therefore, testing the SVM methodology on independent out-of-sample datasets, not used in the cross-validation was critical. Training set data was used for training the classifier, while the testing examples were not exposed to the system during learning and hyper-parameter optimization phases. Use of these optimized values of SVM parameters during testing phase has increased the accuracy of results by many folds.

Conclusion and Future Work

An overall prediction accuracy of 94.5% and other performance evaluation metrics approaching perfect prediction values, for model species *Arabidopsis thaliana* are indicative of the validity of proposed RBF-SVM based classifier for splice site detection. Further, very good predictive performance values obtained for other two species, *Drosophila melanogaster* and *Caenorhabditis elegans*, both in case of intra and inter-species classifier performance analysis studies direct towards the generalization of our classification model over species and establishment of an inter species splice site pattern finding model. These findings suggest that the SVM-based prediction of splice sites might be helpful in identifying potential *exon-intron* boundaries and hence gene annotations. Classifier generated for *Arabidopsis thaliana* can also very efficiently predict the splice site information of other two species in study *Drosophila melanogaster* and *Caenorhabditis elegans;* this shows that there exist an evolutionary conservation among spice site pattern. Similar trend can be found by testing dataset of *Drosophila melanogaster* on model generated by *Caenorhabditis elegans*. The results obtained strongly suggest an interspecies conservation profile among the splice site pattern.

In the process of spliceosome mediated RNA splicing, introns are removed from transcribed pre-mRNA and exons are joined to form the final mRNA, for translation into successive protein. *Spliceosome* recognizes the splice sites (donor and acceptor) on pre-mRNA and it does so for invariably large ranges of pre-mRNA with very accurate intron-exon boundary recognition. This can be only possible if either the intron-exon boundary carries a typical pattern in it or somehow spliceosome can remember the boundary sequence information. But the number of possible proteins in a species and associated splice sites, support only the former theory that; spliceosome does not have memory in it rather it recognizes splice site by typically conserved residual-chemical pattern. The importance of EIIP values as chemical features has already been established by the researchers, we further used 12 residues length sequence which probably carry the chemical environmental effect of splice site surrounding. Seeking a chemical environmental effect is more logical than finding statistical features of the whole sequence and hence this work has an extra edge over other works in the sense that it is probably closer to the phenomenon adopted by *spliceosome* in real life.

This work was started with a simple thought of mathematical modeling of the detection of intron-exon boundary information, the very phenomenon used by *spliceosome* in nature. The results obtained for all three species under study with very high degree of accuracy are quite encouraging for further extension into other species. We also like to extend our work for adopting more robust way of character to numeric sequence conversion, enhancing the accuracy with fine tuning of SVM parameters and window size.

References

1. Uberbacher E.C., Xu, Y. and Mural R.J. (1996) Discovering and understanding genes in human DNA sequence using GRAIL. Methods Enzymol, 266: 259–281.
2. Fickett J.W. and Tung C.S. (1992) Assessment of protein coding measures. Nucleic Acids Res., 20:6441–6450.
3. Fogel, G.B., Chellapilla, K. and Corne, D.W. (2002) dentification of coding regions in DNA sequences using evolved neural networks. In: Evolutionary Computation in Bioinformatics, G. B. Fogel and D. W. Corne, Eds. San Fransisco, CA: Morgan Kaufmann, pp. 195–218.
4. Hebsgaard, S.M., Korning P.G. and Tolstrup N *et al.* (1996) Splice site prediction in Arabidopsis thaliana pre-mRNA by combining local and global sequence information. Nucleic Acids Res., 24(17): 3439–3452.
5. Reese, M.G. (2001) Application of a time-delay neural network to promoter annotation in the Drosophila melanogaster genome. Comput. Chem., 26(1): 51–56.
6. Ranawana, R. and Palade, V. (2005) A neural network based multi-classifier system for gene identification in DNA sequences. Neural Comput. Appl., 14(2):122-131.
7. Sherriff, A. and Ott, J. (2001) Applications of neural networks for gene finding. Adv. Genet., 42: 287–297.

8. Bandyopadhyay, S., Maulik, U. and Roy, D. (2008) Gene identification: classical and computational intelligence approaches. IEEE Transaction on systems, man and cybernatics, 38:1.
9. Rätsch, G., Sonnenburg, S. and Srinivasan, J. *et al.* (2007) Improving the caenorhabditis elegans genome annotation using machine learning. PLoS Computational Biology, 3(2): e20.
10. Jaakkola, T. and Haussler, D. (1999) Exploiting Generative Models in Discriminative Classifiers. *In: Advances in Neural Information Processing Systems* Vol. 11. Edited by: Kearns M, Solla S, Cohn D. Cambridge, MA: MIT Press, pp. 487-493.
11. Zien, A., Rätsch, G. and Mika, S. *et al.* (2000) Engineering support vector machine kernels that recognize translation initiation sites. BioInformatics, 1.
12. Brown, M.P.S., Grundy, W.N. and Lin, D *et al.* (2000) Knowledge-based analysis of microarray gene expression data using support vector machines. PNAS, 97(1): 262-267.
13. Tsuda, K., Kawanabe, M. and Rätsch, G. *et al.* (2002) A New Discriminative Kernel from Probabilistic Models. Advances in Neural information processing systems, 14:977-984.
14. Sonnenburg, S., Rätsch, G. and Jagota, A. *et al.* (2002) New Methods for Splice-Site Recognition. Proc. ICANN.
15. Sonnenburg, S. (2002) New Methods for Splice Site Recognition. *In: Master's thesis Humboldt University* [Supervised by K.R. Müller, H.D. Burkhard and G. Rätsch].
16. Lorena, A. and de Carvalho, A. (2003) Human splice site identifications with multiclass support vector machines and bagging. Artificial Neural Neural Networks and Neural Information Processing – ICANN/ICONIP, LNCS, 2714: 234–241.
17. Yamamura, M. and Gotoh, O. (2003) Detection of the Splicing Sites with Kernel Method Approaches Dealing with Nucleotide Doublets. Genome Informatics, 14: 426-427.
18. Rätsch, G. and Sonnenburg, S. (2004) Accurate splice site detection for caenorhabditis elegans. In: Kernel Methods in Computational Biology Edited by: B Schölkopf KT, Vert JP. MIT Press.
19. Degroeve, S., Saeys, Y. and Baets, B.D. *et al.* (2005) SpliceMachine: predicting splice sites from high-dimensional local context representations. *Bioinformatics*, 21(8): 1332-1338.
20. Huang, J., Li, T. and Chen, K. *et al.* (2006) An approach of encoding for prediction of splice sites using SVM. Biochimie, 88(7): 923-929.
21. Zhang, Y., Chu, C.H. and Chen, Y. *et al.* (2006) Splice site prediction using support vector machines with a Bayes kernel. Expert Systems with Applications, 30:73-81.
22. Baten, A., Chang, B. and Halgamuge, S. *et al.* (2006) Splice site identification using probabilistic parameters and SVM classification. BMC Bioinformatics, 7(Suppl 5): S15.
23. Anastassiou, D. (2001) Genomic signal processing. IEEE Signal Process. Mag, 2001, 18(4)8-20.
24. Zhang, X., Chen, F. and Zhang, Y. *et al.* (2002) Signal processing techniques in genomic engineering. Proc. IEEE, 90(12):1822-1833.
25. Voss, R.F. (1992) Evolution of long-range fractal correlations and 1/f noise in DNA base sequences. Phy. Rev. Lett., 68(25): 3805–3808.
26. Silverman, B.D. and Linsker, R.A. (1986) Measure of DNA periodicity. J. Theor. Biol., (118): 295-300.
27. Ning, J., Moore, C.N. and Nelson J.C. (2003) Preliminary wavelet analysis of genomic sequences. In: Proc. IEEE Bioinformatics Conf., 509–510.
28. Rao, K.D. and Swamy, M.N.S. (2008) Analysis of Genomics and proteomics using DSP Techniques. IEEE Transactions on circuits and systems, 55(1).
29. Akhtar, M., Epps, J. and Ambikairajah, E. (2008) Signal processing in sequence analysis: advances in eukaryotic gene prediction. IEEE Journal of selected topics in signal processing, 2(3).

Inter-Species Conservation of Splice Sites 185

30. Li, W. (1997) The study of correlation structure of DNA sequences: A critical review. Comput. Chem., 21(4) 257–271.
31. Tiwari, S., Ramaswamy, S. and Bhattacharya, A. *et al.* (1997) Prediction of probable genes by Fourier analysis of genomic sequences. Comput. Appl. Biosci., 13: 263–270.
32. Kotlar, D. and Lavner, Y. (2003) Gene prediction by spectral rotation measure: A new method for identifying protein-coding regions. Genome Res., 18:1930–1937.
33. Rao, N. and Shepherd, S.J. (2004) Detection of 3-periodicity for small genomic sequences based on AR techniques. In: Proc. IEEE Int. Conf. Comm., Circuits Syst., 2: 1032–1036.
34. Vaidyanathan, P.P. and Yoon, B.J. (2002) Gene and exon prediction using allpass-based filters. Presented at the IEEE Workshop Genomic Signal Processing and Statistics, Raleigh, NC.
35. Varadwaj, P., Purohit, N. and Arora, B. (2009) Detection of Splice Sites Using Support Vector Machine. Springer Communication in Computer and Information Science, Proceedings of IC3, 40: 493-502.
36. Saxonov, S., Daizadeh, I. and Fedorov, A. *et al.* (2000) An exhaustive database of protein-coding intron-containing genes. Nucleic Acids Res., 28(1): 185-190.
37. Shepelev, V. and Federov, A. (2006) Advances in the Exon-Intron Database EID. Briefings in Bioinformatics, 7(2): 178-185.
38. Cosic, I. (1994) Macromolecular bioactivity: is it resonant interaction between macromolecules?—Theory and applications. IEEE Trans Biomed Eng., 41(12):1101-1114.
39. Burge, C.B. *et al.* (1999) Splicing precursors to mRNAs by the spliceosomes. In: Gesteland, R.F., Cech, T.R., Atkins, J.F., The RNA World, Cold Spring Harbor Lab. Press, 525–560.
40. Cortes, C. (1995) Vapnik, V. Support-Vector Networks. Machine Learning, 20: 273-297.
41. Chih-Chung Chang and Chih-Jen Lin. (2001) LIBSVM : A library for support vector machines, Software available at http://www.csie.ntu.edu.tw/~cjlin/libsvm

Chapter – 9

Applications of Support Vector Machines in Plant Genomes

Shimantika Sharma, Sona Modak and V.K. Jayaraman

Abstract

Support Vector Machine (SVM) is a classification and regression prediction tool, widely employed in different fields of science and engineering. A review on some recent plant bioinformatics applications using SVM is made. Different soft wares for SVM are also explained. There is a growing amount of plant genomes data currently available from different sources. Recent advances made using SVM and Feature selection in plant bioinformatics are reviewed

Introduction

Recent developments in genomic and post-genomic research have generated a vast amount of biological data. This data is growing exponentially with the advancement of research technologies. Several plant genomes have been sequenced recently. For instance, the complete sequence of *Arabidopsis thaliana* and the draft sequence of rice are now available [1]. Also, sequencing of several other plants including maize, tomato, sorghum, *etc* is in progress. Considering the amount and complexity of this data, it is impossible for an expert to analyze and compare the data manually. Thus, there is an increasing need for computational methods that can efficiently store, organize and interpret this large amount of data.

Bioinformatics is one interdisciplinary science which uses computational techniques to solve biological problems. The field of Bioinformatics mainly deals with:

- The development of databases which store, manage and provide easy access to vast amount of biological data.
- The development of novel algorithms to solve several biological problems which involves protein structure and function identification, locating a gene within a sequence and grouping protein sequences into families.

A particular active area of research in bioinformatics is the application of machine learning tools to extract important and useful information from a large pool of biological data. Machine learning algorithms are built in a way such that they can easily recognize complex patterns and further make intelligent decisions based on the data. For solving classification problems, machine learning techniques first obtain information from a set of already classified samples (training set) and then use this information to classify unknown samples (test set).

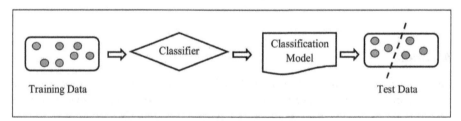

This article focuses on the application of Support Vector Machine (SVM) as a classifier and Feature Selection as a pre-processing step for the identification of plant gene and protein functions. The first section introduces Support Vector Machine for classification. The second section introduces Feature Selection and focuses on some widely used feature selection methods. In the third section, we discuss some of the emerging research topics involving the application of SVM in plant bioinformatics. The fourth section lists various SVM softwares currently in use.

Support Vector Machine

Support Vector Machine (SVM) is a classification and regression prediction tool invented by Vladimir Vapnik. SVM is rigorously based on statistical learning theory. For linearly separable problems SVM employs a maximum margin hyperplane for separating examples belonging to two different classes. For nonlinearly separable problems, SVM first transforms the data into a higher dimensional feature space and subsequently employs a maximum margin linear hyper plane. To deal with the intractability problems SVM judiciously employs appropriate kernel functions so that all computations can be performed in the original input space

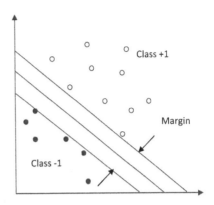

itself. It can be shown that such a methodology requires solution of a convex quadratic optimization problem. This formulation has the beneficial property of possessing a unique global solution. This property along with superior performance capabilities has made SVM a very popular classification tool widely employed in different fields of science and engineering.

Feature Selection

It is very essential to identify the relevant biological features in large and complex datasets in order to understand the molecular mechanisms in plants that regulate biological processes underlying the data. The relevant plant information is provided to the machine learning methods in the form of a set of features that describe the plant data. In most cases however, not all of these features contribute to the classification task. If such features are not eliminated from the dataset, the high dimension of the dataset poses computational difficulties, and introduces unnecessary noise in the process, decreasing the classification performance of the learning algorithm and increasing the computation time. Thus, there is a need to incorporate techniques that search for an "informative" set of features with "increased" classi-fication performance. These dimensionality reduction techniques are often referred to as Feature Selection [2].

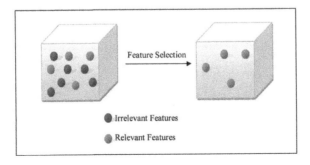

Feature selection algorithms mainly fall into two categories:

- **Wrappers** that estimate the quality of features using the accuracy predicted by the learning algorithm. Methods like Ant Colony Optimization and Genetic Algorithm in combination with a classifier with Support Vector Machines (SVM) fall into this category.

- **Filters** that evaluate the quality of features according to heuristics based on general characteristics of the individual features that is not dependent on the learning algorithm. Methods based on statistical tests and mutual information fall into this category.

Since wrappers employ a learning algorithm for evaluating the quality of the features, they give more accurate results than the filter methods. However, because they need to train the learning algorithm, wrappers are more time consuming than filters. In addition, wrappers need to be re-executed when switching from one learning algorithm to another [3].

Most of the classification algorithms involve the following steps:

- **Feature Extraction:** A pre-processing step in which a compendium of features is extracted which is capable of describing a large data accurately.

- **Feature Selection:** Out of all the domain features extracted in the previous step, only the relevant ones are selected in order to reduce the size of the data and to enhance the quality of prediction.

- **Classification:** The data are then grouped into different classes with the help of a classifier.

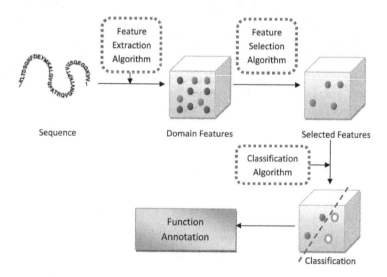

Review of Some Recent Plant Bioinformatics Applications Using SVM

Recently, SVM has been applied to solve a number of Plant Bioinformatics problems. This section highlights some of these applications of SVM in Plant Bioinformatics.

1. Micro - RNA Target Prediction in *Arabidopsis Thaliana*

Micro RNAs (miRNAs) are small non coding RNA, usually 21-23 nucleotides long that play an important role in post transcriptional regulation. They affect development by binding to a target gene and antagonizing various pathways of expression. A prediction mechanism has been presented that is based on two features, antisense transcription and small RNA abundance [4]. These two features are related to each other. A newly discovered phenomenon in antisense strand of the target genes along with small RNAs and a commonly used indicator of binding sites were used in a Support Vector Machine to build a prediction model and were analyzed using the output of the Support Vector Machine. The predicted and validated classifications were compared and the importance of the features was evaluated. It was suggested that the newly discovered feature may be able to identify new miRNA target sites in Arabidopsis and other species based on the accuracy, specificity, sensitivity and precision of the SVM results.

2. Analysis of Antisense Expression by Whole Genome Tiling Microarrays and siRNAs Suggests mis-annotation of Arabidopsis Orphan Protein-coding Genes.

Studies conducted on rice (Oryza sativa) sense and antisense gene expression in publicly available whole genome data and sequenced smRNA libraries found MIRNA genes of transitivity indicated to be similar to that found in Arabidopsis [5]. Statistical analysis of the antisense transcripts, presence of antisense ESTs, and its involvement with smRNAs suggest hundred Arabidopsis 'orphan' hypothetical genes to be non-coding RNAs. Relaying with this hypothesis, newer Arabidopsis homologues of some of MIRNA genes on the antisense strand of already annotated protein-coding genes were found. A Support Vector Machine (SVM) was used to build a prediction model of miRNA targets. miRNA target models were build using thermodynamic energy of binding and new expression features of sense and antisense transcription topology and siRNA. The SVM was trained on targets to predict the most conserved class of validated Arabidopsis MIRNA genes. The prediction accuracy of 84% and 76% was obtained for "new" rapidly evolving MIRNA genes. The results showed that antisense and smRNA expression features employing SVM can identify novel MIRNA genes and other non-coding RNAs in plants, which can be used further for exploring the miRNA evolution and post-transcriptional gene regulation.

3. PlantMiRNAPred: Efficient Classification of Real and Pseudo Plant pre-miRNAs.

MicroRNAs (miRNAs) are short 21-24nt non-coding RNAs that play important role in post-transcription as regulators in animals and plants. Many existing methods use comparative genomics to identify plant precursor miRNAs (pre-miRNAs), others are based on complementary characteristics between miRNAs and target mRNAs sequences. However the limitation is they can only identify the homologous and complementary miRNAs. Furthermore ab-initio methods for animals cannot be used for plants. Therefore, a Support Vector Machine classification method was developed for specifically predicting plant pre-miRNAs [6]. The efficient prediction was done by extracting the pseudo hairpin sequences from protein coding sequences of Arabidopsis thaliana and Glycine max. Further a set of informative features were selected to improve the accuracy. Training samples were selected on the basis of their distributions in high-dimensional sample space. The classifier PlantMiRNAPred achieved more than 90% on plant datasets which were from 8 plant species. The species used included *A.thaliana, Oryza sativa, Populus trichocarpa, Physcomitrella patens, Medicago truncatula, Sorghum bicolor, Zea mays* and *Glycine max.* The accuracy suggested that pseudo pre-miRNAs along with the extracted selected features for training dataset gives better performance. This can be used for discovering new non-homologous plant pre-miRNAs.

4. RSLpred: an Integrative System for Predicting Subcellular Localization of Rice Proteins Combining Compositional and Evolutionary Information.

Oryza sativa was presented in a complete map-based sequence which was a major milestone for researchers. Identification of localization of encoded proteins is the key in understanding the functional characteristics and its purification. In this proposed method, RSLpred, is such an effort in genome-scale subcellular prediction of encoded proteins. SVM based modules have been developed using traditional amino acid, dipeptide and four parts-amino acid composition and achieved an overall accuracy of 81.43, 80.88 and 81.10% respectively [7]. Other SVM based modules have been developed using position-specific iterated-basic local alignment search tool and achieved 68.35% accuracy. Similarly another module was developed using evolutionary information of a protein sequence extracted from position-specific scoring matrix achieved an accuracy of 87.10%. A large number of modules have developed using various schemes like higher-order dipeptide composition, N- and C-terminal splitted amino acid composition and hybrid information. The benchmarking test for RSLpred was done on independent set of rice proteins, performing better than other prediction methods. An online web tool has also been developed.

5. Prediction of Fungicidal Activities of rice Blast disease Based on Least-squares Support Vector Machines and Project Pursuit Regression.

Rice blast disease is frequently studied under agricultural research. Machine learning methods such as genetic algorithm-multilinear regression (GA-MLR), least-squares support vector machine(LS-SVM), and project pursuit regression(PPR), were used to find the relationship between thiazoline derivatives and fungicidal activities with rice blast disease. The GA-MLR method was used to generate molecular descriptors from a large set of descriptors which were used for developing a linear QSAR models. On the basis of the selected descriptors , other models were developed for the same. Linear and nonlinear models gave good prediction results, but the nonlinear models gave better prediction. Thus LS-SVM and PPR methods were able to relate the relationship between structural descriptors and fungicidal activities [8]. The same were used for the study of rice blast, allowing the design and development of new fungicidal compounds to resist such diseases.

6. Prediction of Enological Parameters and Discrimination of Rice Wine age Using least-squares Support Vector Machines and Near Infrared Spectroscopy

Least-squares support vector machines (LS-SVM) combined withnear-infrared (NIR) spectra have been used for prediction of enological parameters and for discriminating wines on the basis of age [9]. Principle Component Analysis (PCA) were used for the scores of first ten principal components (PCs) along with radial basis function (RBF) as kernel function of LS-SVM models. As compared to partial least-squares (PLS) regression, the performance of LS-SVM was slightly better with better and higher determination coefficients for validation (Rval2) and lower root-mean-square error of validation (RMSEP) for alcohol content, titratable acidity, and pH, respectively. LS-SVM gave better results than discriminant analysis (DA) to discriminate rice wine age. LS-SVM together with NIR spectroscopy is a reliable method for rice wine quality estimation.

7. Splicing-site Recognition of Rice (*Oryza sativa L.*) DNA Sequences by Support Vector Machines

Support Vector Machines were used to predict the splicing site in eukaryotic organisms conforming to the principle of GT-AG. Using the same method it was used to test the effect of the test data set of true and pseudo splicing sites in (*Oryza sativa L.*) rice DNA sequences [10] A model was built and used for prediction. The prediction accuracy of 87.53% was obtained at the 5' end splice site and an accuracy of 87.37% at the 3' end splice site. This results suggested that the splicing site recognition of rice DNA sequences can be efficiently done for other plant genomes too.

8. *SpliceIT*: A Hybrid Method for Splice Signal Identification Based on *Probabilistic and Biological Inference*

Splicing sites are defined by the boundaries of exonic regions and order protein synthesis and function. SpliceIT introduces a hybrid method which predicts the splice site that uses a combination of probabilistic modeling and using other discriminative features[11]. Gaussian support vector machine (SVM) is used to train the probabilistic profile that is extracted using two alternative position dependent feature selection methods. Then the extracted predictions are combined with known regulatory elements in order to induce a tree-based modeling. The evaluation was done on *Arabidopsis thaliana* splice site datasets which showed highly accurate, sensitive, specific splice signal identification.

9. Automated Arabidopsis Plant Root Cell Segmentation Based on SVM Classification and Region Merging

Automating any cell processes is important in research to understand the vital processes such as cell division. In this study an automatic cell detection approach for *Arabidopsis thaliana* which is based on segmentation [12]. The segmentation is done by selecting the best cell candidates from a watershed-based image and improves the result by merging adjacent regions. The selection is done by using SVM classifier, which is based on a cell descriptor obtained from the shape and edge strength of the cells contour. Here a newly proposed cell merging criterion was used which is based on edge strength. The results showed that after merging with basic watershed segmentation, gave 1.5% better coverage and up to 27% better precision in correct cell segmentation, which facilitates the study of cells and particularly used in automation of Arabidopsis plant root cell segmentation.

10. Combining Pareto-optimal Clusters Using Supervised Learning for Identifying Co-expressed Genes

Clustering algorithms have been deployed effectively to identify data as optimization problem in case of microarray datasets. The problem with fuzzy clustering in microarray data is that each of these clustering solutions have same amount of information regarding the clustering structure of the input data. Due to this, a newer fuzzy voting approach is used which combines the clustering information from all the solutions in the resultant Pareto-optimal set. This enables to identify the genes which are assigned to some cluster with high degree. The remaining genes were classified using Support Vector Machine (SVM) classifier using the set of genes which were identified and clustered[13]. The proposed clustering technique was carried out on five publicly available benchmark microarray datasets which also contained Arabidopsis thaliana. Comparative studies showed that using different kernel types in SVM have been already done which used microarray datasets. Biological significance for the tests has also been carried using web based gene annotation tools that the proposed clustering method using SVM can be used to identify co-expressed genes efficiently. This has been used to find the genes which belong to same functional groups.

11. Prediction of Regulatory Interactions in Arabidopsis using Gene-Expression Data and Support Vector Machines

Predicting the regulatory interactions between transcription factors (TF) and their targets is widely carried out genomic research. SVM and gene-expression data have been used to predict the regulatory interactions in Arabidopsis[14]. A set containing 125 validated TF target interactions and 750 negative regulatory gene pairs were taken as the training data. Prediction was done using SVM. The overall prediction accuracy of 88.68% was obtained which was used for the correct prediction of regulatory interactions in Arabidopsis which uses expression data and SVM to construct the gene regulatory network in a cost-effective manner.

12. Predicting Genome-wide Redundancy Using Machine Learning

Gene duplication leads to many genetic redundancies, which hide the function of mutated genes. Evolutionary changes due to gene duplication can be studied by identifying the genetic redundancy. Support Vector Machines have been used for classifying gene family members into redundant and non-redundant gene pairs in many species such as *Arabidopsis thaliana* [15]. Combining multiple attributes can be used to predict the genetic redundancy in comparison to single attribute such as BLAST E-values or expression correlation. Employing SVM classification reached the level where most of the redundant cells were correctly labeled. SVM predicts that about half of all genes in Arabidopsis showed the mark of predicted redundancy with at least one but typically less than three other family members. Also showed that large amount of predicted redundant gene pairs were old duplications, providing insights that redundancy or duplication is prominent and constant over long evolution and will exhibit redundancy with very few genes within a same family.

13. PAIR: the Predicted Arabidopsis Interactome Resource

The predicted Arabidopsis interactome resource is the most complete dataset of the Arabidopsis interactome with high reliability. It consists of 5990 predicted interactions in *Arabidopsis thaliana* together with 145,494 predicted interactions. The interactions were fine tuned model that was created using Support Vector Machine (SVM) that integrates gene co-expressions, domain interaction, phylogenetic profile , homologous interactions in other organisms (interlogs) [16]. Predictions were covered to approximately 24% for entire Arabidopsis interactome and reliability estimated to 44%. Validations were done using two independent datasets. PAIR has user friendly interface, rich annotation between two proteins.

14. A New Machine Learning Approach for Protein Phosphorylation Site Prediction in Plants

Protein phosphorylation is important regulatory mechanism in various organisms. Prediction of phosphorylation sites is difficult task. This has always been true in plants, due to limited information on substrate specific phosphorylation data from plant organisms. Also the tools are trained with kinase specific phosphorylation data. So a novel prediction approach using Support Vector Machines have been employed which incorporates protein sequence information and protein disordered regions along with using k-nearest neighbor [17]. Validation on the PhosPhAt dataset of phosphoserines in *Arabidopsis* and TAIR7 non-redundant protein database showed good performance for this method as a tool prediction of Protein phosphorylation site in plants.

15. Computational Identification of Potential Molecular Interactions in Arabidopsis

Protein interaction network is useful in molecular mechanism studies. Protein interaction repositories have been reported that collect and organize protein interactions. Several models for organisms have been created for interactions, yet a very limited number of interactions relating to plants can be found in databases. Thus to facilitate research, a support vector machine (SVM) model was created to predict the potential *Arabidopsis thaliana* protein interactions. [18] A 100 iteration bootstrap evaluation was done to predict the interactions in Arabidopsis which gave an 48.67% estimation and these interactions covered a approx 29.02% of the interactome. The validation was done with independent evaluation dataset which consists of new reported interactions. The model successfully recognized 28.91% of new interactions and with expected sensitivity of 29.02%. This model was applied on Arabidopsis protein pairs which resulted into 224,206 possible interactions. A more detailed annotations and interaction are present in Predicted Arabidopsis Interactome Resource a database freely accessible online. Potential molecular interactions in Arabidopsis provides knowledge of protein network in finding the molecular mechanism of all the Arabidopsis proteins.

16. GolgiP: Prediction of Golgi-resident Proteins in Plants

Golgi apparatus are important in processing of proteins required for the cellular processes and functions. So the prediction of Golgi-resident proteins in plants has been facilitated by a novel Golgi-prediction server, GolgiP for researchers [19]. It predicts both membrane and non-membrane associated Golgi-resident proteins in plants. Support Vector Machines have been used for classification and prediction of such Golgi proteins using three different types of information such as dipeptide composition, transmembrane domain and functional domain of a protein. The functional domain is generated using the Conserved Domains Database, the transmembrane domain includes length and the number of transmembrane domains at the N-terminus of a protein. Using GolgiP, genome scale predictions for Golgi-resident proteins in 18 plant genomes was done. It is publically available.

17. Kernel Based Machine Learning Algorithm for the Efficient Prediction of Type III Polyketide Synthase Family of Proteins

Type III Polyketide Synthases (PKS) are protein family that have been found to have important role in biosynthesis of various polyketides in plants, fungi and bacteria. This protein has also been found to be good for human health. Tool to identify the probability for a sequence to be Type III polyketide synthases have been developed which uses Support Vector Machine classifiers (SVMs) [20]. The tool effectively classifies and predicts the Type III PKS family proteins with high specificity and sensitivity. More work on adding additional and other sequence features are being carried out. PKSIIIpred server is available online. It is based on the limited training dataset.

18. Prediction of Downstream Interaction of Transcription Factors with Mapk3 in *Arabidopsis thaliana* using Protein Sequence Information

Biological processes and functions can be inferred from the Protein-Protein interactions. Prediction of Downstream interaction of transcription factors with Mitogen Activated Protein Kinase3 (MAPK3) in *Arabidopsis thaliana* is done using sequence information. The information was extracted using Support Vector Machine (SVM) which is based on the physicochemical properties of proteins. Prediction of MAPK3 protein interactions with downstream transcription factor proteins was done. The Myb related transcription factor family showed maximum interaction with 71.14% with MAPK3 while minimum percentage is 21.51% shown by NAC transcription factor family. The results obtained clearly give the complexity of MAPK3 interaction with several different transcription factors and its prediction.

19. Prediction of C-to-U RNA Editing Sites in Higher Plant Mitochondria Using only Nucleotide Sequence Features

RNA editing is an important step in post-transcriptional modification which contributes in adding complexity to organism.C-to-U RNA editing is observed in higher plant mitochondira. There are many computational approaches used in predicting the RNA editing sites. But all the methods are sequence based methods. Therefore C-to-U RNA editing sites using only nucleotide sequence is developed using SVM algorithm that uses triplet scoring method [21]. This method gives an overall accuracy of 84%. Many of the triplets were found to be missing from upstream near an edited cytidine, which shows that these triplests play and important shielding for cytidine from being edited. This shows that editing sites in higher plant mitochondria can be predicted only using the nucleotide sequence features thus helping to find editing site recognition mechanism.

20. Identification of TATA and TATA-less Promoters in Plant Genomes by Integrating Diversity Measure, GC-Skew and DNA Geometric Flexibility

Identifying promoter regions is important in understanding the eukaryotic transcription regulation. Promoter prediction of plant genomes was possible by analyzing the conservative patterns, GC composition and TATA box and TATA less promoters in PlantPromDB, a hybrid approach based on SVM [23]. This approach predicts two types of promoters which search for local word content, GC-skew and DNA geometric flexibility. Better prediction results were found especially for TATA-less promoter, for which accuracy was 10% higher

Applications of Support Vector Machines in Plant Genomes   199

against other methods. This can be used to locate and identify the promoter region of plant genome.

21. Plant Noncoding RNA Gene Discovery by "Single-genome Comparative Genomics"

Duplication gives rise to large no of gene families. Plant genomes have such duplications in large number. There is little knowledge about the protein coding genes and their duplications. Most of ncRNA genes have undergone multiple duplications that are conserved in nature. The extent ncRNA gene families' duplications in plants is on large scale. Support vector machine (SVM) was used to develop an SVM model that is used to predict the likely ncRNA candidates among the repeated families in the rice genome [23]. Results showed that among 4000 ncRNA families, only 90 correspond to putative snoRNA or miRNA. Some families were classified as structured RNAs. 89% of ncRNA didn't produce any signal when compared to other genomes. Large fraction of rice ncRNA genes were found to be present in multiple copies and are species specific or of recent origin. Comparison is unique and reliable for discovering plant non-coding RNA genes.

22. Dose Detection of Radiated Rice by Infrared Spectroscopy and Chemometrics

Infrared spectroscopy and chemometrics was used to discriminate different radiations doses of rice. Samples are selected using calibration set randomly and remaining samples were used as prediction set. Using Partial least-squares (PLS) analysis and least-squares support vector machines (LS-SVM) were implemented for calibration models. PLS was used for models with different wavelength bands including near-infrared and mid-infrared regions. Different hidden variables were used as input for LS-SVM to develop models using grid search and RBF kernel [24]. LS-SVM models outperformed PLS models. The overall results indicate that infrared spectroscopy combined with LS-SVM models predict the different radiation doses of rice.

23. Comparison of Calibrations for the Determination of Soluble Solids Content and pH of Rice Vinegars Using Visible and Short-Wave Near Infrared Spectroscopy

Visible and short wave near infrared spectroscopy along with chemometrics have been used to discriminate radiation doses in rice. Similarly determination of soluble solids content (SSC) and pH values of rice vinegars. 225 samples (45 for each variety) were selected randomly for including in calibration set. For the validation set, 75 samples were taken and the remaining 25 sample were used for independent set. Partial least square (PLS) analysis was used for calibration models with different bands. Best PLS models were achieved with Vis/SWNIR region. Different hidden variables were used as input for least square-support vector machine (LS-SVM) models with grid search technique and RBF kernel [25]. Effective wavelengths (EWs) were selected on the basis of regression coefficients. The EW-LS-SVM models were developed and an excellent accuracy was achieved. The overall results indicated Vis/SWNIR spectroscopy combined with LS-SVM gave high precision and efficient way for predicting SSC and pH values of rice vinegars.

24. A Markov Classification Model for Metabolic Pathways

Metabolic pathways can be used to identify the metabolic networks that have relation with biological responses. The model HME3M proposed first identifies frequently traversed network paths by using Markov classification. The performance of HME3M was done with logistic regression and support vector machines (SVM) for both simulated pathways and two metabolic networks, glycolysis and pentose phosphate pathway for *Arabidopsis thaliana* [26]. AltGenExpress microarray data are used and difference between the developmental stages and stress responses of Arabidopsis was studied. Results showed that HME3M is a better method to identify increasing network complexity and pathway noise and confirmed known biological responses of Arabidopsis. This shows that HME3M is an accurate method for classifying metabolic pathways.

25. Detecting Polymorphic Regions in *Arabidopsis thaliana* with Resequencing Microarrays

Detecting single nucleotide polymorphisms (SNPs) in eukaryotic genomes is needed. It is possible to detect isolated SNPs; however it is greatly reduced when other polymorphisms are located near a SNP. Similar to hidden Markov models, method using support vector machines was used for training [27]. This method was applied for resequencing data generated for other genomes including *Arabidopsis thaliana*. Nonredundantly, 27% of the genome was used for polymorphic regions with high specificity (approx 97%). The resulting data provide a fine-scale view of these regions in *Arabidopsis thaliana*. These predictions provide a valuable data for detecting evolution relationship and genetic and functional studies in *Arabidopsis thaliana*.

26. Intragenomic Matching Reveals a Huge Potential for miRNA-Mediated Regulation in Plants

The microRNAs (miRNAs) play important role in post-transcriptional regulators. Intragenomic matching for finding miRNAs was used in algorithm with the use of Support Vector Machines (SVM) for classification of these precursors [27]. Three plant genomes *Arabidopsis thaliana*, *Populus trichocarpa*, and *Oryza sativa* were explored for miRNAs without the requirement of conservation across species. This method was able to approximately 1,200, 2,500, and 2,100 miRNA candidate genes capable of extensive base pairing with potential target mRNAs in *A.thaliana*, *P. trichocarpa*, and *O. sativa* respectively which is more than 5 times the number of currently annotated miRNAs in plants. Conservation analysis showed only a few are conserved between species. Large miRNA mediated regulatory interactions are encoded in the investigated plants.

27. Support Vector Machine Regression for the Prediction of Maize Hybrid Performance

Accurate prediction of phenotypical performance in single-cross hybrids allows rapid genetic pool. Support vector machines based on epsilon-insensitive was used for prediction [29]. Specific kernel functions are commonly used for similarity measures for dominant and co-dominant markers. Three different markers were integrated by using simple kernel functions. Data for grain maize the obtained from private company RAGTR2n was used. Accuracies were best for prediction methods for several combinations of marker types and traits.

28. Prediction of Dual Protein Targeting to Plant Organelles

Many proteins targeting more than one subcellular localization were found in animals, fungi and in plants. In plants N-terminal targeting signals are located in both mitochondria and plastids. Ambiguous Targeting Predictor (ATP) classifies such targeting signals which use Support Vector Machines (SVM) for predicting using 12 different amino acid features [30]. Validation is done using fluorescent protein fusion. Both *in-silico* and *in-vivo* evaluations suggest that target prediction is useful for proteins with dual targeting to organelles. Such proteins are predicted by both ambiguous targeting predictor and a signal targeting prediction tools. Comparison shows that land plant genomes encode an approximately on average, > 400 proteins located in mitochondria and plastids. Thus this prediction tool allows further understanding about the dual protein targeting to plant organelles.

Softwares for Support Vector Machines (SVM)

A number of SVM softwares are developed recently. Following is a list of some of these softwares.

1. SVM^{light}

SVM^{light}, by Joachims, is one of the most widely used SVM classification and regression packages. It has a fast optimization algorithm, can be applied to very large datasets, and has a very efficient implementation of the leave–one–out cross-validation. It is distributed as Cþþ source and binaries for Linux, Windows, Cygwin and Solaris. Kernels available include polynomial, radial basis function, and neural (tanh).

Availability: http://svmlight.joachims.org/

2. SVMstruct

SVMstruct, by Joachims, is an SVM implementation that can model complex (multivariate) output data y, such as trees, sequences, or sets. These complex output SVM models can be applied to natural language pharsing, sequence alignment in protein homology detection and Markov models for part-of-speech tagging. Several implementations exist: SVMmulticlass, for multiclass classification; SVMcfg, which learns a weighted context free grammar from examples; SVMalign, which learns to align protein sequences from training alignments; and SVMhmm, which learns a Markov model from examples. These modules have straightforward applications in bioinformatics, but one can imagine significant implementations for cheminformatics, especially when the chemical structure is represented as trees or sequences.

Availability: http://svmlight.joachims.org/svm_struct.html

3. mySVM

mySVM, by Ru¨ping, is a Cþþ implementation of SVM classification and regression. It is available as Cþþ source code and Windows binaries. Kernels available include linear, polynomial, radial basis function, neural (tanh), and anova. All SVM models presented in this chapter were computed with mySVM. Availability: http://www-ai.cs.uni-dortmund.de/SOFTWARE/MYSVM/index.html

4. mySVM/db

mySVM/db is an efficient extension of mySVM, which is designed to run directly inside a relational database using an internal JAVA engine. It was tested

with an Oracle database, but with small modifications, it should also run on any database offering a JDBC interface. It is especially useful for large datasets available as relational databases.
Availability http://www-ai.cs.uni-dortmund.de/SOFTWARE/MYSVMDB/index.html

5. LIBSVM

LIBSVM (Library for Support Vector Machines) was developed by Chang and Lin and contains C-classification, n-classification, e-regression, and n-regression. Developed in Cþþ and Java, it also supports multiclass classification, weighted SVMs for unbalanced data, cross-validation, and automatic model selection. It has interfaces for Python, R, Splus, MATLAB, Perl, Ruby, and LabVIEW. Kernels available include linear, polynomial, radial basis function, and neural (tanh).
Availability: http://www.csie.ntu.edu.tw/~cjlin/libsvm/

6. SVMTorch

SVMTorch, by Collobert and Bengio,185 is part of the Torch machine learning library (http://www.torch.ch/) and implements SVM classification and regression. It is distributed as Cþþ source code or binaries for Linux and Solaris.
Availability: http://bengio.abracadoudou.com/SVMTorch.html

7. Weka

Weka is a collection of machine learning algorithms for data mining tasks. The algorithms can either 388 Applications of Support Vector Machines in Chemistry be applied directly to a dataset or called from a Java code. It contains an SVM implementation.
Availability: http://www.cs.waikato.ac.nz/ml/weka/

8. BioWeka

BioWeka is an extension library to the data mining framework Weka for knowledge discovery and data analysis tasks in biology, biochemistry and bioinformatics. Includes integration of the Weka LibSVM project.
Availability: http://sourceforge.net/projects/bioweka/

9. Gist

Gist is a C implementation of support vector machine classification and kernel principal components analysis. The SVM part of Gist is available as an interactive Web server at http://svm.sdsc.edu. It is a very convenient server for

204 Agriculture Bioinformatics

users who want to experiment with small datasets (hundreds of patterns). Kernels available include linear, polynomial and radial.
Availability: http://svm.sdsc.edu/cgi-bin/nph-SVMsubmit.cgi

10. MATLAB SVM Toolbox

This SVM toolbox, by Gunn, implements SVM classification and regression with various kernels, including linear, polynomial, Gaussian radial basis function, exponential radial basis function, neural (tanh), Fourier series, spline, and B spline. All figures from this chapter presenting SVM models for various datasets were prepared with a slightly modified version of this MATLAB toolbox.
Availability: http://www.isis.ecs.soton.ac.uk/resources/svminfo/

11. TinySVM

TinySVM is a Cþþ implementation of C-classification and C-regression which use sparse vector representation. It can handle several thousand training examples and feature dimensions. TinySVM is distributed as binary/source for Linux and binary for Windows.
Availability: http://chasen.org/~taku/software/TinySVM/

12. SmartLab

SmartLab provides several support vector machines implementations, including cSVM, a Windows and Linux implementation of two-class classification; mcSVM, a Windows and Linux implementation of multiclass classification; rSVM, a Windows and Linux implementation of regression; and javaSVM1 and javaSVM2, which are Java applets for SVM classification.
Availability: http://www.smartlab.dibe.unige.it/

13. GPDT

GPDT, by Serafini, *et al.*, is a Cþþ implementation for large-scale SVM classification in both scalar and distributed memory parallel environments. It is available as Cþþ source code and Windows binaries.
Availability: http://dm.unife.it/gpdt/

14. Spider

Spider is an object-orientated environment for machine learning in MATLAB. It performs unsupervised, supervised, or semi-supervised machine learning problems and includes training, testing, model selection, cross-validation and statistical tests. Spider implements SVM multiclass classification and regression. Java applets, http://svm.dcs.rhbnc.ac.uk/. These SVM classification and regression Java applets were developed by members of Royal Holloway, University of London and the AT&T Speech and Image Processing Services Research Laboratory. SVM classification is available from http://svm.dcs.rhbnc.ac.uk/pagesnew/ GPat.shtml. SVMregression is available at http://svm.dcs.rhbnc.ac.uk/pagesnew/1D-Reg.shtml.

Availability: http://www.kyb.tuebingen.mpg.de/bs/people/spider/

15. LEARNSC

This site contains MATLAB scripts for the book Learning and Soft Computing by Kecman.16 LEARNSC implements SVM classification and regression.

Availability: http://www.support-vector.ws/html/downloads.html

16. LS-SVMlab

LS-SVMlab, by Suykens, is a MATLAB implementation of least-squares support vector machines (LS–SVMs), a reformulation of the standard SVM that leads to solving linear KKT systems. LS–SVMprimal–dual formulations have been formulated for kernel PCA, kernel CCA, and kernel PLS, thereby extending the class of primal– dual kernel machines. Links between kernel versions of classic pattern recognition algorithms such as kernel Fisher discriminant analysis and extensions to unsupervised learning, recurrent networks, and control are available.

Availability: http://www.esat.kuleuven.ac.be/sista/lssvmlab/

17. LSVM

LSVM (Lagrangian Support Vector Machine) is a very fast SVM implementation in MATLAB by Mangasarian and Musicant. It can classify datasets containing several million patterns.

Availability: http://www.cs.wisc.edu/dmi/lsvm/

18. ASVM

ASVM (Active Support Vector Machine) is a very fast linear SVM script for MATLAB, by Musicant and Mangasarian, developed for large datasets.

Availability: http://www.cs.wisc.edu/dmi/asvm/

19. PSVM

PSVM (Proximal Support Vector Machine) is a MATLAB script by Fung and Mangasarian that classifies patterns by assigning them to the closest of two parallel planes.

Availability: http://www.cs.wisc.edu/dmi/svm/psvm/

20. SVM Toolbox

This fairly complex MATLAB toolbox contains many algorithms, including classification using linear and quadratic penalization, multiclass classification, e-regression, n-regression, wavelet kernel, and SVM feature selection. Availability: http://asi.insa-rouen.fr/%7Earakotom/toolbox/index.html

Conclusion

There is a growing amount of plant genome data which are currently available from many resources and thus a growing number of applications of machine learning techniques in plant bioinformatics. SVM is a simple but very powerful machine learning algorithm for classification. Feature Selection helps in reducing the dimensionality of large plant datasets by selecting only the informative features. In this review, we present some of the recent advances made using Support Vector Machine and Feature Selection in plant bioinformatics. A number of biological problems such as miRNA target prediction, predicting subcellular localization of plant proteins, splicing-site recognition, splice signal identification, regulatory interaction prediction, evolutionary relationship detection, genetic and functional studies, etc. have been solved using these machine learning approaches. These methods can be further enhanced by making use of more complex techniques, such as co-clustering, and high-performing computational systems, such as parallel computers in order to solve complex plant bioinformatics problems.

References

1. D. Vassilev, J., Leunissen, A. Atanassov, A. and Nenov, G. (2005) Dimov, application of bioinformatics in plant breeding, Biotechnol. & Biotechnol. Eq.
2. Hall, M.A. (2000) Correlation based feature selection for discrete and numeric class machine learning, Proc.17th Int'l. Conf., pp. 359-366.
3. Hall, M.A. (1999) Correlation-based Feature Selection for Machine Learning, *PhD Thesis*, Department of Computer Science, University of Waikato, Hamilton, NZ.
4. Viktoria, G. (2007) Feature evaluation of the Support Vector Machine for Micro-RNA target prediction in Arabidopsis thaliana based on antisense transcription and small RNA abundance, Master's Thesis, Texas Tech University.
5. Richardson, C.R., Luo, Q.J., Gontcharova, V., Jiang, Y.W., Samanta, M., Youn, E. and Rock C.D. (2010) Analysis of antisense expression by whole genome tiling microarrays and siRNAs suggests mis-annotation of Arabidopsis orphan protein-coding genes, PLoS One, 26;5(5):e10710.
6. Xuan, P., Guo, M., Liu, X., Huang, Y., Li, W. and Huang, Y. (2011) PlantMiRNAPred: efficient classification of real and pseudo plant pre-miRNAs, Bioinformatics, 27(10): 1368-76. Epub 2011 Mar 26.
7. Kaundal, R. and Raghava, G.P., (2009) RSLpred: an integrative system for predicting subcellular localization of rice proteins combining compositional and evolutionary information, Proteomics, 9(9):2324-42.

Applications of Support Vector Machines in Plant Genomes

8. Du, H., Wang, J., Hu, Z., Yao, X. and Zhang, X. (2008) Prediction of fungicidal activities of rice blast disease based on least-squares support vector machines and project pursuit regression, J. Agric. Food Chem., 56(22):10785-92.
9. Yu, H., Lin, H., Xu, H., Ying, Y., Li, B. and Pan, X. (2008) Prediction of enological parameters and discrimination of rice wine age using least-squares support vector machines and near infrared spectroscopy, J. Agric. Food Chem., 56(2):307-13.
10. Peng, S.H., Fan, L.J., Peng, X.N., Zhuang, S.L., Du, W. and Chen, L.B. (2003) Splicing-site recognition of rice (*Oryza sativa* L.) DNA sequences by support vector machines, J. Zhejiang Univ. Sci., 4(5):573-7.
11. Malousi, A., Chouvarda, I., Koutkias, V., Kouidou, S. and Maglaveras, N. (2010) SpliceIT: a hybrid method for splice signal identification based on probabilistic and biological inference, J. Biomed Inform., 43(2):208-17.
12. Marcuzzo, M., Quelhas, P., Campilho, A., Mendonça, A.M. and Campilho, A. (2009) Automated Arabidopsis plant root cell segmentation based on SVM classification and region merging, Comput. Biol. Med., 39(9):785-93.
13. Maulik, U., Mukhopadhyay, A. and Bandyopadhyay, S. (2009) Combining Pareto-optimal clusters using supervised learning for identifying co-expressed genes, BMC Bioinformatics, 20;10:27.
14. Yu, X., Liu, T., Zheng, X., Yang, Z. and Wang, J. (2011) Prediction of regulatory interactions in Arabidopsis using gene-expression data and support vector machines, Plant Physiol. Biochem., 49(3):280-3.
15. Chen, H.W., Bandyopadhyay, S., Shasha, D.E. and Birnbaum, K.D. (2010) Predicting genome-wide redundancy using machine learning, BMC Evol. Biol., 18;10:357.
16. Lin, M., Shen, X. and Chen, X. (2011) PAIR : the predicted Arabidopsis interactome resource, Nucleic Acids Res., 39(Database issue): D1134-40.
17. Gao, J., Agrawal, G.K., Thelen, J.J., Obradovic, Z., Dunker, A.K. and Xu, D. (2009) A New Machine Learning Approach for Protein Phosphorylation Site Prediction in Plants, Lect. Notes Comput. Sci., 5462/2009:18-29.
18. Lin, M., Hu, B., Chen, L., Sun, P., Fan, Y., Wu, P. and Chen, X. (2009) Source, Computational identification of potential molecular interactions in Arabidopsis, Plant Physiol.,151(1):34-46.
19. Chou, W.C., Yin, Y. and Xu, Y. (2010) GolgiP: prediction of Golgi-resident proteins in plants, Bioinformatics, 1;26(19):2464-5.
20. Mallika, V., Sivakumar, K.C., Jaichand, S. and Soniya, E.V. (2010) Kernel based machine learning algorithm for the efficient prediction of type III polyketide synthase family of proteins, J. Integr Bioinform., 7(1). 10.2390/biecoll-jib-2010-143.
21. Du, P., He, T. and Li, Y. (2007) Prediction of C-to-U RNA editing sites in higher plant mitochondria using only nucleotide sequence features, Biochem. Biophys. Res. Commun., 358(1):336-41.
22. Zuo, Y.C. and Li, Q.Z. (2011) Identification of TATA and TATA-less promoters in plant genomes by integrating diversity measure, GC-Skew and DNA geometric flexibility, Genomics, 97(2):112-20.
23. Chen, C.J., Zhou, H., Chen, Y.Q., Qu, L.H. and Gautheret, D. (2011) Plant noncoding RNA gene discovery by "single-genome comparative genomics", RNA, (3):390-400.
24. Shao, Y., He, Y. and Wu, C. (2008) Dose detection of radiated rice by infrared spectroscopy and chemometrics, J. Agric. Food Chem., 11;56(11):3960-5.
25. Liu, F., He and Y. and Wang, L. (2008) Comparison of calibrations for the determination of soluble solids content and pH of rice vinegars using visible and short-wave near infrared spectroscopy, Anal. Chim. Acta., 610(2):196-204.

26. Hancock, T. and Mamitsuka, H. (2010) A markov classification model for metabolic pathways, Algorithms Mol. Biol., 4:5-10.
27. Zeller, G., Clark, R.M., Schneeberger, K., Bohlen, A., Weigel, D. and Rätsch, G. (2008) Detecting polymorphic regions in *Arabidopsis thaliana* with resequencing microarrays, Genome Res., 18(6):918-29.
28. Lindow, M., Jacobsen, A., Nygaard, S., Mang, Y. and Krogh, A. (2007) Intragenomic matching reveals a huge potential for miRNA-mediated regulation in plants, PLoS Comput. Biol., 3(11):e238.
29. Maenhout, S., De Baets, B., Haesaert, G. and Van Bockstaele, E. (2007) Support vector machine regression for the prediction of maize hybrid performance, Theor. Appl. Genet. Nov.,115(7):1003-13.
30. Mitschke, J., Fuss, J., Blum, T., Höglund, A., Reski, R., Kohlbacher, O. and Rensing, S.A., (2009) Prediction of dual protein targeting to plant organelles, New Phytol., 183(1):224-35.

Chapter – 10

Protein Structure, Prediction and Visualization

Sanjeev Kumar Singh and Sunil Tripathi

Abstract

The field of bioinformatics emerged as a tool to facilitate biological discoveries more than 10 years ago. The ability to capture, manage, process, analyze and interpret data became more important than ever. Bioinformatics in agricultural science has many complex aspects and problems that must be resolved due to the involvement of a wide variety of fields and subjects. Consequently, the establishment of interdisciplinary research areas is required, which should include collaboration among experimental and informatics researchers in different fields within industry, academia, and government. Thus, the Agricultural Bioinformatics Special Interest Group was established to promote interaction among researchers and provide an opportunity to report on bioinformatics studies in agricultural science: design and analysis of functional foods, scientific testing of food safety, breeding of agricultural organisms, development of environmental friendly agricultural chemicals, bioremediation and production of useful chemical materials using microorganisms, measuring and monitoring biodiversity, statistical genetics in resource ecology, and so on. The activities of the Agricultural Bioinformatics Special Interest Group include (1) Studies of applied bioinformatics in the agricultural science (2) Development of new research areas in agricultural science using bioinformatics

(3) Practical application of bioinformatics technologies in the field of agricultural science (4) Development of new bioinformatics technologies characteristic of agricultural science.

Plant life plays important roles in our economy, society, as well as our global environment. In particular, crop is the most important plant to us. The onset of genomics is providing massive information to improve crop phenotypes. The gathering of sequence data allows detailed genome analysis by using open database access. The goals of genome research are the recognition of the sequenced genes and the presumption of their functions by metabolic analysis and reverses genetic screens of gene knockouts. Multiple sequence alignment provides a method to estimate the number of genes in gene families allowing the identification of previously undescribed genes. Crop plant network collections of databases and bioinformatics resources for crop plants genomics have been built to harness the extensive work in genome mapping. This resource facilitates the identification of ergonomically important genes, by comparative analysis between crop plants and model species, allowing the genetic engineering of crop plants selected by the quality of the resulting products. Bioinformatics resources have evolved beyond expectation, developing new nutritional genomics biotechnology tools to genetically modify and improve food supply, for an ever-increasing world population. Bioinformatics can now be leveraged to accelerate the translation of basic discovery to agriculture. The predictive manipulation of plant growth will affect agriculture at a time when food security, diminution of lands available for agricultural use, stewardship of the environment, and climate change are all issues of growing public concern. Bioinformatics and computers can help scientists to solve it. Here are introduced roles of bioinformatics in protein structure prediction and visualization, meanwhile web tools and resources of bioinformatics were reviewed.

Introduction

Proteins play a key role in almost all biological process and emphasize the high importance of protein for life. Proteins take part in maintaining the structural integrity of the cell, transport and storage of small molecules, catalysis, regulation, signaling and the immune system. The linear protein chains fold up into specific three-dimensional (3D) structure and their functional properties depend intricately upon their structure. Protein structure prediction is the prediction of the three-dimensional structure of a protein from its amino acid sequence—that is, the prediction of a protein's tertiary structure from its primary structure (structure prediction is fundamentally different from the inverse, and less difficult, problem of protein design). Protein structure prediction is one of the most important goals pursued by bioinformatics and theoretical chemistry. Protein structure prediction is of high importance in medicine (for example, in drug design) and biotechnology (for example, in

Protein Structure, Prediction and Visualization

211

the design of novel enzymes). Every two years, the performance of current methods is assessed in the CASP experiment.

The practical role of protein structure prediction is now more important than ever. Massive amounts of protein sequence data are produced by modern large-scale DNA sequencing efforts such as the Human Genome Project. Despite community-wide efforts in structural genomics, the output of experimentally determined protein structures-typically by time-consuming and relatively expensive X-ray crystallography or NMR spectroscopy is lagging far behind the output of protein sequences.

A number of factors exist that make protein structure prediction a very difficult task. The two main problems are that the number of possible protein structures is extremely large, and that the physical basis of protein structural stability is not fully understood. As a result, any protein structure prediction method needs a way to explore the space of possible structures efficiently (a search strategy), and a way to identify the most plausible structure (an energy function).

In comparative structure prediction (also called homology modeling), the search space is pruned by the assumption that the protein in question adopts a structure that is reasonably close to the structure of at least one known protein. In *de novo* or *ab initio* structure prediction, no such assumption is made, which results in a much harder search problem. In both cases, an energy function is needed to recognize the native structure, and to guide the search for the native structure. Unfortunately, the construction of such an energy function is to a great extent an open problem.

Direct simulation of protein folding in atomic detail, via methods such as molecular dynamics with a suitable energy function, is typically not tractable due to the high computational cost. Therefore, most *de novo* structure prediction methods rely on simplified representations of the atomic structure of proteins.

The above mentioned issues apply to all proteins, including well-behaving, small, monomeric proteins. In addition, for specific proteins (such as for example multimeric proteins and disordered proteins), the following issues also arise:

- Some proteins require stabilization by additional domains or binding partners to adopt their native structure. This requirement is typically unknown in advance and difficult to handle by a prediction method.

- The tertiary structure of a native protein may not be readily formed without the aid of additional agents. For example, proteins known as chaperones are required for some proteins to properly fold. Other proteins cannot fold properly without modifications such as glycosylation.

- A particular protein may be able to assume multiple conformations depending on its chemical environment.

- The biologically active conformation may not be the most thermodynamically favorable.

Due to the increase in computer power, and especially new algorithms, much progress is being made to overcome these problems. However, routine de novo prediction of protein structures, even for small proteins, is still not achieved.

Methods of Protein Modeling

The prediction of the three-dimensional (3D) structure of a protein from its one dimensional (1D) protein sequence is a much published and debated area of structural bioinformatics. This prediction involves the kind of fold that the given amino acid sequence may adopt; in other words, whether it takes a *new fold* or one of the existing folds. If the sequence takes one of the existing folds, which is the most suitable fold among the known folds (fold recognition)? When fold recognition is clear because of good sequence similarity to one of the known structures, then the question is how best can one model the structure of the given sequence, taking the relevant information from existing homologous structures in the Protein Data Bank (PDB) (comparative modeling).

Despite the obstructions, much progress is being made by many of the research groups. Prediction of protein structure is now a perfectly realistic goal. A wide range of approaches are routinely applied for such predictions. These approaches may be classified into two broad classes (1) comparative protein modeling and (2) *de novo* modeling.

1) Comparative Protein Modeling

Comparative protein modeling uses previously solved structures as starting points, or templates. This is effective because it appears that although the number of actual proteins is vast, there is a limited set of tertiary structural motifs to which most proteins belong. It has been suggested that there are only around 2000 distinct protein folds in nature, though there are many millions of different proteins. When the structure of one protein in a family has been determined by experimentation, the other member of same family can be modeled, based on their alignment to the known structure. These methods may also be split into two groups.

a) **Homology modeling** - Homology modeling is based on the reasonable assumption that two homologous proteins will share very similar structures. Because a protein's fold is more evolutionarily conserved than its amino acid sequence, a target sequence can be modeled with reasonable

Protein Structure, Prediction and Visualization

accuracy on a very distantly related template, provided that the relationship between target and template can be discerned through sequence alignment. It has been suggested that the primary bottleneck in comparative modeling arises from difficulties in alignment rather than from errors in structure prediction given a known-good alignment. Unsurprisingly, homology modeling is most accurate when the target and template have similar sequences.

b) **Protein threading-** Protein threading scans the amino acid sequence of an unknown structure against a database of solved structures. In each case, a scoring function is used to assess the compatibility of the sequence to the structure, thus yielding possible 3D models.

2) *de novo* Protein Modeling

De novo or Ab initio, protein modeling methods seek to build 3D protein models 'from scratch'. There are many possible procedures that either attempt to mimic protein folding or apply some stochastic method to search possible solutions (i.e., global optimization of a suitable energy function). These procedures tend to require vast computational resources, and have thus only been carried out for tiny proteins. To predict protein structure *de novo* for larger proteins will require better algorithms and larger computational resources like those afforded by either powerful supercomputers (such as Blue Gene) or distributed computing. Although these computational barriers are vast, the potential benefits of structural genomics (by predicted or experimental methods) make *ab initio* structure prediction an active research field.

a) Homology Modeling

Homology modeling or comparative modeling is the most used and consistent theoretical method for predicting protein structures from its sequence. The homology modeling conveys precisely that this method is about modeling a structure using a homologous model as template (usually a correct X-ray or NMR structure). Basically homology modeling is usually based on sequence-similarity between the input sequence of the protein (target) and the homologue structure that is used as template.

By definition "Homology" or comparative modeling is the prediction of three-dimensional (3D) structure of a target protein from the amino acid sequence of a homologous (template) protein for which an X-ray or NMR determined structure is available. Due to following facts the comparative approach to protein structure prediction is possible, which is as follows:

- A small change in the protein sequence generally outcome in a small change in its 3D structure.

- It is also one of the facts that the protein's 3D structure from the same family is more conserved than their amino acid sequences. As a result, if likeness between the proteins is noticeable at the sequence level, structural similarity may be presume.

- During evolution, the structure is more stable and changes much slower than the associated sequence, so that similar sequences adopt practically identical structures and distantly related sequences still fold into similar structures.

Why Homology Model?

Homology models of proteins are of great interest for planning and analyzing biological experiments when no experimental 3D structures are available. Homology modeling combines sequence analysis and molecular modeling to predict 3D structures. Homology protein structure modeling builds a 3D model for a protein of unknown structure (the target) based on one or more related proteins of known structure (template, X-ray or NMR-determined structures). This method usually provides models that are comparable to low-resolution X-ray crystallography or medium resolution NMR solution structures. This is the most used and reliable theoretical method for predicting protein structures out of a sequence. We can also choose a distant homologue of the desire protein which does not have an experimentally solved structure and use the appropriate modeling software such as Swiss-Model, Modeller or Homology (InsightII) resource to sequence with an accuracy that is comparable to the best results achieved experimentally. This would allow users to use rapidly generated *in silico* protein models in all the contexts with proper reliabilities. The NMR and X-ray protein crystal structures are classified into families based on a limited set of folding patterns. In general, functional sites maintain identical structural folds, and proteins of the same function generally have similar structure (trypsin and subtilisin are exceptions case). Such observations became the foundation of homology modeling. The key to its success is thorough understanding of the dataset. Among the three major approaches (as has been discussed briefly earlier) to 3D structure prediction, homology modeling is the easiest. There are several softwares and Web servers (Table 10.1) that automate the comparative modeling process. These servers accept a sequence from the user and return a comparative mode. In addition to modeling there are different databases which are useful in retrieval of target sequence and for many more purpose (Table 10.2).

Table 10.1: List of useful software and server for protein structure prediction

Steps	Server	Websites
Databases	NCBI	http://www.ncbi.nlm.nih.govhttp://
	GeneBank	www.ncbi.nlm.nih.gov/Genbank/
	TrEMBL	GenbankSearch.htmlhttp://
	Pfam	www.srs.ebi.ac.ukhttp://
	PDB	www.sanger.ac.uk/Software/Pamhttp://
	MSD	www.rcsb.orghttp://www.rcsb.org/
	CATH	databases.htmlhttp://
	SCOP	www.biochem.ucl.ac.uk/bsm/cathhttp://
		www.scop.mrc-lmb.cam.ac.uk/scop
Plant related Database	AGBASE	http://www.agbase.msstate.edu/
	ARABINET	http://weeds.mgh.harvard.edu/atlinks.html
	AraCyc	http://www.arabidopsis.org/biocyc/index.jsp
		http://www.athamap.de/
	AthaMap	http://atidb.org/cgi-perl/index
	ATIDB	http://www.ncbi.nlm.nih.gov/mapview/
		map_search.cgi?chr=barley.inf
	BARLEY GENOME (NCBI)	http://bioinf.scri.ac.uk/barley_snpdb/
	BARLEY SNP	http://www.sigmaaldrich.com/Area_of_Interest/
		Life_Science/Nutrition_Research/Key_Resources/
	BIOACTIVE NUTRIENT	BN_Explorer.html
	EXPLORER	http://www.bio.indiana.edu/~nsflegume/
		http://dendrome.ucdavis.edu/
	COMPARATIVE ANALYSIS OF	http://193.51.165.9/projects/FLAGdb++/HTML/
	LEGUME GENOMES	index.shtml
	DENDROME: FOREST TREE	http://wheat.pw.usda.gov/GG2/index.shtml

(Contd.)

Steps	Server	Websites
	GENOME DATABASE	
	FLAGdb++	http://www.gramene.org/
	GRAINGENES	http://www.gramene.org/pathway/
	GRAMENE	http://www.ipni.org/index.html
	GRAMENE PATHWAY	http://www.maizegdb.org/
	IPNI: THE INTERNATIONAL	
	PLANT NAMES INDEX	http://www.hort.purdue.edu/newcrop/
	MAIZEGDB: MAIZE GENETICS	http://www.pathoplant.de/
	AND GENOMICS DATABASE	http://pgrc-35.ipk-gatersleben.de/portal/page/portal/
	PG_BICGH/P_BICGH	
	NewCROP™	http://pgrc.ipk-gatersleben.de/
	PathoPlant®	http://www.ncbi.nlm.nih.gov/genomes/PLANTS/
	PlantList.html	
	PLANT BIOINFORMATICS PORTAL	http://mpss.udel.edu/
		http://bioinf.scri.sari.ac.uk/cgi-bin/plant_snorna/home
	PLANT GENOME RESOURCES CENTER	http://bioinformatics.psb.ugent.be/webtools/plantcare/html/
		http://www.plantgdb.org/
	PLANT GENOMES CENTRAL (NCBI)	http://markers.btk.fi/
	PLANT MPSS DATABASES	http://www.ba.itb.cnr.it/PLANT-PIs/
	PLANT snoRNA DATABASE	http://plantsp.genomics.purdue.edu/
	PlantCARE: Cis-ACTING	http://plantst.genomics.purdue.edu/
	REGULATORY ELEMENT	https://gabi.rzpd.de/PoMaMo.html
	PlantGDB: PLANT GENOME	http://golgi.gs.dna.affrc.go.jp/SY-1102/rad/index.html
	DATABASE	http://soybase.agron.iastate.edu/
	PLANTMARKERS	
	PLANT-Pis: PROTEASE	
	INHIBITORS	
	PlantsP: FUNCTIONAL GENOMICS	

(Contd.)

Steps	Server	Websites
	OF PLANT PHOSPHORYLATION	
	PlantsT: FUNCTIONAL GENOMICS	
	OF PLANT TRANSPORTERS	
	POMAMO	
	RICE ANNOTATION DATABASE	http://www.tigr.org/tdb/e2k1/plant.repeats/index.shtml
	SOYBASE	http://www.bioinfo.no/tools/TAED
	TAED: THE ADAPTIVE	http://phytophthora.vbi.vt.edu/
	EVOLUTION DATABASE	
	TIGR PLANT REPEAT DATABASE	http://tropgenedb.cirad.fr/
	TropGENE	http://rgp.dna.affrc.go.jp/whoga/index.html.en
	VBI MICROBIAL DATABASE	
	WhoGA: WHOLE GENOME	
	ANNOTATION	
Template Search	ModBase	http://www.salilab.org/modbase
	TargetDB	http://www.targetdb.pdb.org
	PDB-Blast	http://www.bioinforamtics.burnham-inst.org/pdb_blast
	BLAST	http://www.ncbi.nlm.nih.gov/BLAST
	Fasta	http://www.ebi.ac.uk/fasta33
	DALITHREADER	http://www2.ebi.ac.uk/dali
	123D	http://www.bioinf.cs.ucl.ac.uk/threader
	PROFIT	http://www.123d.ncifcrf.gov
	3D-PSSM	http://www.lore.came.sbg.ac.at
	LOOPP	http://www.sbg.bio.ic.ac.uk/~3dpssm
	FASS03	http://www.ser-loopp.tc.cornell.edu/loopp.html
	Prospect	http://www.ffas.ljcrf.edu/ffas-cgi/cgi/ffas.pl
		http://www.compbio.ornl.gov/structure/prospect

(Contd.)

Steps	Server	Websites
Alignments	Superfamily	http://www.supfam.mrc.cam.ac.uk/SUPERFAMILY
	BLAST	http://www.ncbi.nlm.nih.gov/BLAST
	PDB-Blast	http://www.bioinformatics.burnham-inst.org/pdb_blast
	ClustalW	http://www.ebi.ac.uk/clustalW/
		CMsearchlauncher.bcm.tmc.edu/multi-align
	SAM	http://www.csc.ucsc.edu/research/compbio/sam.html
	Smith-Waterman	http://www-hto.usc.edu/software/seqaln/seqaln-query.html
Comparative, loop and	T-coffee	http://www.ch.embnet.org/software/TCoffee.html
side-chain modelling	3D-JIGSAW	http://www.bmm.icnet.uk/servers/3dijigsaw
	CPH-Models	http://www.cbs.dtu.dk/services/CPHmodels
	COMPOSER	http://www-cryst.bioc.cam.ac.uk
	FAMS	http://www.phychem.pharm.kitasato-u.ac.jp/FAMS/fams.html
	Swiss-Model	http://www.expasy.ch/swissmod/SWISS-MODEL.html
	WHATIF	http://www.cmbi.kun.nl/whatif
	ICM	http://molsoft.com
	Modeller	http://www.salilab.org/modeller/modeller.html
Model evaluation	PrISM	http://www.honiglab.cpmc.columbia.edu
	SCWRL	http://www.dunbrack.fccc.edu/SCWRL3.php
	MODLOOP	http://www.salilab.org/moloop
	SHOTGUN5	http://www.bioinfo.pl/meta
	PROCHECK	http://www.biochem.ucl.ac.uk/~roman/procheck/procheck.html
	WHATCHECK	http://www.cmbikun.nl/gv/server/WIWWWI
	ProsaII	http://www.camesbg.ac.at
	BIOTECH	http://www.biotech.embl-ebi.ac.uk:8400
	VERYFY_3D	http://www.doe-mbi.ucla.edu/Services/Veryfy_3D
	ERRAT	http://www.doe-mbi.ucla.edu/Services/Errat.
	ANOLEA	htmlhttp://www.protein.bio.puc.cl/cardex/servers/anolea
	AQUA	http://www.urchin.bmrb.wisc.edu/~jurgen/Aqua/server
	PROVE	http://www.ucmb.ulb.ac.be/UCMB/PROVE

Postulation of Homology Model

The following are the postulation assumptions of homology modeling:

- Regions of homologous sequence have similar structure.

- The overall 3D structure of the target protein is not dissimilar to that of the related proteins.

- Homologous residues throughout a family of proteins are conserved structurally.

- Residues involved in biological activity have similar topology throughout the protein family.

- Loop regions (non-conserved residues) allow insertions and deletions without disrupting the overall structure of the protein. Loop regions are flexible and therefore need to be constructed as strictly as the conserved regions, assuming that they play no role in biological activity.

Many proteins are simply too large for NMR analysis and cannot be crystallized for X-ray diffraction. Protein modeling is the only way to obtain structural information if experimental techniques fail. At almost all the steps choices have to be made. The modeler can never be sure to make the best ones, and thus a large part of the modeling process consists of serious thought about how to speculate between multiple seemingly similar choices. A lot of research has been spent on teaching the computer how to make these decisions, so that homology models can be built fully automatically.

The practice homology modeling is a multi-step process (Fig. 10.1) and can be summarized in the following steps:

- Template identification and alignment the target sequence with the template

- Correction of alignment

- Model building

- Backbone generation

- Side-chain modeling

- Refinement of Model

- Model validation

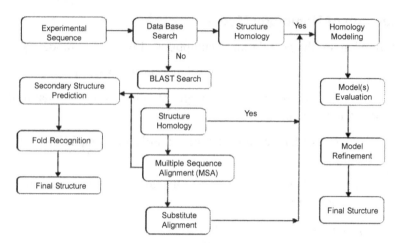

Fig. 10.1: Flowchart illustrating steps of homology modeling

Template Identification and Alignment the Target Sequence with the Template

The comparative protein modeling requires safe modeling zone, the percentage identity between the sequence of interest and a possible template is high enough to be detected with simple sequence alignment programs (BLAST or FASTA). For comparative protein modeling requires at least one sequence of known 3D structure with significant similarity to the target sequence. This is usually achieved by searching the PDB of know protein structures using the target sequence as the query and this search is generally done by comparing the target sequence with the sequence of each of the structures in the database. Generally protein comparison depends mainly three methods that are useful in fold identification:

1. The first methods compare the target sequence (query) with each of the database sequences independently, using pair-wise sequence-sequence comparison. The most popular programs in the class include FASTA and BLAST.

2. The second methods rely on the most well known program is PSI-BLAST which perform multiple sequence comparisons to improve the sensitivity of the search. In this series another similar approach that appears to execute even slightly better than PSI-BLAST has been implemented in the program PDB-BLAST, which begins by finding all sequences in a sequence database that are clearly related to the target and easily aligned with it. The templates are then found by comparing the target sequence profile with each of the template sequence profiles, using a local dynamic programming method that relies on the common BLOSUM62 residue

Protein Structure, Prediction and Visualization

substitution matrix. Such sensitive fold identification techniques based on sequence profiles are especially useful for finding significant structural relationships when sequence identity between the target and the template.

3. The third method is based on a pair-wise comparison of a protein sequence and a protein structure. It means structural information is used for one of the two proteins that are being compared, and the target sequence is matched against a library of 3D profiles or threaded through a library of 3D folds. Those methods are especially useful when it is not possible to construct the sequence profiles because of the absence of sufficient known sequences that are clearly related to the target or potential templates.

If once a list of potential templates is obtained using searching methods, now it is necessary to select best one or more templates that are suitable for the particular modeling problem. Having known one or more possible modeling templates using the fast methods described earlier, let us now consider more sophisticated methods to arrive at a better alignment. Sometimes it may be difficult to align two sequences in a region where the percentage sequence identity is very low. One can then use other sequences from homologous proteins to find a solution. Generally Multiple Sequence Alignments (MSA) is nevertheless useful in homology modeling for example, to place deletions (missing residues in the model) or insertions (additional residues in the model) only in areas where the sequences are strongly divergent. Although a simple sequence alignment (BLAST or FASTA) gives the highest score for the wrong (Fig. 10.2. alignment 1), a look at the structure of the template reveals that alignment 2 is correct, because it leads to a small gap, compared to a huge hole associated with alignment 1.

	1	2	3	4	5	6	7	8	9	10	11	12	13
Template	PHE	ASP	ILE	CYS	ARG	LEU	PRO	GLY	SER	ALA	GLU	ALA	VAL
Model (bad) 1	PHE	ASP	ILE	CYS	ARG	LEU	PRO	---	---	---	GLU	ALA	VAL
Model (good) 2	PHE	ASP	ILE	CYS	ARG	---	---	---	SER	ALA	GLU	ALA	

Fig. 10.2: Showing the example of sequence alignment in which three-residue must be modelled (Krieger *et al.*, 2003)

Correction of Alignment

Having identified one or more possible modeling templates using the fast methods described above, it is time to consider more sophisticated methods to arrive at a better alignment. Sometimes it may be difficult to align two sequences in a region where the percentage sequence identity is very low. One can then use other sequences from homologous proteins to find a solution. For the above problem there is a very powerful concept called "Multiple Sequence Alignment (MSA)." There are many programs available to align a number of related sequences, for example ClustalW, and the resulting alignment contains a lot of additional information. The MSA implicitly contains information about this structural context. If at a certain position only exchanges between hydrophobic residues are observed, it is highly likely that this residue is buried. To consider this knowledge during the alignment, one uses the multiple sequence alignment to derive position specific scoring matrices, also called profiles. When building a homology model, we are in the fortunate situation of having an almost perfect profile—the known structure of the template. Generally, MSA are nevertheless useful in homology modeling, for example, to place deletions (missing residues in the model) or insertions (additional residues in the model) only in areas where the sequences are strongly divergent. A example for correcting an alignment with the help of the template is shown in Fig. 10.2. Although a simple sequence alignment gives the highest score for the wrong answer (see alignment 1 in Fig. 10.2), a easy look at the structure of the template reveals that alignment 2 is correct, due to a small gap, compared to a huge hole associated with alignment 1.

Model Building

The model building process involves the following steps:

Construction of framework: The construction of the protein to be modeled takes into account the following aspects:

- Builds the Cα for only the conserved regions in the alignments.

- Averages the position of conserved Cα atoms in the target sequence from the positions of the corresponding atoms in the templates.

- Where multiple templates are used, the position is based on the amount of local sequence identity.

Construction of loops: Loops (non-conserved region) often play an important role in defining the functional specificity of a given protein framework, forming the active and binding sites. The accuracy of loop modeling is a major factor determining the usefulness of comparative models in applications such as ligand docking. In majority of cases, the alignment between model and template sequence contains gaps or inserts. The gaps are present in the target (model) sequence and inserts are in the template sequence. In the case of gaps, we just skip residues from the template, creating a hole in the model that must be closed. On the other hand in the case of inserts, we take the continuous backbone from the template, cut it, and insert the missing residues. In both cases imply a

Protein Structure, Prediction and Visualization 223

conformational change of the backbone and these conformational changes cannot happen within regular secondary structure elements. It is therefore safe to shift all insertions or deletions in the alignment out of helices and strands, placing them in loops and turns. The main obstruction here is that these changes in loop conformation are difficult to predict. To make things worse, even without insertions or deletions we often find quite different loop conformations in template and target.

Here three core reasons identify are the following:

1. Surface loops tend to be involved in crystal contacts, which lead to a significant conformational change between template and target.

2. The exchange of small to bulky side chains underneath the loop pushes it aside.

3. The mutations of a loop residue to proline are from glycine to any other residue. In both cases, the new residue must fit into a more restricted area in the Ramachandran plot, which most of the time requires conformation changes of the loop.

Loop Modeling Method

There are two main classes of loop modeling methods:

a) Database Search Approaches

The database or knowledge-based search approach to loop modeling is correct and proficient when a database of specific loops is created to address the modeling of the some class of loops, such as β-hairpins, or loops on a specific fold, such as the hyper variable regions in the immunoglobulin fold. For example, an analysis of the hypervariable immunoglobulin regions resulted in a series of rules that allowed a very high accuracy of loop prediction in other members of family. These rules were based on the small number of conformations for each loop, and the dependence of the loop conformation on its length and certain key residues. Attempts are still being made to classify loop conformations into more general categories, thus extending the applicability of the database search approach to far more cases. Conversely, the database methods are limited by the fact that the number of possible conformations increases exponentially with the length of a loop. The situation is made even worse by the requirement for an overlap of at least one residue between the database fragment and the anchor core regions, which means that the modeling of a five-residue insertion requires at least seven-residue fragment from the database. A recent study reported a more favorable coverage of loop conformations in PDB. Therefore, it is possible that the database approaches are rather limited by the ability to recognize suitable fragments, and not by the lack of these segments.

b) Conformational Search Approaches

The **Conformational** search (energy-based) strategies include the minimum perturbation method, molecular dynamics simulations, genetic algorithms, Monte Carlo and simulated annealing, multiple-copy simultaneous search, self-consistent field optimization, and an enumeration based on the graph theory. Loop prediction by optimization is applicable to both simultaneous modeling of several loops and loops interacting with ligands, either of which is not straightforward for the database search approaches, where fragments are collected from the unrelated structures with different environments. One of the module, Modloop in Modeller has been developed recently, which implementing the optimization-based approach. Optimization of loop (Modloop) relies on conjugate gradients and molecular dynamics with simulated annealing. The main reasons are the generality and conceptual simplicity of energy minimization, as well as the limitations on the database approach imposed by a relatively small number of known protein structures. Loop prediction by optimization is applicable to simultaneous modeling of several loops and loops interacting with ligands, which is not straightforward for the database search approaches. To simulate comparative modeling problems comparative modeling problems, the loop modeling procedure was evaluated by predicting loops of known structure in only approximately correct environments. It is possible to estimate whether a given loop prediction is correct or not, based on structural variability of the independently derived lowest-energy loop conformations.

Backbone Generation

Since the loop building only adds Cα atoms, the backbone carbonyl and nitrogens must be completed in these regions. This step can be performed by using a library of pentapeptide backbone fragments derived from the PDB entries determined with a resolution better than 2.0 Å. Creating the backbone is trivial for most of the model: One simply copies the coordinates of those template residues that show up in the alignment with the model sequence. If two aligned residues differ, only the backbone coordinates (N, Cα, C and O) can be copied. If they are the same, one can also include the side chain (at least the more rigid side chains, since rotamers tend to be conserved). Experimentally determined protein structures are not perfect (but still better than models in most cases). There are countless sources of errors, ranging from poor electron density in the X-ray diffraction map to simple human errors when preparing the PDB file for submission. A lot of work has been spent on writing software to detect these errors (correcting them is even more difficult). It is obvious that a straightforward way to build a good model is to choose the template with the fewest errors. For this the PDBREPORT database (www.cmbi.nl/gv/pdbreport) can be very helpful. But what will be happen if two templates are available, and each has a poorly determined region, but these regions are not the same? One should clearly combine the good parts of both templates in one model— an approach known as multiple template modeling. (The same applies if the alignments between the model sequence and possible templates show good matches in different regions). Although in principle multiple template modeling is a nice idea (and can be done by modeling servers for example Swiss-Model), it is difficult in practice to achieve results that are really closer to the true structure than all the templates. Yet, it is possible and has been done by Salis's group.

Side-Chain Modeling

For many of the protein side chains, there is no structural information available in the templates. Therefore, these cannot be built during the framework generation and must be added later. The number of side chains that need to be built is dictated by the degree of sequence identity between target and template sequences. To this end one uses a table of the most probable rotamers for each amino acid side chain, depending on their backbone conformation. All the permissible rotamers of the residues missing from the structure are analyzed to see if they are acceptable by a VDW exclusion test. The most favored rotamer is added to the model. The atoms defining the $\div1$ (torsion angle of $C\alpha$-$C\beta$ bond) and $\div2$ angles of incomplete side chains can be used to restrict the choice of rotamers to those that fit in these angles. If some side chains cannot be rebuilt in a first attempt, they will be assigned initially in a second pass. This allows some side chains to be rebuilt even if the most probable permissible rotamer of a neighboring residue already occupies some of this portion of space. The latter may then switch to a less probable but permissible rotamer. In case if all the side chains cannot be added, an additional tolerance of 0.15 Å can be introduced in the VDW exclusion test and the procedure repeated.

Practically all successful approaches to side-chain placement are at least partly knowledge based. They use libraries of common rotamers extracted from high resolution X-ray structures. The various rotamers are tried successively and scored with a variety of energy functions. More research has been done on the development of methods to make this enormous search space tractable. Beside the small fact that copying conserved rotamers from the template often splits up the protein into distinct regions where rotamers can be predicted independently, the key to handling the combinatorial explosion lies in the protein backbone. Certain backbone conformations strongly favor certain rotamers (a hydrogen bond between side chain and backbone) and thus to a great extent reduce the search space. For a given backbone conformation, there may be only one strongly populated rotamer that can be modeled right away, thereby providing an anchor for surrounding, more flexible side chains. To predict a rotamer, the corresponding backbone stretch in the template is superposed on all the collected examples, and the possible side-chain conformations are selected from the best backbone matches.

Refinement of Model

Energy minimization (EM) with force fields such as CHARMM, AMBER, or GROMOS can be performed for idealization of bond geometry and removal of unfavorable non-bonded contacts. EM will judge these features in the model and try, as far as possible, to make them close to the ideal values. The refinement (optimization) of a primary model should be performed by not more than 100 steps of steepest descent, followed by 200-300 steps conjugate gradient energy minimization. At every minimization step, a few major errors (too short atomic distances) are removed, while several minor errors are introduced. When the major errors have disappeared, the minor ones start accumulating and the model moves away from the target. Therefore, current modeling programs either restrain the atom positions and/or apply only a few hundred steps of energy minimization. Constraining the positions of selected atoms in each residue generally helps in avoiding the excessive structural drift during force field computations.

Validation of Model

After homology modeling every model contains errors. The overall accuracy of comparative models spans a wide range. At the low end of the spectrum are the low-resolution models, whose only indispensable correct feature is their fold. At the high end of the spectrum are the models with accuracy comparable with medium resolution crystallographic structures. Low-resolutions models are often useful to address biological questions because function can often be predicted from only coarse structural features of a model. An essential step in the homology modeling process is therefore the confirmation of the model.

The errors in the homology models can be divided into different categories which are as follows.

- Change of a region that does not have an equivalent segment in any of the template structures.

- Errors in side-chain packing.

- Change of a region that is aligned incorrectly with the template structures.

- Change of a region that is aligned correctly with the template structures.

- A misfolded structure resulting from using an incorrect template.

There are two different methods for estimation of errors in a structure, which is as follows.

1. Model's energy calculation based on force field: This method checks whether the bond lengths and bond angles are within standard ranges or not. One of the important questions such as 'Is the model folded correctly or not?' still in this way cannot be answered, because completely misfolded but well-minimized models often reach the same force field energy as the target structure. This result is mainly due to the fact that molecular dynamics force fields do not explicitly contain entropic terms (Such as the hydrophobic effect), but relay on the simulation to generate them. Although this problem can be addressed by extending the force fields and adding, e.g., salvation parameters, the major drawback is that always a single number is obtained for the entire protein and thus it is not easy to trace problems down to individual residues.

2. Determination of normality indices that describe how well a given characteristic of the model resembles the same characteristic in real structures. Many features of protein structures are well suited for normality analysis. Most of them are directly or indirectly based on the analysis of inter-atomic distances and contacts.

Model Evaluation

There are two methods of evaluation for homology model which is as follows:

1. The internal evaluation method, used for self-consistency checks in which a model satisfies the restraints used to calculate it or not. It deals with evaluation of the stereochemistry property of a model such as bonds, bond angles, dihedral angles, and non bonded distances. This is done by programs like PROCHECK and WHATCHECK. While errors in stereochemistry are rare and less revealing than errors detected by the methods for external evaluation., group of a errors may specify that the corresponding regions also contains other larger errors (e.g., alignment errors).

2. External evaluation method is based on information that was not used in the calculation of the model. At a low resolution, external evaluations

test whether a correct template was used. A more challenging task for the scoring functions is the prediction of unreliable regions in the model. One way to approach this problem is to calculate a pseudo-energy profile of a model, such as that produced by Verify3D. The profile reports the energy for each position in the model. Peaks in the profile frequently correspond to errors in the model. There are several pitfalls in the use of energy profiles for local error detection. For example, a region can be identified as unreliable just because it interacts with an incorrectly modeled region.

Procheck suite for Stereochemistry

The PROCHECK (www.biochem.ucl.ac.uk/~roman/procheck/procheck.html) suite of programs provides a detailed check on the stereochemical quality of a given protein structure. The aim of PROCHECK is to assess how normal, or conversely how unusual, the geometry of the residues in a given protein structure is, as compared with stereochemical parameters derived from well-refined, high-resolution structures.

Its output comprises a number of plots in post-script format and a comprehensive residue-by-residue listing (Fig. 10.3a & b). These give an assessment of overall quality of the structure, as compared with well-refined structures of the same resolution, and also highlight regions that may need further investigation. The PROCHECK programs are useful for assessing the quality not only protein structures in the process of being solved, but also of existing ones and those being modeled on known structure.

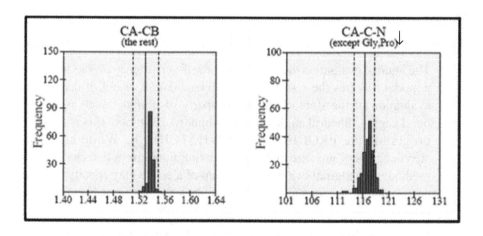

Fig. 10.3a: An output of PROCHECK showing bond length and bond angle

Protein Structure, Prediction and Visualization

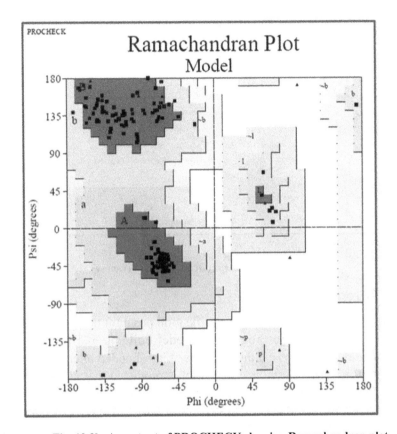

Fig. 10.3b: An output of PROCHECK showing Ramachandran plot

VERIFY_3D Server

VERIFY 3D structure evaluation server is a tool designed to help in the refinement of crystallographic structure. It will provide a visual analysis of the quality of a putative crystal structure for a protein. Verify 3D expects this crystal structure to be submitted in PDB format. This VERIFY 3D works best on proteins with at least 100 residues. (Fig. 10.4)

This program analyzes the compatibility of an atomic model (3D) with its own amino acid sequence (1D). Each residue is assigned a structural class based on its location and environment (alpha, beta, loop, polar, apolar etc). Then a database generated from vetted good structures is used to obtain a score for each of the 20 amino acids in this structural class. For each residue, the scores of a sliding 21-residue window (from -10 to +10) are added and plotted. The returned 3D-1D profile should for the most part stay above 0.2 and never really dip below 0.

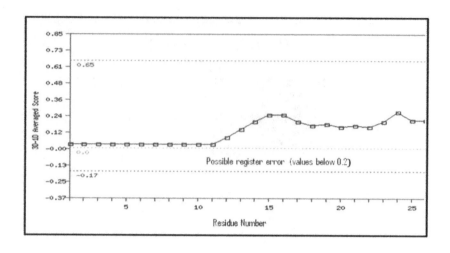

Fig. 10.4: An output of VERYFY_3D server

ERRAT Program

The ERRAT is a protein structure verification program that is especially well suited for evaluating the progress of model building and refinement. The program works by analyzing the statistics of non-bonded interactions between different atom types. A single output plot is produced that gives the value of the error function versus position of a 9-residue sliding window. By comparison with statistics from highly refined structures, the error values are calibrated to provide confidence limit (Fig. 10.5). Generally, the method is sensitive to smaller errors than 3-D Profile analysis, but is more forgiving than Procheck.

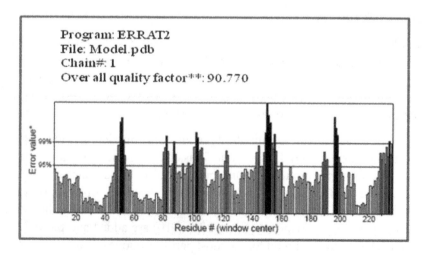

Fig. 10.5: An output of Program ERRAT2

PROVE (PROtein Volume Evaluation)

The volume-based structure validations procedures are implemented in the program PROVE, which is accessible through the World Wide Web. It calculates the volumes of atoms in macromolecules using an algorithm which treats the atoms like hard spheres and calculates a statistical Z-score deviation for the model from highly resolved (2.0 Å or better) and refined (R-factor of 0.2 or better) PDB-deposited structures (Fig. 10.6).

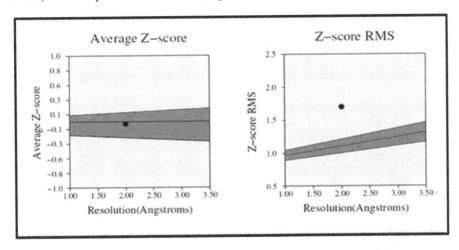

Fig. 10.6: An output of PROVE program showing average Z-score and Z-score RMS

Utility of Homology Modeling

Homology modeling is often an efficient way to obtain useful information about the proteins of interest. Comparative models can be helpful in the following ways:

- Designing mutants to test hypotheses about the function of a protein.
- Protein-protein interaction prediction, protein-protein docking, molecular docking, and functional annotation of genes.
- Identify subtle differences between related proteins that have not all been solved structurally (For example, cation binding sites on the Na^+/K^+ ATPase).
- Identification of active and binding sites.
- Searching for, designing, and improving ligands for a given binding site.
- Modeling substrate specificity.

Homology Modeling

Name	Method	Description	Web link
LOMETS	Local Meta threading server	Meta-server combining 9 other threading programs with final models built by Modeller	http://zhang.bioinformatics.ku.edu LOMETS/
3D-JIGSAW	Fragment assembly	Automated webserver	http://bmm.cancerresearchuk.org/~3djigsaw/
Biskit	wraps external programs into automated workflow	BLAST search, T-Coffee alignment, and MODELLER construction	http://biskit.pasteur.fr/
CABS	Reduced modeling tool	Downloadable program	http://www.biocomp.chem.uw.edu.pl/services.php
CPHModel	Fragment assembly	Automated web server	http://www.cbs.dtu.dk/services/CPHmodels/
ESyPred3D	Template detection, alignment, 3D modeling	Automated webserver	http://www.fundp.ac.be/sciences/biologie urbm/bioinfo/esypred/
GeneSilico	Consensus template search/fragment assembly	Webserver	https://genesilico.pl/meta2/
Geno3D	Satisfaction of spatial restraints	Automated webserver	http://geno3d-pbil.ibcp.fr/cgi-bin/geno3d_automat.pl?page=/GENO3D/geno3d_home.html
HHpred	Template detection, alignment, 3D modeling	Interactive webserver with help facility	http://toolkit.tuebingen.mpg.de/hhpred
LIBRA I	LIght Balance for Remote Analogous proteins, ver. I	Webserver	http://libra.ddbj.nig.ac.jp/top-e.html
MODELLER	Satisfaction of spatial restraints	Standalone program in Python	http://salilab.org/modeller/
EasyModeller	GUI to MODELLER	Standalone windows executable	http://www.uohyd.ernet.in/modellergui/
ROBETTA	Rosetta homology modeling and ab initio fragment assembly with Ginzu domain prediction	Webserver	http://robetta.org/

(Contd.)

Name	Method	Description	Web link
Selvita Protein Modeling Platform	Package of tools for protein modeling	Free demo, interactive webserver and standalone program including: BLAST search, CABS modeling, 3D threading, Psi-Pred secondary structure prediction	http://www.selvita.com/
SWISS-MODEL	Local similarity/fragment assembly	Automated web server (based on ProModII)	http://swissmodel.expasy.org/
TIP-STRUCTFAST	Automated Comparative Modeling	Webserver and knowledgebase	https://tip.eidogen-sertanty.com/Login.po
WHAT IF	Position specific rotamers	Webserver	http://swift.cmbi.ru.nl/servers/html/index.html
Commercial software			
Modeller	Homology protein structure modeling by satisfaction of special restrains	Academic/Commercial	http://www.salilab.org/modeller/
InsightII	Molecular modeling and simulation programs	Commercial	http://www.accelrys.com/
SYBYL	General molecular modeling program	Commercial	http://www.tripos.com/
TRITON	Graphical tool for modeling protein mutant and assessment of their activities	Commercial	http://www.ncbr.chemi.muni.cz/trito triton.html/
Threading/Fold recognition			
Selvita Protein Modeling Platform	Package of tools for protein modeling	Free demo, interactive webserver and standalone program including: 3D threading and flexible 3d threading	http://www.selvita.com/selvita-protein-modeling-platform.html
RAPTOR	Integer programming based fold recognition	Free demo	http://www.bioinformaticssolutions.com/products/raptor/index.php

(Contd.)

Name	Method	Description	Web link
HHpred	Template detection, alignment, 3D modeling	Interactive webserver with help facility	http://toolkit.tuebingen.mpg.de/hhpred
3D-PSSM	3D-1D sequence profiling	Webserver	http://www.sbg.bio.ic.ac.uk/~3dpssm/
I-TASSER	Combination of several methods	Webserver	http://zhang.bioinformatics.ku.edu/I-TASSER/
SUPERFAMILY	Hidden Markov modeling	Webserver/standalone	http://supfam.org/SUPERFAMILY/
Bioingbu	Evolutionary information recognition	Webserver	http://www.cs.bgu.ac.il/~bioinbgu/
mGen THREADER/ GenTHREADER	Sequence profile and predicted secondary structure	Webserver	http://bioinf.cs.ucl.ac.uk/psipred/
LOOPP	Multiple methods	Webserver	http://cbsuapps.tc.cornell.edu/loopp.aspx
***Ab initio* structure prediction**			
I-TASSER	Threading fragment structure reassembly	On-line server for protein modeling (best in CASP)	http://zhang.bioinformatics.ku.edu/I-TASSER/
Selvita Protein Modeling Platform	Package of tools for protein modeling	Interactive webserver and standalone program including: CABS ab initio modeling	http://www.selvita.com/selvita-protein-modeling-platform.html
ROBETTA	Rosetta homology modeling and ab initio fragment assembly with Ginzu domain prediction	Webserver	http://robetta.org/
Rosetta@home	Distributed-computing implementation of Rosetta algorithm	Downloadable program	http://boinc.bakerlab.org/rosetta/
CABS	Reduced modeling tool	Downloadable program	http://www.biocomp.chem.uw.edu.pl/services.php

(Contd.)

Name	Method	Description	Web link
Bhageerath	A computational protocol for modeling and predicting protein structures at the atomic level.	Webserver	http://www.scfbio-iitd.res.in/bhageerath/index.jsp
Abalone	Molecular Dynamics folding	Program	http://www.agilemolecule.com/Abalone/Protein-folding.html
Secondary structure prediction			
NetSurfP	Profile-based neural network	Webserver	http://www.cbs.dtu.dk/services/NetSurfP/
GOR	Information theory/Bayesian inference	Many implementations	http://gor.bb.iastate.edu/http://abs.cit.nih.gov/gor/
Jpred	Neural network assignment	Webserver	http://www.compbio.dundee.ac.uk/www-jpred/
Meta-PP	Consensus prediction of other servers	Webserver	http://www.predictprotein.org/404page.php
PREDATOR	Knowledge-based database comparison	Webserver	http://mobyle.pasteur.fr/cgi-bin portal.py?form=predator
Predict Protein (PHD)	Profile-based neural network	Webserver	http://www.predictprotein.org/
PSIpred	two feed-forward neural networks which perform an analysis on output obtained from PSI-BLAST	Webserver	http://bioinf.cs.ucl.ac.uk/psipred/
YASSPP	Cascaded SVM-based predictor using PSI-BLAST profiles	Webserver	http://glaros.dtc.umn.edu/yasspp/
SSPRED	Multiple alignment-based program	Webserver	http://www.embl-heidelberg.de/sspred sspmul.html
SOPM	Multiple alignment-based program	Webserver	http://www.ibcp.fr/predict.html

(Contd.)

Name	Method	Description	Web link
Zpred	Multiple alignment- based method	Webserver	http://www.kestrel.ludwig.ucl.ac.uk zpred.html
HNN	Accessible through the NSP server	Webserver	http://www.npsa-pbil.ibcp.fr/NPSA npsa_mlr.html
Target99	Perform iterated search against a library of HMMs of protein with known 3D structure	Webserver	http://www.cse.ucsc.edu/research /compbio/HMM-apps.html
PHDsec	Sequence analysis, structure and function prediction	Webserver	http://www.cubic.bioc.columbia.edu predictprotein

Coiled-coils structure prediction

COILS	Single sequence based prediction for coiled-coil protein in database	Webserver	http://www.ulrec3.unil.ch/software COILS_form.html
Paircoil	Predicts the location of coiled-coil regions in amino acid sequences	Webserver	http://www.gropups.csail.mit.edu/cb paircoil/paircoil.html

Transmembrane helix and signal peptide prediction

HMMTOP	Hidden Markov Model	Webserver/standalone	http://www.enzim.hu/hmmtop/
MEMSAT	Neural networks and SVMs	Webserver/standalone	http://bioinf.cs.ucl.ac.uk/psipred/
PHDhtm	Multiple alignment-based neural network system	Webserver/standalone	http://www.predictprotein.org/
Phobius	Homology supported predictions	Webserver/standalone	http://phobius.sbc.su.se/
TMHMM	Prediction of transmembrane helices in proteins	Webserver/standalon	http://www.cbs.dtu.dk/services TMHMM/
Signalp	Neural network prediction of presence and location of signal peptide cleavage sites in amino acid sequences from different organisms	Webserver	http://www.cbs.dtu.dk/services/SignalP/

Protein Structure, Prediction and Visualization

- Used in conjunction with molecular dynamics simulations.
- Predicting antigenic epitopes.
- Inferring function from calculated electrostatic potential around the protein.
- Facilitating molecular replacement in X-ray structure determination.
- Refining models based on NMR constraints.
- Testing and improving a sequence structure alignment.
- Confirming a remote structural relationship.

Comparative, loop and side-chain modelling T-coffee3D-JIGSAWCPH-ModelsCOMPOSERFAMSSwiss-ModelWHATIFICMModeller http://www.ch.embnet.org/software/TCoffee.htmlhttp://www.bmm.icnet.uk/servers/3dijigsawhttp://www.cbs.dtu.dk/services/CPHmodelshttp://www-cryst.bioc.cam.ac.ukhttp://www.phychem.pharm.kitasato-u.ac.jp/FAMS/fams.htmlhttp://www.expasy.ch/swissmod/SWISS-MODEL.htmlhttp://www.cmbi.kun.nl/whatifhttp://molsoft.comhttp://www.salilab.org/modeller/modeller.html

Model Evaluation

PrISMSCWRLMODLOOPSHOTGUN5PROCHECKWHATCHECKProsaIIBIOTECHVERYFY_3DERRATANOLEAAQUAPROVE http://www.honiglab.cpmc.columbia.eduhttp://www.dunbrack.fccc.edu/SCWRL3.phphttp://www.salilab.org/moloophttp://www.bioinfo.pl/metahttp://www.biochem.ucl.ac.uk/~roman/procheck/procheck.htmlhttp://www.cmbikun.nl/gv/server/WIWWWIhttp://www.camesbg.ac.athttp://www.biotech.embl-ebi.ac.uk:8400http://www.doe-mbi.ucla.edu/Services/Veryfy_3Dhttp://www.doe-mbi.ucla.edu/Services/Errat.htmlhttp://www.protein.bio.puc.cl/cardex/servers/anoleahttp://www.urchin.bmrb.wisc.edu/~jurgen/Aqua/serverhttp://www.ucmb.ulb.ac.be/UCMB/PROVE

Protein Threading

Protein threading, also known as fold recognition, is a method of computational protein structure prediction used for protein sequences which have the same fold as proteins of known structures but do not have homologous proteins with known structure. Protein threading predicts protein structures by using statistical knowledge of the relationship between the structure and the sequence. The prediction is made by "threading" (aligning) each amino acid contained in the target sequence to a position in the template structure, and evaluating how well the target fits the template. After the best-fit template is selected, the structural model of the sequence is built based on the alignment with the chosen template. The protein threading method is based on two basic interpretations. One is that the number of different folds in nature is fairly small (approximately 1000) i.e. the sequence has little or no primary sequence similarity to any sequence with a known structure, and the other is that some model from the structure library (PDB) represents the fold of the sequence.

There are many reasons for preferring threading modelling which is as follows.

1. The secondary structure is more conserved than primary structure.

2. The tertiary structure is more conserved than secondary structure.

3. As a result, very distant associations can be better detected through 2D or 3D structural homology instead of sequence homology.

Threading is the method by which a library of unique or representative structures is searched for structure analogs to the target sequences, and is based on the theory that there may be only a limited number of distinct protein folds.

Methods of Threading Modelling

Prediction-based methods (PBM)/2D Threading

- Predict secondary structure.

- Estimate the structure on the basis of secondary structure.

Distance based methods (DBM)/3D Threading

- Create a 3D model of the structure.

- Evaluate using a distance-based 'hydrophobicity' or pseudo-thermodynamic (empirical) potential.

There are some basic components which involve in threading modelling which are as follows.

- **Representation of the query sequence.**

- **Representation of Protein Structure.**

- **Objective function for scoring sequence-structure alignment.**

- **Method of aligning a sequence to the model.**

- **Method of selecting a model from the library.**

Representation of the Query Sequence

In naturally occurring proteins, sequences that are similar to the query sequence carry useful information about its 3D structure. It is extensively accepted that significantly similar protein sequences also assume a similar 3D structure. A multiple sequence alignment centred on the query sequence reflects sequence inconsistency within the protein family to which the query sequence belongs. Generally most threading algorithms use this actuality.

Representation of Protein Structure

For threading, the 3D coordinates are reduced to more abstract representations of protein structure. Protein structure is entirely determined by the 3D coordinates of all non-bonding atoms. Usually, structural core elements are defined by the secondary structure elements, α-helices and β-strands, usually with side chains removed. Among proteins with similar structures, large variations occur in the loop regions connecting the structural elements. In consequence, loop lengths, loop conformations, and loop residue interactions are rarely conserved, and often the loop residues are not represented explicitly in the structural models.

The main distinction among threading approaches is the choice of the structure mode representation. Threading algorithms fall into two main categories. They use (1) the protein structure is represented as a linear model. In a linear representation, protein structure is modelled as a chain of residue positions that do not interact. In a second-order representation, the model also includes interacting pairs of residue positions, e.g., to account for hydrophobic packing, salt bridges, or hydrogen bonding. Approaches that represent protein structure as a linear model consider each structural position in the model independently, neglecting spatial interactions between amino acids in the sequence. This allows very fast alignment algorithms, but loses structural information that may be present in amino acid interactions. (2) A protein structure is represented as a higher-order model. The higher-order models have been considered to represent triples and higher multiples of interacting residue positions, but are less common. Approaches that use higher-order models explicitly consider spatial interactions between amino acids that are distant in the sequence but are brought into close proximity in the model. This potentially allows for more realistic and informative structural models.

Objective Function for Scoring Sequence Structure Alignment

Generally threading modelling approaches do not use the physical full-atom free energy functions normally used by macromolecular modelling software.

As a substitute, most threading objective functions are determined empirically by statistical analysis of the 3D data deposited in the PDB. As a result, they are often referred to as empirical potentials or knowledge-based potentials. In the case of non-linear structural models one more general name is contact potentials, which is reflecting their origin in analysis of contacts between atoms or residues in crystal structures. Numerous approaches enlarge empirical potentials with other terms thought to be important.

Method of Aligning a Sequence to the Model

The objective of a threading alignment algorithm is to find a best possible match between the query sequence and a structural model among all possible alignments. The possibility of the match is defined by the objective function. One key difference among threading algorithm is determined by the objective function, and fall into three broad categories.

- 1D algorithms that use only the information associated with the 1D features of the structure.

- 1D/2D algorithms that apply the 1D search logic (objective function representation) for the optimal alignment but at various steps use the information inferred from the 2D features or redefine the objective function.

- 2D algorithm that uses full 2D representation of the problem, and deals with the higher complexity of the search space.

Method of Selecting a Model from the Library

Finding the best possible (optimal) score and alignment of a sequence to a structure depend open the possibility that what is the most likely structure of the query sequence (target). There are largely two approaches for this. The first approach selects the structure based on the best alignment score, usually after normalizing the scores in some way so that the scores from different models are comparable. This first approach is more popular and intuitive. The second approach includes the total probability of a model across all alignments of the sequence to the model. This approach is better grounded with the probability theory.

Unlike sequence-sequence alignment, there is no straightforward method to determine the statistical significance of the optimal score. The statistical significance of the score indicates how likely it is to obtain a given optimal score by chance. The distribution of scores for a given sequence and structural model depends on both the length of the query sequence and structural model depends on both the length of the sequence and the size of the model. Currently, there is no generic analytical description of the shape of the distribution of threading scores across different models, and sequences, though it is well understood that the distribution of optimal scores is not normal. For gapped local alignment of the two sequences, or a sequence and a sequence profile, the distribution of optimal alignment scores can be approximated by an extreme value distribution. Fitting the observed distribution of scores to the extreme value distribution function has been applied by the profile threading method FFAS. Some profile based methods approximate the distribution of scores by a normal distribution and calculate the Z-scores. The Z-scores are calculated with the mean and standard deviation of the scores of a query sequence with the library of all structural models. Similarly, many threading approaches with the quadratic pair-wise objective function use the optimal raw score as the primary measure of structure and sequence compatibility and estimate the statistical significance of the score assuming a normal distribution of the sequence scores threaded to a library of available models.

The neural network score approach implemented in GenThreader, determines the compatibility of the structure with the model. The input to the neural network is a set of values of different scores: sequence similarity scores, the solvent accessibility score and the pair-wise interaction scores. The Gibbs-sampling threading approach estimates the significance of the optimal score by comparison to the distribution of scores generated by threading a shuffled query sequence to the same structural model. The distribution of shuffled scores is assumed to be normal.

Threading modelling recognition involves similar steps as comparative modelling (Fig. 10.7). Depending on the algorithm to align the target sequence with the folds and the energy functions to determine the best fits, the threading methods can be roughly divided into four classes, which is as follows.

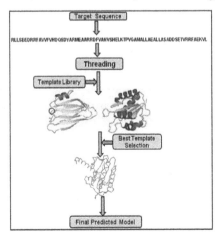

- The initial methods used the situation of each residue in the structure in the energy function and dynamica programming to evaluate and the alignment.

- As an alternative of using overlay simplified residual environment as the energy function, statically derived pair-wise interaction potential between residue pairs or atom pairs can be used to evaluate the best possible fit between the target sequence and library folds.

Fig. 10.7: Schematic representation of protein threading modelling

- This method does not use any explicit energy function at all. Instead, secondary structure and accessibility of each residue are predicted first and target sequence and library folds are encoded into string for the purpose of sequence-structure alignment.

- Finally, sequence similarity and threading can be obtained for fold recognition. For large-scale genome-wise protein structure predication, sequence similarity can be first used for the initial alignments and the alignments can be evaluated by threading methods.

The threading methods are limited by the high computational cost since each entry in the whole library of thousands of possible folds needs to be aligned in all possible ways to select the fold(s). Another major bottleneck is the energy function used for the evaluation of the alignment. As these functions are drastically simplified for efficient evaluation, it is not a reasonable expectation to be able to find the correct folds in all cases with a single form of energy function. Prospect is designed to find the globally optimal sequence-structure alignments for the given form of energy function. The divide-and-conquer algorithm is used to speed up the calculation by explicitly avoiding the conformation search space that is shown not to contain the optimal alignment. In numerous cases that have sequence identity very less (e.g.17%), perfect sequence-structure alignment is still achieved for the portions that can be aligned between the target and template structures. Yet in cases where no fold templates exist for the target sequence, important features of the structure are still recognized through threading the target sequence to the structures.

CASP (Critical Assessment of Techniques for Protein Structure Prediction) is a community-wide experiment for protein structure prediction taking place every two years since 1994. CASP provides users and research groups with an opportunity to assess the quality of available methods and automatic servers for protein structure prediction. Targets for prediction are proteins that are about to have their structures solved. The CASP contest provides an opportunity to evaluate methods on the same sample set and in the context of truly 'blind' predictions. This setting allows for performance comparison across methods using varied self-evaluation criteria applied by authors. Many threading approaches have been the subject of evaluation by past CASP contests. Examples of prediction methods that have been successful are: GenThreader, FFAS, and PROSPECTOR. Over earlier period CASP experiments, fold recognition methods proved capable to recognize distant sequence and structure similarities, untraceable by sequence comparison methods.

De Novo Protein Modeling

De novo or *Ab initio* protein modeling methods try to find to build 3D protein models "from scratch", i.e., based on physical principles rather than (directly) on previously solved structures (Fig 10.8). There are many possible procedures that either attempt to mimic protein folding or apply some stochastic method to search possible solutions (i.e., global optimization of a suitable energy function). These procedures tend to require vast computational resources, and have thus only been carried out for tiny proteins. To predict protein structure *de novo* for larger proteins will require better algorithms and larger computational resources like those afforded by either powerful supercomputers (Blue Gene) or distributed computing (Folding@home, the Human Proteome Folding Project). The *de novo* methods assume that the native structure corresponds to the global free energy minimum, accessible during the lifespan of the protein and attempts to find this minimum by an exploration of many conceivable protein conformations. The two key components of *de novo* methods are the procedure for efficiently carrying the conformational search, and the free energy function used for evaluating possible conformations.

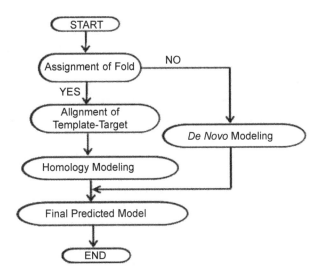

Fig. 10.8: Schematic representation of *De Novo* protein modeling

De novo protein structure prediction methods attempt to predict tertiary structures from sequences based on general principles that govern protein folding energetic and/or statistical tendencies of conformational features that native structures acquire, without the use of explicit templates. A general paradigm for de novo prediction involves sampling conformation space, guided by scoring functions and other sequence-dependent biases such that a large set of candidate ("decoy") structures are generated. Native-like conformations are then selected from these decoys using scoring functions as well as conformer clustering. High-resolution refinement is sometimes used as a final step to fine-tune native-like structures. There are two major classes of scoring functions. Physics-based functions are based on mathematical models describing aspects of the known physics of molecular interaction. Knowledge-based functions are formed with statistical models capturing aspects of the properties of native protein conformations.

When no suitable structure templates can be found, *ab initio* methods can be used to predict the protein structure form the sequence information only. Common to all *ab initio* methods are:

- The properly defined protein representation and corresponding protein conformation space in that representation
- Energy functions compatible with the protein representation
- The proficient and reliable algorithms to search the conformational space to minimize the energy function.

The conformations which minimize the energy function are considered to be the structures that the protein is likely to adopt at native conditions. The folding of the protein sequence is ultimately dictated by the physical forces acting on the atom of the protein and thus the most accurate way of formulating the protein folding or structure prediction problem is in terms of an all-atom model subject to the physical forces. Unluckily, the complexity of such a representation makes the solution simply possible with the currently available computational capacity. For reasonable reasons, most *de novo* or *ab initio* methods use reduced representations of the protein to limit the conformational space to a manageable size and use empirical energy functions that capture the most important interactions that drive the folding of the protein sequence towards the native structures. At present, many *de novo* methods can predict large contiguous segments of the protein to accuracy within six of RMSD.

A problem essential to the reduced representation of the protein and the simplified empirical potential is that the energy function is not sensitive enough to differentiate the correct native structures from conformations that are structurally close to the native state. The energy landscape calculated from such energy functions will not be correctly funneled but flattened and caldera -like around the native structure. In reality, as the native state is approached, the correlation between the calculated energy and the measure of similarity between predicted and native structures are not valid any longer. The usual practice is then to produce a large number of decoy structures and then use

Protein Structure, Prediction and Visualization | 245

various filtering and clustering techniques to pick up the more native-like structures. Filters can be used or eliminate structures with poorly formed secondary structures and low contact orders compared with that for sequences with compatible length. The other important technique is to use multiple sequences similar to the target sequence to generate decoy structures. Structures thus generated usually form dense clusters that are more compatible to the native structures of protein families of similar sequences than those obtained from a single sequence only.

Benefit and Drawback of *De Novo* modeling

De novo approach can be applied to model any sequence. On the other hand there is a lot of draw of this approach which is as follows.

- This method, usually results in relatively low-resolution models because of the complexity and limited understanding of the protein folding problem.

- This method remains applicable to a limited number of sequences of less than about 100 residues.

- Insufficient knowledge of force fields during energy minimization is the major drawback of this method.

Applications of *De Novo* Methods

Genome functional annotation and structural genomics initiatives are two main areas of research where *de novo* protein structure predication could make important contributions.

Genome Annotation

Though usually genome annotation has been accomplished using sequence-similarity search tools, many factors reduce the capacity of sequences homology to identify distant homologs. Circular permutations, domain insertions, exchange of secondary structure elements, and the genetic drift all give to the divergence of functionally related proteins over time. Thus, the annotation of open reading frames lacking detectable sequence homology to proteins of known function represents a promising application for *de novo* models. Low resolution *de novo* predicted structures may be able to reveal structural and functional relationships between proteins not apparent from sequence similarity only. This insight is well illustrated by some examples of predictions from CASP4. For examples (1), the predicted structures were each found to be structurally related to a protein with a similar function, but no significant sequence similarity.

(2) The functionally important residues were found clustered in the predicted structures.

De novo structures could be probed for the presence of residues adopting conserved geometric motifs. Although this approach has been applied to *de novo* model with a few success, it remains indistinct how to apply the technique more efficiently to low-resolution structures. In particular, some questions remain as to how indistinct structural motifs must be in order to detect homologies in low-resolution models. The Previous work has shown that weak matches to functional motif patterns may be filtered effectively by requiring similarity between the structures of pattern matches and the known structural environments of particular motifs. Therefore, it seems possible that *de novo* models could provide this structural information when high-resolution structures are unavailable.

Structural Genomics Initiatives

De novo structure prediction can help direct target selection by focusing experimental structures determination on those proteins likely to adopt novel folds or else to be of particular biological significance. While comparative modeling methods have been applied on a genomic extent, these approaches are inherently limited by their need for at least one homologous of known structure with good coverage and sufficient sequence similarity to be structurally the same.

Homologues of this quality are not always available, and therefore homology methods tend to leave significant fractions of both sequences and genomes improperly modeled. *De novo* techniques do not face this limitation, and thus may be precious adjunct to homology methods, filling in structural gaps and producing much more complete set of models than what could be obtained by either technique alone. However, small amounts of experimental data can noticeably improve the quality and reliability of *de novo* structure prediction with the application of spatial constraints. Apart from that, other sources of experimental data such as chemical cross-linking experiments could be used, allowing rapid structure determination for proteins not readily amenable to X-ray or NMR analysis. *De novo* structure prediction may therefore be useful for increasing the speed of structure determination, which is particularly important for structural genomics.

Successful structure prediction requires a free energy function sufficiently close to the true potential for the native state to be at one of the lowest free energy minima, as well as a method for searching conformational space for low energy minima. *De novo* structure prediction is challenging because current potential functions have limited accuracy, and the conformational space to be searched

Protein Structure, Prediction and Visualization | 247

is vase. Many methods use reduced representations, simplified potentials, and coarse search strategies in recognition of this resolution limit. Encouragingly, these simplified methods are starting to show some success in protein structure predication and have advanced to the point where genome scale modeling may become useful. Recently, however, there have been significant advancements in the field. There is hope that *de novo or ab initio methods* will continue to improve, and that this improvement will provide both fundamental insights into the physics underlying protein folding and a valuable, practical resource for genome analysis.

Conclusion

The 3D structure of a protein is of great importance to design experiments aimed towards understanding the function of a protein. It is often very difficult to experimentally determine 3D structure of a protein due to different barriers. This ground has initiated the need for homology modeling of proteins. Supplementary protein from different sources and sometimes of diverse biological function can have similar sequences. High sequences similarity implies distinct structural similarity. All these have served as the premise for the development of protein modeling. There are many possible procedures that either attempt to mimic protein folding or apply some stochastic method to search possible solutions. These procedures tend to require vast computational resources. Apart from vast computational barriers, the potential benefits of structural genomics make *de vono* or *ab initio* structure prediction an active field of research.

Protein Structure Visualization

An apparent visual representation of a macromolecular structure, be it a single image, a pair of stereo images, or a full-blown, interactive three-dimensional (3D) view, remains probably the most expressive way to describe the very significant amount of data that is encapsulated in the atomic coordinates of a model. There are different ways to examine the macromolecular structures and many packages for macromolecular visualization that are currently available. The programs and packages mentioned here can be divided roughly into three classes: the first and possibly the least visually demanding visualization task surrounding macromolecular models is the construction of an atomic model. In beginning point for model building may be a blank screen and a target sequence, in which case the model must be built largely from scratch, or it may be an existing structure that can be modified and molded to fit the data for the target structure. Either way, the process is highly interactive, and packages that deal with this specialized area of visualization are generally large and complex but at the same time rather limited in the styles of representation that are available, the emphasis being on the data themselves rather than their representation. Once a structure is built, refined, and available for wider use, the challenge becomes that of obtaining useful information from the structure. Extracting detailed information from a model requires tools that allow interactive manipulation and query of atomic coordinates, from measuring distances and angles to displaying a series of overlaid multiple structures. Although there are more possibilities for using different styles of representation at this stage, the emphasis at this point is again on clarity and visual simplicity, and more importance is generally attached to tools that enable a user to interrogate a structure than to those that provide a range of visual styles. Finally, once the structure is ready for publication, software is required to generate clear, informative, and attractive representations of atomic data, most often in the form of static images, but possibly also as 2D animations or 3D, interactive scenes.

Many researchers trying to determine the first molecular structures were faced with not only the theoretical challenge of calculating electron density values, but also the daunting task of somehow interpreting this electron density and obtaining the atomic coordinates that constitute a theoretical model of a macromolecular structure. The earliest molecular models, such as that of myoglobin, were built from masses of rods, wires, and spheres, so complex that the molecule itself was often lost in the web of supporting metalwork that was required to maintain its structure. Peering into the dim image the crystallographer could manually build the model of the protein structure by joining together small fragments of molecule and adjusting them to fit the faint outlines of electron density by eye. As with many other areas of science, macromolecular structure determination truly took off with the advent of electronic computing, and, as graphics technologies developed, so did the field of macromolecular visualization. Until relatively cheap and plentiful computing power became available, any calculation of the kind required for molecular structure determination was a painful manual undertaking, while visualization still involved either a hand-built physical model or a computer-generated, two-dimensional representation, formed by plotting electron density values or atomic coordinates on paper. The earliest attempts at electronic representations of molecular models used a computer-controlled oscilloscope to display a rotating image of a protein structure, with the speed and direction of rotation being controlled by the user. The system had many shortcomings, but it was an essential proof of concept and undoubtedly a herald of things to come.

The early graphics systems were milestones in molecular visualization, as crystallographers were finally able to view their models as truly three-dimensional objects, but they were mostly ad hoc constructions, difficult to build and almost as difficult to use. In the late 1960s, the head of the computer science department of the University of Utah and a professor from Harvard founded a company with the goal of developing and advancing the new field of computer graphics. Evans and Sutherland in 1969, produced one of the first commercial vector graphics systems, the crude and very expensive Line Drawing System (LDS1). While it was distant from a financial success, LDS1 proved that computer graphics was an invaluable way to display complex data, and with high success company develop a range of graphics systems and workstations that were the workhorses of the field of macromolecular visualization for many years.

Vector graphics systems were limited in the range of styles of representation that were available, and they gradually gave way to far more flexible systems that used raster-based displays. FRODO was one of the earliest and most widely used programs for constructing and manipulating molecular structures. It provided high-quality, interactive, color images of electron density maps and structures, and gave the user the ability to fit a model into displayed density by moving fragments of the model or even individual atoms. Till 1980s, program remained in widespread use before eventually being replaced by O, allowing more complex features to be added and giving more flexibility for the user and remains one of the most popular crystallographic model-building packages. In the early 1990s, one of the programs was RasMol, developed by Roger Sayle and gradually became what is now one of the most popular and widely used general-purpose molecular visualization programs. KINEMAGE was designed for visual description of the molecule that was created by the author of the structure, and provides the user with views and descriptions of the model.

The next key change in molecular visualization was once again the result of new developments and progress in computer technology. In 1980s, University of Utah (computer science department) had a reputation for computer graphics, and given a number of young researchers a leader start in the computer graphics industry. Jim Clark (1982) founded Silicon Graphics, with the aim of making the most powerful graphics platform in the world and making it affordable. Finally in 1987 Silicon Graphics had achieved these goals and had produced the standard in computer graphics. In late 1990s the revolution in Personal Computer (PC) hardware brought major changes to all aspects of structural biology, as cheap and extremely powerful desktop PCs largely outpaced all but the most expensive workstations in both raw power and graphics capability. PC graphics technology has quickly caught up with the more established graphics platforms from established vendors such as Silicon Graphics. It is promising for any user to generate views of macromolecular structures that are

250   Agriculture Bioinformatics

both visually appealing and graphics technologies continue to develop, the potential for macromolecular visualization can only increase.

Visualization Software

When trying to solve a structure by crystallography, the goal of the crystallographer is to turn a sequence into a three-dimensional atomic model of a structure, and representation styles necessarily focus on the constituent atoms of the molecule. Several visualization packages are used for crystallographic model building, and common to all of them are the ability to display electron density, and tools for manipulating atomic coordinates to fit that density. The XFit is part of the XtalView crystallography package, is an alternative model building program which is entirely mouse-driven, and all functions can be accessed from a graphical user interface. QUANTA is a commercial package for macromolecular crystallography from MSI (now part of Accelrys) that provides a similar range of tools to the other model-building programs. Probably the most useful style for displaying atomic coordinates for manipulation is the simple wire-frame bonds representation, with the lines being colored according to the type of the atoms being linked (blue for nitrogen, red for oxygen, yellow for sulfur etc.). Basic lines can usually be drawn very quickly by most graphics systems, so even a large structural ensemble can generally be represented in this fashion without bringing the computer to a complete halt. Many different programs can generate these relatively simple representations of molecules, but probably the most widely used is RasMol. It is simple to use, yet flexible, and can be used for display and interrogation of atomic models. RasMol runs well on every platform and is probably the most useful general-purpose structure viewer. RasTop is a only Windows based program that adds a more comprehensive user interface and extensions to the features of the original RasMol. The next level of detail at which molecules are commonly represented utilizes somewhat abstract views of macromolecular (protein) structures. The propensity of proteins to form well-defined secondary structural elements- α-helices and β-strands-is a fundamental property of protein structure. By abstracting away the atomic coordinates and representing a structure according to secondary structure alone, this schematic style aptly describes both the arrangement of individual atoms in adjacent residues along the chain, as well as the interaction between more widely separated atoms through hydrogen bonding. Practically all programs that can be used to view macromolecular structures can also generate interactive, Richardson-style, three-dimensional representations of protein structures. A noninteractive program for generating attractive images of molecular structures is MolScript with representation styles for atomic coordinates include wire-frame, CPK, or

Protein Structure, Prediction and Visualization

ball-and-stick styles, while secondary structure elements may be drawn as solid spirals and arrows for α-helices and β-strands, respectively. A widely used modification of the original version of MolScript, known as bobscript, adds enhanced coloring capabilities and, most significantly, the ability to display electron density maps as solid surfaces or meshes, although there is no facility for previewing a scene as in MolScript v2. Along with electron density, many visualization programs can display various other kinds of 3D data. In some cases it may be appropriate to display the data in the form of meshes, as for electron density, but another common style of representation is a projection of the data values onto molecular surfaces. Grasp was one of the earliest programs to be able to display surfaces interactively and it remains one of the most commonly used programs for looking at the properties of macromolecules. Grasp is capable of coloring a solid surface according to the density and polarity of charge surrounding surface vertices, and the resulting patches of positive and negative charge dramatically illustrate the nature of the charge on the surface of a molecule. One of the package is Chimera, which can display both electron density and various kinds of surface as meshes and opaque or semitransparent solid surfaces. The core features of Chimera include a flexible interactive viewer for displaying molecular structures in a wide variety of styles, as well as basic structure editing tools, but the program also has a modular design that allows new features to be added easily. On the other hand, PyMol (Fig. 10.9), another flexible, extensible package for molecular visualization, sports many of the same features as Chimera, but is freely available and open source.

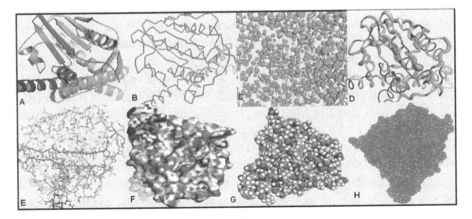

Fig. 10.9: Different types of molecular visualization of modeled protein through PyMol, namely A-Ribbon (Cartoon), B-Simple, C-Ball and Stick, D- b Factor Putty, E- Technical, F- Surface, G- Sphere, H- Mesh

The program supports a wide range of representation styles, from simple line drawings through chicken wire or solid molecular surfaces and density maps.. Finally, one of the most powerful packages that can be used for visualizing molecular surfaces is AVS, from Advanced Visual Systems. AVS is a completely general-purpose visualization tool that uses a novel graphical editor to design networks that link together many separate modules, each of which performs a single task. It is notoriously difficult to master, but, with the right modules and careful design of networks, it can generate good results and provides an extremely flexible environment for investigation of molecular properties and structure and for all types of interactive visualization.

Web-Based Visualizing Software

Traditionally, in order to run a particular program, the user had to download and install that program locally. With the introduction of the World Wide Web and the explosion of interest in server-based rather than locally installed software, this scenario is no longer always the case. Technologies such as Java, Active X, JavaScript, and so forth mean that users can run increasingly complex applications without having to explicitly download them, and rely instead on having the application delivered automatically along with a Web page.

Java applets are probably the most commonly used form of Web-deliverable, platform- independent software, allowing developers to write a single program that can be deployed on many different kinds of computer, from PC to different operating system. For low resolution visualization tasks, Java is a very capable solution and is becoming popular with developers because of its power and ease of use. WebMol is a trivial Java applet that can display and query a molecular structure. In this such as live, interactive Ramachandran plots make WebMol a useful tool for assessing the quality of a model. One of another Java applet, QuickPDB is one of very few visualization programs, which provide tools for interrogating both the structure and sequence of a protein structure simultaneously. QuickPDB is one of the visualization options from the Research Collaboratory for Structural Bioinformatics (RCSB) Protein Data Bank (PDB) site. Apart from that the Molecular Interactive Collaborative Environment (MICE) is another Java applet for viewing molecular structures, but it provides less support for querying the structure and places more emphasis on the representation of the structure. The key feature of MICE is that it allows users to generate interactive views of any structure in the PDB and to then share that view in real time with other users across a network. It is also available as one of the visualization options in the RCSB PDB site. Web browsers began life as simple tools for displaying simple pages, but as the potential of the browser was realized, it became clear that there was no way that developers of the

Protein Structure, Prediction and Visualization 253

browser itself could hope to keep up with the mass of applications, technologies, and services that would soon be available on the Web. Netscape got around this problem by introducing a "plug-in" architecture to their browser that would allow third-party software developers to create modular applications that would be installed inside Netscape and used to cope with suitably enhanced Web pages and sites. Chime, which is a commercial derivative of RasMol, packaged in the form of a browser plug-in but having all of the same capabilities and flexibility as its stand-alone cousin. Once installed in a Web browser, Chime is invoked to handle PDB format files, creating a RasMol-style view of the structure. Chime also forms the kernel of the Protein, a Web application that provides a framework for examining and manipulating structures. Table 10.2 lists the programs discussed in this chapter along with their capabilities as well as their website address.

Table 10.2: Lists the useful database, alignment, comparative, loop, side-chain modeling, model evaluation programs and servers for the comparative protein structure modeling

Visualizing software	Capabilities	Website
Protein Explorer	Protein Structure Explore Tool	http://www.umass.edu/microbio/chime/pe_beta/pe/protexpl/
QuikPDB	Interactive Sequence-Structure Viewer	http://xtal1.sdsc.edu/misha/QuickPDB.html
WebMol	Java applet for structure display	http://www.cmpharm.ucsf.edu/cgi-bin/webmol.pl
Chime	Visualization Tool Browser-plug in version of RasMol	http://hub.opensolaris.org/bin/view/Project+dtrace-chime/
Cn3D	3-D Structure Viewer	http://www.ncbi.nlm.nih.gov/Structure/CN3D/cn3d.shtml
RasMol	Structure interrogation Presentation	http://www.umass.edu/microbio/rasmol/
MAGE	Kinemage Image Viewing	http://kinemage.biochem.duke.edu/kinemage/magepage.php
AVS	Structure query, presentation	http://www.avs.com/
Bobscript	Presentation	http://www.csb.yale.edu/userguides/graphics/bobscript/bobscript.html
Chimera	Basic model building, model query, presentation	www.cgl.ucsf.edu/chimera/
Dino	Presentation	www.dino3d.org/
Grasp	Electrostatics calculation andrepresentation	http://wiki.c2b2.columbia.edu/honiglab_public/index.php/Software:GRASP
MICE	Java applet for structure display	http://mice.sdsc.edu/
MolScript	Presentation	http://www.avatar.se/molscript/
Model builder ModBldReview.htm	Model-building	http://www.housatonicrr.com/

(Contd.)

Visualizing software	Capabilities	Website
PyMOL	Presentation	http://pymol.sourceforge.net/
Quanta	Model building Structureinterrogation	http://accelrys.com/products/quanta/
RasTop	Enhanced RasMol	http://www.geneinfinity.org/rastop/
VMD	Structure analysis and presentation, with an emphasis onMolecular dynamics	http://www.ks.uiuc.edu/Research/vmd/
XTALVIEW	Electron density map inter pretation Model building Presentation	www.sdsc.edu/CCMS/Packages/XTALVIEW/

This session of chapter gives only the briefest of overviews of a few of the packages that are available to look at the structures of macromolecules. The reader will be able to assess the usefulness and usability of these and other packages for themselves, and will be able to find the most appropriate one for any given task.

Further Reading

1. Gregoret, L.M. and Cohen, F.E. (1990) Novel method for the rapid evaluation of packing in protein structures. J. Mol. Biol., 211:959–74.
2. Holm, L. and Sander, C. (1992) Evaluation of protein models by atomic salvation preference. J. Mol. Biol., 225:93–105.
3. Xue J, et al. (2008) (In) IFIP International federation for information processing, 259; Computer and Computing Technologies in Agriculture, Vol. 2; Daoliang Li; (Boston: Springer), pp. 977–982.
4. Ghosh, Z. and Mallick, B. (2008) Bioinformatics: principles and applications, (Ist Edition) Oxford University Press, New Delhi.
5. Krieger, E., Naburus, S.B. and Vriend, G. (2003) 'Homology modelling', in: structural bioinformatics', Bourne P.E, Weissig H., (eds), Wiley Liss, Inc, Hoboken, New Jersey, 512-513.
6. CASP4 (Forthcoming) Results from the comparative assessment of techniques for protein structure prediction. Proteins 45(S5):98–162.
7. Chothia, C. (1984) Principles that determine the structure of proteins. Ann. Rev. Biochem., 53:537–72.
8. Kabsch, W. and Sander, C. (1984) On the use of sequence homologies to predict protein structure: identical pentapeptides can have completely different conformations. Proc. Natl. Acad. Sci., USA 81:1075–8.
9. Lazaridis, T. and Karplus, M. (2000) Effective energy functions for protein structure prediction. Curr. Opin. Struct. Biol., 10:139–45.
10. Simons, K.T., Strauss, C. and Baker, D. (2001) Prospects for ab initio protein structural genomics. J. Mol. Biol., 306:1191–9.

References

1. Altschul, S.F. et al. (1990) Basic local alignment search tool. J. Mol. Biol., 215:403–10.
2. Bates, P.A. and Sternberg, M.J.E. (1999) Model building by comparison at CASP3: using expert knowledge and computer automation. Proteins, (Suppl. 3):47–54.

Protein Structure, Prediction and Visualization

3. Chinea, G., *et al.* (1995) The use of position specific rotamers in model building by homology. Proteins, 23:415–21.
4. Chothia, C. and Lesk, A.M. (1986) The relation between the divergence of sequence and structure in proteins. EMBO J., 5:823–36.
5. Dayringer, H.E. *et al.* (1986) Interactive program for visualization and modelling of proteins, nucleic acids and small molecules. J. Mol. Graph., 4:82–7.
6. de Filippis, V., *et al.* (1994) Predicting local structural changes that result from point mutations. Protein. Eng., 7:1203–8.
7. Dean, C.M. and Blundell, T.L. (2001) 'CODA: A combined algorithm for predicting the structurally variable regions of protein models', Protein. Sci., 10: 599-612.
8. Desmet, J., (1992) The dead-end elimination theorem and its use in protein side-chain positioning. Nature, 356:539–42.
9. Dodge, C. *et al.* (1998) The HSSP database of protein structure–sequence alignments and family profiles. Nucleic Acids Res., 26:313–5.
10. Dunbrack, R.L. Jr and Karplus, M. (1994) Conformational analysis of the backbone dependent rotamer preferences of protein side chains. Nat. Struct. Biol., 5:334–40.
11. Eisenberg, D., Luthy, R. and Bowie, J.U. (1997) VERIFY3D: Assessment of protein models with three-dimensional profiles, Methods Enzymol., 277: 396-404.
12. Epstein, C.J. *et al.* (1963) The genetic control of tertiary protein structure: studies with model systems. Cold Spring Harb Symp. Quant. Biol., 28:439.
13. Fiser, A. *et al.* (2000) Modeling of loops in protein structures. Protein, Sci., 9:1753–73.
14. Esnouf, R.M. (1997) An extensively modified version of MolScript that includes greatly enhanced coloring capabilities. J. Molec. Graphics, 15(2):132–4, 112–3.
15. Frishman, D. and Argos, P. (1997) Seventy-five percent accuracy in protein secondary structure prediction, Proteins, 27:329-335.
16. Garnier, J., Gibrat, J.F. and Robson, B. (1996) GOR secondary structure prediction method version IV, Methods Enzymol., 266:540-553.
17. Gerstein, M. (1998) Patterns of protein-fold usage in eight microbial genomes: A comprehensive structural census, Proteins, 33: 518-534.
18. Go, N. (1983) Theoretical studies of protein folding, Ann. Rev. Biophys Bioeng, 12:183.
19. Huang, C.C. *et al.* (1996) Chimera: An Extensible Molecular Modeling Application Constructed Using Standard Components, Pacific Symposium on Biocomputing, 1:724.
20. Jones, D.T. (1999) GenTHREADER: Efficient and reliable proteins fold recognition method for genomic sequences, J. Mol. Biol., 287(4):797-815.
21. Jones, D.T. *et al.* (1992) A new approach to protein folds recognition, Nature, 358(6381): 86-89.
22. Jones, D.T. and Moody, C.M. (1996) Towards meeting the Paracelsus challenge: The design, synthesis, and characterization of paracelsin-43, an alpha-helical protein with over 50% sequence identity to an all-beta protein, Proteins, 24(4): 502-513.
23. Jones, D.T. (1999) Protein secondary structure prediction based on position-specific scoring matrices, J. Mol. Biol., 292: 195-202.
24. Jones, T.A. (1978) A graphics model building and refinement system for macromolecules, J. Applied Crystallogr, 11:268–272.
25. Jones, T.A. *et al.* (1991) Improved methods for the building of protein models in electron density maps and the location of errors in these models, Acta Crystallogr, A47:110–119.
26. Jones, T.A. and Thirup, S. (1986) Using known substructures in protein model building and crystallography, MBOJ, 5: 819-822.
27. Kendrew, J.C. *et al.* (1958) A three dimensional model of the myoglobin molecule obtained by x-ray analysis, Nature, 181:662–666.

28. Kraulis, P.J. (1991) MOLSCRIPT: a program to produce both detailed and schematic plots of protein structures, J. Applied Crystallogr, 24:946–950.
29. King, R.D., Sternberg, M.J.E. (1996) Identification and application of the concepts Important for accurate and reliable protein secondary structure prediction, Protein. Sci., 5: 2298-2310.
30. Laskowski, R.A., *et al.* (1993) Procheck: a program to check the stereochemical quality of protein structures. J. Appl. Cryst., 26: 283-291.
31. Kouranov, A. *et al.* (2006) The RCSB PDB information portal for structural genomics, Nucleic Acid Res., 34: D302-D305.
32. Levin, J.M. *et al.* (1986) An algorithm for secondary structure determination in proteins based on sequence similarity, FEBS Lett., 205(2): 303-308.
33. Levinthal, C. *et al.* (1968) Computer graphics in macromolecular Chemistry. *Emerging Concepts in Computer Graphics* (Nievergelt J, Secrest D, editors). New York W.A. Benjamin, pp. 231–253.
34. McRee, D.E. (1999) XtalView/Xfit-A versatile program for manipulating atomic coordinates and electron density, J. Struct. Biol., 125(2–3):156–165.
35. Moult, J. and James, M.N. (1986) An algorithm for determining the conformation of polypeptide segments in proteins by systematic search, Proteins, 1:146-163.
36. Nicholls, A. (1993) GRASP: graphical representation and analysis of surface properties, Biophys J., 64:A166.
37. Novotny, J. *et al.* (1988) Criteria that discriminate between native proteins and incorrectly folded models. Proteins, 4:19–30.
38. Oliva, B. *et al.* (1997) An automated classification of the structure of protein loops, J. Mol. Biol., 266:814-830.
39. Ortiz, A.R. *et al.* (1998) Fold assembly of small proteins using Monte Carlo simulations driven by restraints derived from multiple sequence alignments, J. Mol. Biol., 277:419-448.
40. Pauling, L. and Corey, R.B. (1951) Configurations of polypeptide chains with favored orientations around single bonds: Two new pleatd sheets, Proc. Natl. Acad. Sci. USA, 37: 729-740.
41. Pauling, L. *et al.* (1951) The structure of proteins: Two hydrogen bonded helical conformations of the polypeptide chain, Proc. Natl. Acad. Sci. USA, 37: 205-211.
42. Peitcsch, M.C. *et al.* (2000) Automated protein modeling-the proteome in 3D', Pharmacogenomics, 1: 257-266
43. Pillardy, J. *et al.* (2001) Conforamtion-family Monte Carlo: A new method for crystal structure prediction, Proc. Natl. Acad. Sci. USA, 98(22): 12,351-12,356.
44. Rost, B. and Sander, C. (1993) 'Prediction of protein secondary structure at better than 70% accuracy, J. Mol. Biol., 232: 584-599.
45. Rychlewski, L. *et al.* (2000) Comparison of sequence profiles: Strategies for structural predictions using sequence information, Protein Sci., 9(2): 232-241.
46. Salamov, A.A. and Solovyev, V.V. (1995) Prediction of protein secondary structure by combining nearest-neighbor algorithms and multiply sequence alignments, J. Mol. Biol., 247: 11-51.
47. Sali, A. and Blundell, T.L. (1993) Comparative protein modelling by satisfaction of spatial restraints. J. Mol. Biol., 234:779–815.
48. Skolnick, J. and Kihara, D. (2001) Defrosting the frozen approximation: Prospector-a new approach to threading, Proteins, 42(3): 319-331.
49. Sanchez, R. and Sali, A. (1997) Evaluation of comparative protein structure modeling by MODELLER-3. Proteins, (Suppl. 1):50–8.

Protein Structure, Prediction and Visualization

50. Sanchez, R. and Sali, A. (1999) ModBase: a database of comparative protein structure models. Bioinformatics, 15:1060–1.
51. Sander, C. and Schneider, R. (1991) Database of homology-derived protein structures and the structural meaning of sequence alignment. Proteins, 9:56–68.
52. Simons, K.T., Bonneau, R., Ruczinski, I. and Baker, D. (1999) Ab initio structure prediction of CASP III targets using ROSETTA. Proteins, (Suppl. 3):171–6.
53. Sippl, M.J. (1993) Recognition of errors in three dimensional structures of proteins. Proteins, 17:355–62.
54. Stites, W.E., Meeker, A.K. and Shortle, D. (1994) Evidence for strained interactions between side-chains and the polypeptide backbone. J. Mol. Biol., 235:27–32.
55. Tate, J.G. *et al.* (2001) Design and implementation of a collaborative molecular graphics environment, J. Molec. Graphics, 19:280–287.
56. Tappura, K. (2001) Influence of rotational energy barriers to the conformational search of protein loops in molecular dynamics and ranking the conformations. Proteins, 44:167-79.
57. Taylor, W.R. (1986) Identification of protein sequence homology by consensus template alignment. J. Mol. Biol., 188:233–58.
58. Thompson, J.D., Higgins, D.G. and Gibson, T.J. (1994) ClustalW: improving the sensitivity of progressive multiple sequence alignments through sequence weighting, position-specific gap penalties and weight matrix choice. Nucleic Acids Res., 22:4673–80.
59. Vriend, G. (1990) WHAT IF-A molecular modeling and drug design program. J. Molec. Graphics, 8:52–56.

Chapter – 11

Mitogenomics: Mitochondrial Gene Rearrangements, its Implications and Applications

Tiratha Raj Singh

Abstract

Mitochondrial (mt) genomic study may reveal significant insight into the molecular evolution and several other aspects of genome evolution such as gene rearrangements evolution, gene regulation, and replication mechanisms. Other questions such as patterns of gene expression, mechanism of evolution, genomic variation and its correlation with physiology, and other molecular and biochemical mechanisms can be addressed by the mt genomics. Rare genomic changes have attracted evolutionary biology community for providing homoplasy free evidence of phylogenetic relationships. Gene rearrangements are considered to be rare evolutionary events and are being used to reconstruct the phylogeny of diverse group of organisms. Mitochondrial gene rearrangements have been established as a hotspot for the phylogenetic and evolutionary analysis of closely as well as distantly related organisms. This evolutionary episode has significant contribution among animals as well as plants and serves as an important genomic evolutionary event for the analysis of mitogenomes on large scale.

Introduction

Mitochondria are the power house of the cell. They are present in virtually every cell in body. They play a central role in metabolism, apoptosis, disease and aging. They are the site of oxidative phosphorylation, essential for the production of ATP, as well as for other biochemical functions. Mitochondria have a genome separate from the nuclear genome referred to as mitochondrial DNA (mt DNA). Animal mt DNA is a small (~16 kb), compact, economically-organized circular molecule, composed of 37 genes coding for 22 tRNAs, 2 rRNAs and 13 proteins, with few exceptions specifically in invertebrates (Boore 1999). The thirteen proteins are mainly involved in electron transport and oxidative phosphorylation of the mitochondria.

Recent advances in sequencing techniques have made available a great deal of data on whole genome basis. Complete mt genome sequences are available for thousands of organisms. The order of the genes in the mitochondrial DNA molecule in a wide variety of organisms has begun to be disclosed during last two decades. The gene order is highly conserved in vertebrates except for the region around the control region (D-loop), which is more prone to gene rearrangement. Maximum variability has been found in the gene order in invertebrates and plants. The control region of the mitochondrial genome is frequently used in population studies due to the high variability in its nucleotide sequence, while protein-coding genes, such as cytochrome b (Cyt b), are generally used for phylogenetic analysis of taxa above the species level.

Mitogenomics and Phylogenetic Markers

Genome level character comparison can address a large number of evolutionary branch points. A small number of of such comparisons have provided strong resolution of some evolutionary relationships which were controversial earlier. This suggests the reliability and confidence in their usage as markers [3]. Mitochondrial gene rearrangements are considered to be rare evolutionary events. In principle, 'rare genomic changes' [2], such as changes in gene order, can retain phylogenetic information for long periods of time, even when primary sequence data must have become randomized due to the involvement of long time periods.

Differences in mitochondrial gene order have been useful for phylogenetic resolution of some groups of species, for example Arthropoda being monophyletic, and within this Crustacea grouping with Hexapoda to the exclusion of Myriapoda and Onychophora. The 'universal' gene order for living vertebrates (Fig.11.1) is not followed by birds. In some recent studies based on complete mt genomes in birds, it has been suggested that gene orders evolved independently more than once. Also mapping of these gene orders on the avian phylogenetic tree suggested that one of these gene orders is ancestral while the other is a derived form. The most acceptable model for the mt gene rearrangements till date is the tandem duplication followed by the deletion.[3,8,25]

Genes and genomes are the products of complex processes of evolution, influenced by mutation, random drift, and natural selection [30,29]. The inference of genome rearrangement events such as duplication, inversion, and translocation, is crucial in multiple genome comparisons. During the past three decades, DNA and protein sequence data have increasingly become the main asset to infer the general phylogeny; however, certain difficulties and limits have been found with the use of nucleotide sequences in phylogeny, in particular evolutionary rate differences between taxa, non-homogenous base composition, and multiple substitutions. "Rare genomic changes" have attracted great interest because of their potential to provide homoplasy-free evidence of phylogenetic relationships [24]. Gene rearrangements are considered to be rare evolutionary events, and as such the existence of a shared derived gene order between taxa is often indicative of common ancestry [14,3].

An enormous number of phylogenetic studies using mitochondrial (mt) gene sequences have been carried out and reported [20]. The success of mt DNA in molecular systematics lies in its prevalent characteristics, such as maternal inheritance, rapid rate of evolution, and haploid nature [24]. Different parts of mt genome with different functional constraints are expected to evolve at different rates. Thus, comparative mt genomics promises to offer a comprehensive study of distinct patterns and processes of molecular evolution [18,22].

Mitogenomes and their Diversity

Despite their monophyletic origin, animal and plant mt genomes have been described as exhibiting different modes of evolution. Plant mt genomes feature a larger size, a lower mutation rate and more rearrangements than their animal counterparts. Gene order variation in animal mt genomes is often described as being due to translocation and inversion events, but tandem duplication followed by loss has also been proposed as an alternative process. In plant mitochondrial genomes, at the species level, gene shuffling and duplicate occurrence are such that no clear phylogeny has ever been identified, when considering genome structure variation.

Most animal mitogenomes are circular and compact, share the same gene content and have a size that does not exceed 20 kb. Only the control region (CR) and its surroundings are more prone to variations (Fig. 11.1). In contrast, plant mitogenomes exhibit larger size (most are from 200 to 700 kb) and are less compact than their animal counterparts due to the occurrence of non- coding sequences and duplicated fragments. Plant mitogenomes are known to evolve rapidly in structure and slowly in sequence. The plant mitogenomes exhibited a large variation in size mainly due to large duplicated fragments, and gene shuffling was such that no clear evolutionary scenario could be pictured. However, on the basis of nucleotide divergence, groups of related mitogenomes could be defined and qualified as ancestral or derived though no phylogeny could be established [5].

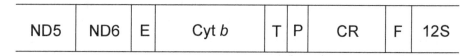

Fig. 11.1: Universal vertebrate mt gene order around control region (CR). Cyt b, Cytochrome b; T, Threonine tRNA; P, Proline tRNA; ND5 and ND6, nicotinamide adenine dinucleotide dehydrogenase subunit 5 and subunit 6 respectively; E, Glutamic acid tRNA; F, Phenylalanine tRNA.

The first complete mt DNA sequence determined was that of human, and was described by the phrase "small is beautiful" (Borst and Grivell, 1981). Vertebrate mt DNA is a small (~16 kb), compact, economically-organized circular molecule, composed of 37 genes coding for 22 tRNAs, 2 rRNAs and 13 proteins [3]. All the thirteen proteins are mainly involved in electron transport and oxidative phosphorylation of the mitochondria (Fig.11.2).

Plant Mitogenomes: An Evolutionary Insight

Plant mitogenomes have larger size and are less compact than their animal counterparts. The reason is the occurrence of non-coding sequences and duplicated fragments. The mitochondrial genomes of higher plants show several organizational features that distinguish them from those of other organisms. These are exemplified by the mitochondrial genome of *maize*: at 570 kb, which is 7 times as big as the yeast (*Saccharomyces cerevisiae*) mitogenome and 35 times the size of mammalian mitogenomes. The large size, presence of large direct repeats, and consequent multipartite structure are common features of higher plant mitochondrial genomes (Lonsdale *et al.*, 1984; Palmer and Herbon, 1988).

Mitogenomics

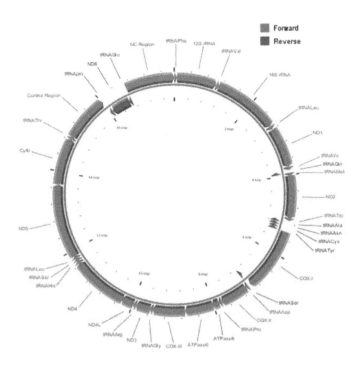

Fig. 11.2: An indicative circular genome of mitochondria with gene positions and both forward and reverse fragments in red and blue and their orientations respectively.

Angiosperms have the largest mitogenome ranging from 200kb to over 2000kb. Angiosperm mitogenome is least dense also among eukaryotes also because its intergenic regions are expanded. Five general characteristic features of angiosperm mitogenomes are:

- Mitogenomes consist of a mixture of DNA species.
- Angiosperm mitogenome contains sequences not descended from the ancestral mitogenomes, but apparently acquired from other genetic compartments, such as plastids, the nucleus, and through horizontal gene transfer (HGT).
- Next two features are related to the expression system. Most angiosperm mitochondrial genes require the posttranscriptional conversion of cytidine residues to uridines through RNA editing [26].
- Presence of group II introns in atleast 10 angiosperm mito genes is another feature of these mitogenomes.
- Another remarkable feature of angiosperm mitogenomes is they are the least gene dense mitogenomes based on genome size [28,13].

However with the availability of wide range of genome sizes (222 to 773 kb among the fully sequenced genomes, the numbers of genes in angiosperm mitochondria (about 50 to 60) do not vary greatly. Angiosperm mitochondria contain more genes than those of humans (37 genes in a 16.6 kb genome) but less than those of the liverwort Marchantia polymorpha (71 genes in 186 kb) and the moss Physcomitrella patens (67 genes in 105 kb) [27]. The mitochondrial genome of the gymnosperm *Cycas taitungensis* (415 kb) contains 64 genes, which is slightly higher than the highest number found in angiosperm mitochondria.

Land plants analysis indicates that some genes have been lost from the angiosperm mitochondria, but the genome size has not been reduced accordingly. In some cases the lost genes appear to have migrated into the nuclear genome, and their translation products are imported into the mitochondria [2]. In other cases, the gene loss is compensated by paralogues in the nuclear genome. For example, the gene rps13 is not present in the Arabidopsis mitochondrion and no migrated copy occurs in the nuclear genome. However, a nuclear rps13 copy of plastid origin appears to encode a mitochondrial RPS13 polypeptide [1]. The genes for tRNALeu and tRNA Arg are present in Cycas mitochondria [27] but lost from those of angiosperms, and results from an analysis of potato indicate that tRNA molecules charged with Leu and Ara are imported from the cytosol into the mitochondria [16].

The amounts of DNA that are known to be derived from external sources such as plastids and the nucleus are insufficient to explain the origin of the whole intergenic regions, since more than 60% of the intergenic regions show no homology with any other sequences. It is possible that some of the intergenic regions of angiosperm mitochondria may have been horizontally transferred from sexually incompatible species including nonplant organisms. Evidence of horizontal transfer into angiosperm mitochondria is accumulating. For example, viral sequences and a group I intron presumably transferred from fungi have been found [7]. Additionally, the plasmid-like DNA elements in some angiosperm mitochondria may be derived from fungi [12].

Cox2, present in the mitochondrion of virtually all plants, was transferred to the nucleus during recent legume evolution. Examination of nuclear and mt Cox2 presence and expression in over 25 legume genera has revealed a wide range of intermediate stages in the process of mt gene transfer, providing a portrayal of the gene transfer process in action. Cox2 was transferred to the nucleus via an edited RNA intermediate. Once nuclear Cox2 was activated, a state of dual intact and expressed genes of trans-compartmental functional redundancy was established; this transition stage persists most fully (i.e., with both compartments' Cox2 genes highly expressed in terms of steady-stated, properly processed RNAs; Cox2 protein levels have not been assayed) only in Dumasia among the many studied legumes. Four other, phylogenetically disparate legumes also retain intact and expressed copies of Cox2 in both compartments, but with only one of the two genes expressed at a high enough level, in the one tissue type examined thus far, to presumably support respiration. Silencing of either nuclear or mt Cox2 has occurred multiple times and in a variety of ways, including disruptive insertions or deletions, cessation of transcription or RNA editing, and partial to complete gene loss [13].

There has been debate on why are a few protein genes preferentially retained by mt genomes across all or most eukaryotes? One view is that the products of these genes, all of which function in respiration, are highly hydrophobic and difficult to both import into mitochondria and properly insert (post-translationally) into the inner mt membrane [21,19].

Evidence for this includes experiments in which cytoplasmically synthesized cytochrome b, a highly hydrophobic protein with eight transmembrane helices, could not be imported in its entirety, with successful import limited to regions comprising only three to four transmembrane domains [4]. In general, genes whose products have many hydrophobic transmembrane domains are usually located in the mitochondrion whereas genes whose products have a few such domains are more often transferred to the nucleus [10, 27]. Indeed, the only two protein genes contained in all of the many completely sequenced mt genomes encode what are by some criteria [4] the two most hydrophobic proteins present in the mitochondrion, cytochrome b and subunit 1 of cytochrome oxidase [9,10]. Although the *hydrophobicity hypothesis* cannot account for the distribution of every mt gene in every eukaryote, it seems likely to be a factor favoring the retention of certain respiratory genes. A second hypothesis for retention of certain genes in organelles is that their products are toxic when present in the cytosol or in some other, inappropriate cellular compartment to which they might be misrouted after cytosolic synthesis [17,9].

Mitogenomes: Gene Rearrangement Events

Extensive gene rearrangements have been observed at intra-genus level in invertebrates and vertebrates both such as in genus Dendropoma in marine gastropods and in genus Sylvia in birds [25]. When a rearrangement occurs, the evolutionary constraints in the regions involved in the new organization may also change (Kumazawa and Nishida, 1993). Gene order may be used to study the constraints on the evolution of the mitochondrial genome, the mechanism responsible for new gene rearrangements and deep-branch phylogeny. Gene order is applied to resolve the phylogeny because of the unlikely possibility of rearrangement convergence and the certainty of mtDNA gene homology [6,25].

Recent study on maize mitogenome showed that rearrangements could result from a mechanism similar to that found in animals, i.e. tandem duplication, but where some duplicates were partially lost [5]. Using this evolution model, a methodology was developed to reconstruct a phylogeny based on rearrangement events that integrated most duplicates, and ended up with an evolutionary scenario of the mitochondrial genome in maize (Fig.11.3). Despite important structural shuffling among genomes, even at the species level, we were able to build a phylogenetic tree using rearrangement events between plant mitochondrial genomes that were congruent with a sequence based tree. Under the hypothesis of structure evolution through inversions and tandem duplications with loss, an evolutionary path could be drawn for each genome. There is a need to develop new tools in order to automatically look for signatures of tandem duplication events in other plant mitogenomes and evaluate the occurrence of this process on a larger scale [5].

The number of possible new mitochondrial genome organizations is enormous. Few of the possible rearrangements have been found in vertebrates and invertebrates. In many cases an intermediate hypothetical stage has been proposed to resolve the mysteries of gene rearrangements and its role in evolution. This is not indicative that other gene organization has not occurred but that the organization as seen today may have functional constraints. Stable structural gene rearrangement may be due to constraints imposed on gene movement, selection on potential differences in transcription of genes, and low probability of transposition to specific gene junctions. Due to their paucity, it is not conceivable that different groups will have the same rearrangement for example birds have some specific gene organization, different from other vertebrates [3,25]. Similar kind of hypothesis has also been proposed for plant mt genome rearrangements (Fig.11.3) [5].

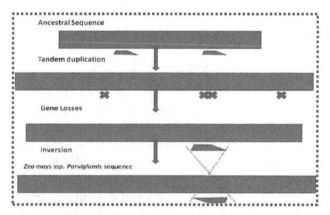

Fig. 11.3: Tandem duplication hypothesis. A hypothetical scenario of evolution from an ancestral sequence to *Zea mays* ssp. *Parviglumis* through tandem duplication, followed by deletions and inversions. Redrawn after Darracq *et al.*, 2010.

Concluding Remarks and Future Challenges

With the increasing amount of mt and nuclear sequence data there will be new outcomes for phylogenetic studies at several levels and will soon emerge as the knowledge base for molecular evolution. As the understanding of the secondary structure of the control region, non coding region and ribosomal genes and their role in the mt genome will improve, these sequences will be helpful for use in phylogeny. Several studies and debate are still in progress on the structure of tRNAs and their specific role in gene regulation and evolution. Gene regulation may also serve as a substrate for evolutionary change, since control of the timing, location, and amount of gene expression can have an insightful effect on the functions and existence of the genes in the organism. Mt gene rearrangements are likely to reveal structural, functional and evolutionary aspects of the biology distinct from those evident from homologous sequence comparisons. It will also help in the prediction and reconstruction of homoplasy free phylogenetic relationships, and will also be used to develop models for the phylogenetic analysis of remotely related organisms and to better understand the evolutionary history of animal and plant kingdoms.

Despite sequential and structural annotation at mitogenome level, there is a strong need to develop methods, and tools for the automatic annotation of gene rearrangement events among available or upcoming mitogenomic sequences. There is a need to analyze duplication and gene loss events in mitogenomic scenario on broad scale so as to decipher some new hypothesis for the analysis of such rare evolutionary events. Such kind of annotative analyses will definitely help molecular and evolutionary biologists, biotechnologists, and agricultural scientists to establish gold standard methods for gene rearrangement events.

References

1. Adams, K.L., Daley, D.O., Whelan J. and Palmer, J.D. (2002) Genes for two mitochondrial ribosomal proteins in flowering plants are derived from their chloroplast or cytosolic counterparts, Plant Cell, 14(4): 931-943.
2. Adams, K.L. and Palmer, J. D. (2003) Evolution of mitochondrial gene content: gene loss and transfer to the nucleus, Molecular Phylogenetics and Evolution, 29(3):380-395.
3. Boore, J.L. (1999) Animal mitochondrial genomes, Nucleic Acids Res., 27:1767-1780.
4. Claros, M.G., Perea, J., Shu, Y., Samatey, F.A., Popot, J.L. and Jacq, C. (1995) Genome structure and gene content in protist mitochondrial DNAs, Eur. J. Biochem., 228:762-771.
5. Darracq *et al.* (2010) A scenario of mitochondrial genome evolution in maize based on rearrangement events, BMC Genomics, 11:233.
6. Dowton, M., Castro, L.R., Campbell, S.L., Bargon, S.D. and Austin, A.D. (2003) Frequent mitochondrial gene rearrangements at the hymenopteran nad3-nad5 junction, J. Mol. Evol., 56(5):517-26.

7. Goremykin, V.V., Salamini F., Velasco R. and Viola R. (2009) Mitochondrial DNA of *Vitis vinifera* and the issue of rampant horizontal gene transfer, Mol. Biol. Evol., 26(1): 99-110.

8. Graur D. and Li, W.H. (2000) Fundamentals of molecular evolution, Sinauer associates, Sunderland, MA.

9. Gray, M.W. (1999) Evolution of organellar genomes Curr. Opin. Genet. Dev., 9:678-687.

10. Gray, M.W., Lang, B.F., Cedergren, R., Golding, G. B., Lemieux, C., Sankoff, D., Turmel, M., Brossard, N., Delage, E., Littlejohn, T.G., *et al.* (1998) Nucleic Acids Res., 26:865–878.

11. Gupta P.K., (2001) Elements of biotechnology, Rastogi Publications, India.

12. Handa H. (2008) Linear plasmids in plant mitochondria: peaceful coexistences or malicious invasions?" Mitochondrion, 8:1,15–25.

13. Kitazaki K. and Kubo T. (2010) Cost of Having the LargestMitochondrial Genome: Evolutionary Mechanism of Plant Mitochondrial Genome, Journal of Botany, Article ID 620137, 12 pages.

14. Liò P., and Goldman N. (1998) Models of molecular evolution and phylogeny, Genome Res., 8/12:1233-1244.

15. Lonsdale, D.M., Hodge, T.P. and Fauron, C.M.R. (1984) The physical map and organization of the mitochondrial genome from the fertile cytoplasm of maize. Nucleic Acids Res., 12(24):9249-9261.

16. Marechal-Drouard L., Guillemaut P., Cosset A., *et al.* (1990) Transfer RNAs of potato (*Solanum tuberosum*) mitochondria have different genetic origins, Nucleic Acids Research, 18(13): 3689-3696.

17. Martin, W. and Schnarrenberger, C. (1997) The evolution of the Calvin cycle from prokaryotic to eukaryotic chromosomes: a case study of functional redundancy in ancient pathways through endosymbiosis. Curr. Genet., 32:1-18.

18. Page, R.D.M. and Holmes, E.C. (2000) Molecular evolution: A phylogenetic approach, Blackwell Science Ltd., U.K.

19. Palmer, J.D. (1997) Organelle genomes: going, going, gone. Science, 275:790–791.

20. Palmer, J.D., Herbon, L.A. (1988) Plant mitochondrial DNA evolved rapidly in structure, but slowly in sequence. J. Mol. Evol., 28(1):87-97.

21. Phillips M.J., and Penny D. (2003) The root of the mammalian tree inferred from whole mitochondrial genomes, Mol. Phylogenet. Evol., 28/2:171-185.

22. Popot, J.L. and de Vitry, C. (1990) On the microassembly of integral membrane proteins Annu. Rev. Biophys. Biophys. Chem., 19: 369-403.

23. Rand, D. (2001) Mitochondrial genomics flies high, Trends Ecol. Evol., 16:2-4.

24. Rawlings, T.A., Collins, T.M. and Bieler, R. (2001) A major mitochondrial gene rearrangement among closely related species, Mol. Biol. Evol., 18(8):1604-9.

25. Rokas A., and Holland P.W.H. (2000) Rare genomic changes as a tool for phylogenetics, Trends Ecol. Evol., 15:454-459.

26. Singh, T.R. (2008) Mitochondrial Gene Rearrangements: new paradigm in evolutionary biology and systematics, Bioinformation, 3/2:95-97.

27. Takenaka, M. Verbitskiy, D., van der Merwe, J.A., Zehrmann A. and Brennicke A. (2008) The process of RNA editing in plant mitochondria, Mitochondrion, 8(1): 35–46.

28. Terasawa, K., Odahara, M. and Kabeya, Y., *et al.* (2007) The mitochondrial genome of the moss physcomitrella patens sheds new light on mitochondrial evolution in land plants, Molecular Biology and Evolution, 24(3): 699–709.

29. Unseld, M., Marienfeld, J.R., Brandt, P. and Brennicke, A. (1997) The mitochondrial genome of Arabidopsis thaliana contains 57 genes in 366, 924 nucleotides, Nature Genetics, 15:1 57–61.

30. Yang, Z. (2006) Computational molecular evolution, Oxford University Press, New York.
31. Zuckerkandal, E. and Pauling, L. (1965) Evolutionary divergence and convergence in proteins, in Evolving genes and proteins (Ed. V. Bryson and h. J. vogel), 97- 166, Academic Press, New York.

Chapter – 12

Spice Bioinformatics

Santhosh J. Eapen

Abstract

Spices are aromatic substances of plant origin which are commonly used for flavouring, seasoning and imparting aroma in foodstuffs. They are cultivated in an area of 5.98 million ha globally with a total production of 7.31 million tons. India is considered as the home of majority of these spices from ancient times and sizable area of the country is under spice cultivation. The latest statistics shows that the total cultivated area in India under spices is 2.89 million ha with a total production of 3.33 million tonnes. About 109 plants belonging to 31 families are recognized as spices useful as ingredients in food. Many of them have been demonstrated to mediate therapeutic benefits for wide spectra of diseases ranging from multiple sclerosis to colorectal cancer.

Introduction

Advances in molecular genetics in the last 20 years like rapid DNA amplification techniques using PCR, rapid sequencing methods and computational software programs have enabled the placement of molecular markers on to the maps of chromosomes of most major crop species and the subsequent tagging of genes of interest by their placement near those markers [38]. Unfortunately very little or scanty molecular data of spices are available in the public domain (Table 12.1). 'Orphan crops' like spices which have not yet received the investment of research effort or funding can exploit comparative genomics and bioinformatics to assist research and crop improvement provided they are related to a well-characterized model plant species.

Table 12.1: Genomic information on common spices available in GenBank

Sl. No.	Common name	Scientific name	Nucleotide sequence	Protein sequence	EST	Protein structure
1	Aniseed	*Pimpinella anisum* L.	5	19	0	0
2	Basil	*Ocimum basilicum* L.	66	73	0	7
3	Bay leaf	*Laurus nobilis* L.	226	117	0	0
4	Black pepper	*Piper nigrum* L.	88	27	63	0
5	Caper	*Capparis spinosa* L.	11	8	0	0
6	Capsicum, Paprika	*Capsicum annuum* L	1676	1299	118060	2
7	Cardamom (small)	*Elettaria cardamomum* Maton	9	6	0	0
8	Celery	*Apium graveolens* L.	158	134	2224	2
9	Chilli, Bird's Eye	*Capsicum frutescens* L.	118	58	33	0
10	Cinnamon	*Cinnamomum verum*	27	15	27	0
11	Cinnamomum	*Cinnamomum aromaticum*	23	4	0	0
12	Coriander	*Coriandrum sativum* L.	54	32	0	0
13	Cumin	*Cuminum cyminum* L.	10	1	0	0
14	Dill	*Anethum graveolens* L.	116	101	0	0
15	Fennel	*Foeniculum vulgare* Mill.	43	14	0	0
16	Fenugreek	*Trigonella foenum-graecum* L.	45	20	0	0
17	Garlic	*Allium sativum* L.	213	185	42	5
18	Ginger	*Zingiber officinale* Rosc.	27	68	38115	1
19	Greater galanga	*Alpinia galanga* Willd.	49	29	0	0
20	Horse radish	*Armoracia rusticana* Gaertn.	67	97	0	20
21	Juniper berry	*Juniperus communis* L.	143	67	0	0
22	Mint	*Mentha piperita* L.	171	95	1316	7
23	Mustard	*Brassica juncea* L.Czern	998	759	5175	9
24	Nutmeg	*Myristica fragrans* Houtt.	24	19	0	0
25	Oregano	*Origanum vulgare* L.	46	22	0	0
26	Parsley	*Petroselinum crispum* Mill.	378	315	0	4
27	Pepper long	*Piper longum* L.	16	4	11	0
28	Pomegranate	*Punica granatum* L.	254	31	0	0
29	Poppy seed	*Papaver somniferum* L.	202	132	0	1
30	Rosemary	*Rosmarinus officinalis* L.	41	11	0	0
31	Saffron	*Crocus sativus* L.	105	82	6767	0
32	Sage	*Salvia officinalis* L.	45	32	0	7
33	Star anise	*Illicium verum* Hook.	11	6	0	0
34	Sweet flag	*Acorus calamus* L.	176	355	0	0
35	Tamarind	*Tamarindus indica* L.	27	17	0	0
36	Tarragon	*Artemisia dracunculus* L.	15	3	0	0
37	Thyme	*Thymus vulgaris* L.	17	6	0	0
38	Turmeric	*Curcuma longa* L.	122	45	12593	2
39	Vanilla	*Vanilla planifolia* Andr.	71	42	31	0

Genetic Markers

DNA markers based on PCR amplification are the choice now-a-days, as they offer several advantages over morphological and biochemical markers. These markers are highly useful for the assessment of genetic diversity in endangered species, for studying the genetic relationships between different cultivars or for comparison of the molecular marker analysis to the chemical composition of the plants. RAPDs, AFLPs, ISSRs, ITSs, SSRs and SNPs are the most popular markers. RFLP linkage maps have been constructed for many economically important crops but lacking in the case of spices. SSRs and SNPs are highly informative and important molecular marker tools for the near future and bioinformatics play a major role in development of these tools. SSR based tool has been employed only in a few spice crops like *Capsicum* [19] and turmeric [30]. An EST based in silico approach was recently adopted to find out SNPs in ginger and turmeric [6,7].

Genome Sequencing

The genome size of perennial spices is quite big because of polyploidy and chromosomal duplication events. The expansion of genomes has mainly been the result of multiplication of retro-transposon repeat sequences that are problematic within the context of complete genome sequencing. Although the main emphasis of plant genome sequencing is currently on discovering and characterizing the range of protein-coding genes present within the genome, thousands of copies of large repeats yield no information on the proteome. Such repeats additionally generate statistical issues that prevent the assembly of individual sequence reads into meaningful contigs. The larger plant genomes are therefore currently precluded from complete sequencing. However, sequencing of several model and crops species has been completed and the next phase of plant genomics will necessarily build on new phylogenies assisted by molecular techniques. Unfortunately the genomic information of model plants is not suitable for studying secondary metabolism in spices because the levels of compounds such as volatiles is extremely low and the phenotypes are unfortunately quite subtle and difficult to identify because of the volatile nature of the substances involved [33].

EST's are a suitable alternative to complete-genome sequencing. This 'poor man's genome' resource forms the core foundations for various genome-scale experiments within the plant genomes that are yet to be sequenced [27]. Databases like dbEST and EMBL database contain over 22 million sequences from approximately 820 plant species. Sizeable number of ESTs from spices like ginger, capsicum, turmeric and mustard are available in these databases (Table 12.1). Several new tools are available for the clustering and annotation of these

ESTs. ESTs provide a robust sequence resource that can be exploited for gene discovery, genome annotation and comparative genomics. A database of ESTs of spices and their annotation has been developed recently [5].

Comparative Genomics

Over the last two decades, comparative genetics has shown that the organization of genes within plant genomes has remained more conserved over the evolutionary periods than previously thought. The complete sequencing of the *Arabidopsis thaliana* genome, a landmark in plant sciences, has boosted developments in agricultural science because of the insights from the *Arabidopsis* sequence. However, the bigger challenge today is to use this information and to extend it to food, feed and fiber crops. RFLP analysis of genes among various plant species has led to the elucidation of extensive genome comparison and conservation of gene order within homologous chromosomal segments. These findings suggest that improvement of other important food crops can benefit from information obtained in these model crop systems, though complete co-linearity between even closely related species is rare. However, the agricultural research community need not wait to reap the benefit as they can make better-informed decisions based on current knowledge from the model crops.

Information available from *Arabidopsis* genome on resistance (RRS1) gene for *Ralstonia solanacearum* [16,17] can be used for tagging and isolating genes for *Ralstonia* resistance in ginger. Candidate genes responsible for pathogenesis can also be identified from sequence information available on *R. solanacearum* [28]. Similarly information available on mapping of heterologous loci for example, Ph-2 locus controlling partial resistance to *Phytophthora infestans* in tomato [29], molecular markers linked to the *Phytophthora* resistance gene, Rps 1-k, in soybean [31] can be used in black pepper also. Intra specific comparative genomics to identify avirulence genes from *Phytophthora* was reported [14]. Whole genome sequencing of *P. capsici*, the causal agent of foot rot of black pepper, is in progress.

Comparative genomics which has already made much headway for solanaceous crops can greatly contribute to research on paprika and other capsicums [13,38]. Similarly Global Musa Genome Consortium has assembled BAC genomic libraries, EST libraries, short insert DNA libraries, repetitive DNA clones, and molecular cytogenetic markers. The information generated will be of immense help in better understanding of other related sub families like Zingiberaceae to which important spices like cardamom, ginger and turmeric belong.

Functional Genomics

In spices, one of the most studied species is pepper, *Capsicum* spp., which belongs to the family Solanaceae. In this family there is a comprehensive collection of related information and tools available to the public: the Solanaceae Genomics Network (SGN;http://sng.cornell.edu) [20].

In black pepper (*Piper nigrum* L.), preliminary work on isolation of genes responsible for agronomically important characters, especially for biotic and a biotic stresses was done. A few putative genomic and cDNA fragments associated with resistance related genes are isolated and programmes on isolating, cloning and validating of full length genes is in progress. Several attempts to isolate resistance genes in black pepper were made by various workers like β-1,3-glucanase in *P. nigrum* and methyl glutaryl CoA reductase in *P. colubrinum* [11,12]; PR5 gene homologue in *P. nigrum* [21] and *P. colubrinum* [9]. Reported Amplification, isolation and partial sequencing of a putative gene imparting resistance to *P. capsici* and a chitinase gene from *P. colubrinum* were reported 29.1. Novel genes such as pea lectin gene and tomato protease inhibitor gene were identified in low copy numbers from black pepper using heterologous probes [1]. In ginger, a pool of candidate genes responsible for the defence mechanisms in *Z. zerumbet* against *Pythium* infection was identified [2,15]. A novel form of type III polyketide synthase was isolated from *Z. officinale* [24]. Targeted cloning of WRKY transcription factor genes from *P. colubrinum* has yielded a 143 bp gene fragment similar to WRKY sequences already identified in different plant species. Preliminary transcriptome analysis of transcripts expressed within leaf tissues challenged with *Phytophthora* revealed expression of many stress induced genes as well as genes related to secondary metabolism. A variety of transcription factors and genes involved in primary metabolism with significant similarity to those characterized in other plants were also identified.

Proteomics and Cheminformatics

Over many years of crop agriculture, the focus has been on agronomic traits such as yield, productivity and pest resistance, while nutrient content has largely been ignored. This field of nutritional genomics takes advantage of the many genes that have been cloned for vitamin pathways and for the synthesis of many other "nonessential" compounds and macronutrients [8].

Proteomics provides a promising approach for studying secondary metabolism in plants and plant cells. The complicated biochemistry of the biosynthesis of these compounds is poorly understood, since there are many steps involved that are tightly regulated [33]. Studying plant proteomes involves a huge amount of work and data, due to the large amount of information. In future, it should possible to directly manipulate the content and composition of many chemical constituents in spices that may add their medicinal or nutraceutical properties. Realizing the importance of this area, IISR has initiated some efforts to organize the chemical information on an array of compounds present in spices and to subject them to *in silico* analysis to predict their biological properties. Drug ability and pharmacological fitness of various compounds in spices was recently worked out using such tools [25,26,35]. The curcumin pathway in turmeric is elucidated using a similar approach [22]. *In silico* tools can also help in predicting the deleterious effects of some chemical compounds [3].

Bioinformatics with the aid of biotechnology can help to address and solve problems of chemical pesticide reduction, improved quality food products, biological controls of pests, and other related issues. Improved quality of food products and biological controls of pests can be achieved through employing bioinformatics techniques like gene identification, promoter prediction and primer design to amplify the beneficial genes and embedding in required crops of interest to make transgenic crops with highest yield. Chemical pesticide reduction is also possible by adopting cheminformatics methods to identify naturally occurring chemical compounds in crops which acts against pests.

Biodiversity Informatics

Biodiversity, including genetic resources, and species and ecosystem diversity, is of great importance to agriculture. As a natural-resource-based industry, agriculture depends on a healthy diversity of organisms and ecosystems that are its foundation. It is because of this close relationship that natural-resource-based industries, such as agriculture, have a direct impact on biodiversity. Integrating ecologically pertinent data into the chain of information from the gene to the biosphere will significantly enhance our understanding of the natural world and promote wise management strategies for natural resources [13]. IISR has made some pioneering attempts to organize the biodiversity information available on major spices and some pathogens [10]. Biodiversity, including genetic resources, and species and ecosystem diversity, is of great importance to spices. As a natural-resource-based industry, agriculture depends on a healthy diversity of organisms and ecosystems that are its foundation. Bioinformatics can contribute to promote databases that will result in preserving the biological data including biodiversity details of important species and habitat digitally. Efforts in this direction in spice sector are summarized in Table 12.2.

Spice Bioinformatics   277

Table 12.2: List of major databases and software tools developed at IISR, Calicut

Name of the database/tool	Description	URL
Database		
Spice genes	A database of spice germplasm resources available at IISR	http://www.spices.res.in/germplasm index.htm
Piperbase	A database of *Piper* species of India (Vinod *et al.*, 2008)	Offline
Chitinase database	An online database on plant chitinases	http://www.spices.res.in /bioinformatics/project/index.htm
PAL database	An enzyme database of Phenylalanine Ammonia Lyases	http://www.spices.res.in /bioinformatics/project/index.htm
CardCC& Mpbase	A cheminformatics database on chemical compounds in cardamom	http://www.spices.res.in/ spicebioinfo/project/cardamom/ index.htm
PASSCOM	A database of predicted activity spectra of spice compounds	http://www.spices.res.in/passcom
PLASBID	A database for identification of plant associated bacteria	http://www.spices.res.in/plasbid
PhytoWeb	A comprehensive web resource for the *Phytophthora* species diseases of major horticultural crops and their management	http://220.227.138.213/phytofura /phydish/index.php
SpicEST	A database of ESTs, SNPs, mi RNAs and microsatellites of various spice crops (Chandrasekar *et al.*, 2009a)	http://www.spices.res.in/spicest
Spice Prop	A database of micro propagation protocols developed at IISR, Calicut for various spices	http://www.spices.res.in/spice prop/ index.php
Spicepat	A database on all patent information related to spices developed using Visual Studio dot Net	Offline
Spice bibliography	A database of world literature on 32 spices published in various scientific journals	http://www.spices.res.in/biblio/ index.php
Software tools		
Phytfinder	An expert system for identification of *Phytophthora* species based on their morphological characters	http://220.227.138.213/phytofura phydish/ index.php?centre=finder_body.php
Rapid	A software program to identify unique bands from PAGE data	http://spices.res.in/rapid/
Sign-O-Bacteria	A tool for identifying the species specific signatures in plant associated bacteria	http://spices.res.in/signobact/

(Contd.)

Name of the database/tool	Description	URL
Readthru	A tool to search for termination codons in sequences	http://spices.res.in/readthro/
Juzbox	A tool to search for Prosite patterns and biomedical words in amino acid sequences [4].	http://spices.res.in/juzbox/

Future Thrusts

IISR has developed an excellent base to take further the spice bioinformatics research. The leads obtained will be utilized to develop a comprehensive online spice information base to cater to the needs of researchers, farmers, traders and policy makers. Another thrust area is establishing a *Phytophthora* e-lab to support the genomic studies on *Phytophthora*, the most important pathogen of horticultural crops in the country. The cheminformatics initiatives will be further intensified so as to promote nontraditional uses of spices. Training and support to research personnel will continue to be one of the core areas.

Genomics and bioinformatics are approaches that will be an essential part of plant research and every plant researcher has to incorporate more such tools in their projects. Plant biology is going to witness a lot of integration of basic and applied research, of other related biological research, from microbe to human. Bioinformatics will provide the glue with which all of these types of integration will occur. Biologists will spend more time on the computer and internet to generate and describe data. To reach such a stage, increased educational and research efforts are needed by government agencies, and R&D organizations for development, enhancement and/or adoption of genomics, computational biology and bioinformatics tools in agriculture, food and healthcare delivery systems. Furthermore, partnerships need to be forged between Indian research institutions and international, public and private sector genomics and breeding efforts. Finally, it has to be hoped that the enormous financial resources being spent annually on reckless biotechnology research be channeled into useful research to address the urgent needs facing the horticulture section.

References

1. Alex S.M., Dicto J., Purushothama M.G. and Manjula S. (2008) Differential expression of metallothionein type-2 homologues in leaves and roots of black pepper (*Piper nigrum* L). Gen. Mol. Biol., 31: 551-554.
2. Aswati Nair, R., Kiran, A.G., Sivakumar, K.C. and Thomas, G. (2010) Molecular characterization of an oomycete-responsive PR-5 protein gene from *Zingiber zerumbet*. Plant Mol. Biol. Reptr., 28:128–135.
3. Balaji, S. and Chempakam, B. (2008) Mutagenicity and carcinogenicity prediction of compounds from cardamom (*Elettaria cardamomum* Maton). Ethnobotanical leaflets, 12: 682-689.
4. Bobby Paul, Balaji, S., Sathyanath, V. and Eapen, S.J. (2009) JUZBOX: A web server for extracting biomedical words from the protein sequence. Bioinformation, 4(5): 179-181.
5. Chandrasekar, A., Riju, A., Sathyanath, V. and Eapen, S.J. (2009a) SpicEST – An annotated database on expressed sequence tags of spices. Genes, Genomes and Genomics 3, (Special Issue 1): 50-53.
6. Chandrasekar, A., Riju, A., Sithara, K., Anoop, S. and Eapen, S.J. (2009b) Identification of single nucleotide polymorphism in ginger using expressed sequence tags, Bioinformation, 4(3):119-122.
7. Chandrasekar, A., Riju, A., Sithara, K., Anoop, S. and Eapen, S.J. (2009c) Single nucleotide polymorphisms (SNPs) and indels in expressed sequence tag libraries of turmeric (*Curcuma longa* L.) Online J. Bioinform., 10 (2): 224-232.
8. Dellapena, D. (1999) Nutritional genomics: manipulating plant micronutrients to improve human health. Science, 285: 375-379.
9. Dicto, J. and Manjula, S. (2005) Identification of elicitor-induced PR5 gene homologue in Piper colubrinum Link by suppression subtractive hybridization. Curr. Sci., 88: 624-627.
10. Eapen, S.J. and Riju, A. (2008) Agro-biodiversity informatics with special reference to spices. Biobytes, 3: 14-18.
11. George, M.R., Nazeem, P. A. and Girija, D. (2006) Isolation and characterization of β-1, 3-glucanase gene from *Piper* spp. J. Plant. Crops, 34: 562-567.
12. Girija, D., Beena, P.S., Nazeem, P.A. and Puroshothama, M.G. (2005) Molecular cloning of cDNA fragment encoding hydroxy methyl glutaryl CoA reductase in *Piper colubrinum*. In Proc. National Symposium on Biotechnological Interventions for Improvement of Horticultural Crops: Issues and Strategies. pp: 303-306. Kerala Agricultural University, Thrissur, Kerala.
13. Jones, M.B., Schildhauer, M.P., Reichman, O.J. and Bowers, S. (2006) The new Bioinformatics: Integrating ecological data from the gene to the biosphere. Annu. Rev. Ecol. Evol. Syst., 37: 519–544.
14. Jorunn, I.B.B., Miles, A., Stephen. C.W., Trudy. A.T., Mildred. O., Paul. R.J.B. and Sophien, K. (2003) Intra specific comparative genomics to identity avirulence genes from *Phytophthora*. New Phytologist, 159: 63-72.
15. Kavitha, P.G. and George Thomas (2008) Defence transcriptome profiling of *Zingiber zerumbet* (L.) Smith by mRNA differential display. J. Biosciences, 33: 81-90.
16. Laurent, D., Frederic, P., Laurence, L., Sylvie, C., Canan, C., Kevin, W., Eric, H., Jim, B., Matthieu, A. and Yves, M. (1998) Genetic characterization of *RRS1*, a recessive locus in *Arabidopsis thaliana* that confers resistance to the bacterial soil borne pathogen *Ralstonia solanacearum*, MPMI, 11(7): 659-667.

17. Laurent, D., Jocelyne, O., Frederic, T., Judith, H., Dong Xin, F., Peter Bitter-Eddy, Jim, B. and Yves, M. (2002) Resistence to *Ralstonia solanacearum* in *Arabidopsis thaliana* is conferred by the recessive RRSi-R gene, a member of a novel family of resistance genes. PNAS, 99(4): 2404-2409.

18. Livingstone, K.D., Lackney, V.K., Blauth, J.R., van Wijk, R. and Jahn, M.K. (1999) Genome mapping in *Capsicum* and the evolution of genome structure in the Solanaceae. Genetics, 152: 1183-1202.

19. Minamiyama, Y., Tsuro, M. and Hirai, M. (2006) An SSR-based linkage map of *Capsicum annuum*. Mol. Breed., 18: 157-169.

20. Mueller, L.A., Solow, T.H., Taylor, N., Skwarecki, B., Buels, R., Binns, J., Lin, C., Wright, M.H., Ahrens, R., Wang, Y., Herbst, E.V., Keyder, E.R., Menda, N., Zamir, D. and Tanksley, S.D. (2005) The SOL Genomics Network: a comparative resource for Solanaceae biology and beyond. Plant Physiol., 138:1310-7.

21. Nazeem, P. A., Achuthan, C.R., Babu, T.D., Parab, G.V., Girija, D., Keshavachandran, R. and Samiyappan, R. (2008) Expression of pathogenesis related proteins in black pepper (*Piper nigrum* L.) in relation to *Phytophthora* foot rot disease. J. Trop. Agric., 46: 45-51.

22. Neema Antony (2005) Investigations on the biosynthesis of curcumin in turmeric *(Curcuma longa L.)*. PhD Thesis, Calicut University, Kerala, India, pp 171.

23. Philippe, M., Philippe, T., Jocelyne, O., Henri, L. and Nigel, G. (1998) Genetic mapping of Ph-2, a single locus controlling partial resistance to Phytophthora infestans. MPMI, 11(4): 259-269.

24. Radhakrishnan, E.K. and Soniya E.V. (2009) Molecular characterization of novel form of type III polyketide synthase from *Zingiber officinale* Rosc. and its analysis using bioinformatics method. J. Proteomics Bioinform., 2(7): 310-315.

25. Riju, A., Sithara, K., Suja, S. Nair, Shamina, A. and Eapen, S.J. (2009) *In silico* screening of major spice phytochemicals for their novel biological activity and pharmacological fitness, J. Bioequivalence & Bioavailability, 1(2): 063-073.

26. Riju, A., Sithara, K., Suja S. Nair and Eapen, S.J. (2010) Prediction of toxicity and pharmacological potential of selected spice compounds. In: Proc. Int. Symp. on Biocomputing, ACM Digital Library, http://doi.acm.org/10.1145/1722024.1722060

27. Rudd, S. (2003) Expressed sequence tags: alternative or complement to whole genome sequences? Trends Plant Sci., 8: 321-329.

28. Salanoubat, M., Genin, S., Artiguenave, F., Gouzy, J., Mangenot, S., Arlat, M., Billaultk, A., Brottier, P., Camus, J.C., Cattolico, L., Chandler, M., Choisen, N., Claudel-Renardl, C., Cunnac, S., Demange, N., Gaspin, C., Lavie, M., Moisan, A., Robert, C., Saurin, W., Schiex, T., Siguier, P., Thebault, P., Whal, en M., Wincker, P., Levy, M., Weissenbach, J. and Boucha, C.A.(2002) Genome sequence of the plant pathogen *Ralstonia solanacearum*. Nature, 415: 497-502.

29. Sandeep Varma, R., George, J.K., Balaji, S. and Parthasarathy, V.A. (2009) Diûerential induction of chitinase in *Piper colubrinum* in response to inoculation with *Phytophthora capsici*, the cause of foot rot in black pepper. Saudi J. Biol. Sci., 16: 11–16.

30. Siju, S., Dhanya, K., Syamkumar, S., Sheeja, T.E., Sasikumar, B., Bhat, A.I. and Parthasarathy, V. A. (2010) Development, characterization and utilization of genomic microsatellite markers in turmeric (*Curcuma longa* L.). Biochem. System. Ecol., 38: 641-646.

31. Takao, K., Shanmukhaswami, S.S., Jinrui, S., Mark, G., Richard, I.B. and Madan, K.B. (1997). High resolution genetic and physical mapping of molecular markers linked to the *Phytophthora* resistants gene *Rps 1*-k in soybean. MPMI, 10(9): 1035-1044.

32. Tanksley, S.D., Bernatzky, R., Lapitan, N.L. and Prince, J.P. (1988). Conservation of gene repertoire but not gene order in pepper and tomato. Proc. Natl. Sci. USA, 85: 6419-6423.

33. Trindade, H. (2009) Molecular biology of aromatic plants and spices. A review. Flavour Fragrance J., DOI 10.1002/ ffj.1974.
34. Vinod, T.K., Labeena, P., Saji, K.V. and Eapen, S.J. (2008) PiperBase – an information bank on *Piper* species in India. In: National Seminar on Piperaceae, 21-22 November 2008, IISR, Calicut.
35. Wilson, M., Balaji, S. and Eapen, S.J. (2007) Druggability of lead compounds from turmeric (*Curcuma longa* L). JMAPS, 29(1):384-390.

Chapter – 13

Application of Bioinformatics in Palm and Cocoa Research

R. Manimekalai, K.P. Manju and S. Naganeeswaran

Abstract

The study of palm genomics can aid to understand the genetic and molecular know-how of all biological processes in palms that are vital tothe palm species.This knowledge will allow the path for the fruitful exploitation of palms as biological resources in the development of new cultivars of improved quality and with shortened economic and environmental costs. This knowledge is also vital for the development of new diagnostic tools. Traits considered of primary interest are resistance to pathogens and abiotic stress, fruit quality and yield.Genomics and bioinformatics approaches helped in unraveling the genomes of economicpalsm like coconut, oilpalm and datepalm. The genome characteristics and tools for genome analysis are provided in the chapter. Due to the economic importance of cocoa, many scientific research studies (genome, EST, transcriptome) have been carried out in the past years in this crop also.

Introduction

Plantation crops constitute a large group of crops. The major plantation crops include coconut, arecanut, oil palm, cashew, tea, coffee and rubber; and the minor plantation crops include cocoa (Prapulla and Indira, 2014). India is the third largest producer of coconut and goes in front of the ninety coconut-producing countries of the world (Richard 2013). India is the largest producer and consumer of cashew nuts (Mohod *et al.*, 2011.). India also occupies number one position in arecanut production (Dwivedi *et al.*, 2008) besides being the largest tea producers in the world. While Coffee is the second largest traded commodity in the world, it has, in recent times become an extremely important foreign exchange (Prapulla and Indira, 2014).

Over the past decades, the genomic and gene/EST researchhas generated an exponential amount of data that will require storage,processing and analysis.The introduction of bioinformatics leads to the easier handling and analysis of these bulk data. The dataincludes not only sequence information, but also information on mutations, structures, markers, mapsand functional discoveries. The storage, analysis, interpretation and conversion of these data into knowledge to aid the scientific community could rely on bioinformatics, the branch of life science that can deal and resolve with the multiple branches of science.It encompasses various databases to store and exchange various information related to the field. So far there have been several databasesthat have been aiding to create new discoveries using the existing knowledge/data.

Bioinformatics for Perennial Plantations and Palms – Role of Model Plants

As it is well known that most of the biological processes are similar in much of the plants, comparing the revealed information with those of the model plants with that of fresh palm members plays a very significant role (Dimitar et al., 2005). Relatively perennial plantation crops are with less biology invasion and a large amount of data can be derived from model plants. On sequence levels, comparison over regions of gene regulation, genetic diseases etc are widely possible and reliable.

The genomes of date palm and oil palm has only recently been released to the public domain in the past few years. So, earlier much of the comparisons were done using the standard model plants such as arabidopsis, *grape*, maize, rice etc. As all flowering plants are closely related, the complete sequencing of all the genes in a single representative plant species will provide a lot of knowledge about all higher plants. Also, discovering the functions of the proteins produced by a model species will provide a lot of information on protein function in all higher plants.

Comparing Genome Sequences

The economic importance of the palm family plants includes staple food, beverges, ornamental, building wood and industrial materials (Ibrahim *et al.*, 2013). With the advance in the technology and web world, the study of plant biology in the 21st century is, vastly different from that in the 20th century to reveal the genetic blueprints. Genomics technology has progressed significantly since publication of the first plant genome sequence (*Arabidopsis thaliana*, 2000). Recently, in palms the increasing number of chloroplast and mitochondrial genomes and the vast deposition of EST/nucleotides in public domain open the wide space for the area on comparative sequence analysis. The annotated genes of known chloroplast and mitochondrial genomes of the related members within the family are found similar and can serve as model genes to discover novel genic regions of the yet to reveal members. Many more perennial genome sequences are about to unwind in the near future. The genome comparisons, comparative analysis and relationships can be obtained with whole genome web based informatics tools. Also the inference of relationships from proteins of known function to proteins of unknown function that are structurally similar can be accomplished through comparative analysis (Conte *et al.*, 2008; Sanchez *et al.*, 2011)

Genomic Release of Palms

List of the genome releases in palm family

Palm	Organelles	Size
Phoenix dactylifera L.	Nuclear	605.4Mb
Phoenix dactylifera L.	Chloroplast	158 Kb
Phoenix dactylifera L.	Mitochondria	715Kb
Elaeisguineensis	Nuclear	1.8Gb
Elaeisguineensis	Chloroplast	156 Kb
CocosnuciferaL.	Chloroplast	154 Kb
Pseudophoenixvinifera	Chloroplast	157 Kb
Bismarckianobilis	Chloroplast	158 Kb
Calamuscaryotoides	Chloroplast	157 Kb

- Major areas of sequence analysis includes, Sequence annotation, phylogenetic analysis, microRNA prediction, repeat finding, sequence filtering, SNP prediction, multiple sequence analysis and RNA editing analysis and transcriptome analysis.

Major Tools/Programs for Sequence Analysis

Sequence annotation- It is the process of identifying the locations/segments of genes, coding regions and other important specific locations and associating relevant information with those locations/segments

Major sequence annotation programs are

- BLAST (Altschul *et al.*, 1990) - General Sequence annotation.

- ORF finder- ORF prediction.

- TRNAscan-SE search server - Used for identifying tRNA genes using BLAST search.

- SNAP gene finder - Used for gene finding.

- MAUVE, Artemics - Genome sequence annotation and comparisons.

- CGAP- A new interactive web-based platform for the comparative analysis of chloroplast genomes CGAP integrated genome collection, visualization, content comparison, phylogeny analysis and annotation functions together.

- DOGMA - a web-based annotation tool for chloroplast and mitochondrial genomes.

- GeneOrder and BADGER - For comparative analysis of gene arrangements in small genomes.

Phylogenetic analysis- Comparing DNA sequences infer evolutionary relationships between the sequences with respect to knowledge of the evolutionary events themselves. Major phylogenetic analysis programs are GRAPPA (http://www.cs.unm.edu/~moret/GRAPPA/) (Tang and Moret, 2004) and MGR (http://grimm.ucsd.edu/MGR/) (Sankoff and Blanchette, 1998) perform phylogenetic analysis based on gene order changes, PAUP4.0b10 - Phylogenetic tree construction using the maximum parsimony method, MEGA- Can construct trees based on Maximum likelihood, neighbor-joining, minimum evolution, UPGMA, maximum parsimony methods.

MicroRNA prediction- MicroRNAs (miRNAs) are a class of non-coding RNAs of ~22 nucleotides that regulates the gene expression by binding to their mRNA.

- Mfold web server- MicroRNAs can be predicted with reference to the known miRNAs from mirBase and can be structured using mfold web server.

- For plants, microRNA targets can be predicted using psRNAtarget web server Sequence alignment methods and its applications.

Repeat finding- Simple Sequence Repeats (SSRs) or microsatellites are short DNA repeats, evenly dispersed across the genome,more frequently in non-coding regions of the genome, of both eukaryotes and prokaryotes. Repeats can be mined using a variety of tools such as MISA, Repeat finder, REPuter and tandem repeat finder.

Sequence filtering- Sequences might have been contaminated with ambiguous sequences, distal oligoN series from either the 5' or 3' end and poly A and poly T repeats. EST trimmer and SeqClean are widely used programs.

SNP Prediction- Single Nucleotide Polymorphisms (SNPs) are single DNA sequence variation in the genome, usually bi-allelic in nature, so can be easily assayed. Available programs for SNP prediction are AutoSNP, SSAHA and BioScope.

Multiple Sequence Analysis: It is the arrangement of two or more amino acid or nucleotide sequences from one or more organisms so that the sequences sharing common properties are aligned. The degree of relatedness or homology between the sequences is predicted computationally or statistically based on weights assigned to the elements aligned between the sequences. Major programs to the field are, MUSCLE, Clustalx, clustal W and MAFFT (version 5)

RNA Editing Analysis- Process edits C to U sites in organellar protein-coding genes by comparing its predicted protein sequence to homologous proteins from other plants.Eg. PREP (Predictive RNA Editor for Plants) suite of programs.

Transcriptome Analysis: Transcriptomes studyhelps to determine when and where genes are turned on or off in various types of cells and tissues. The number of transcripts can be quantified to get some idea of the amount of gene activity or expression in a cell. Furthermore, by considering the transcriptome, it is possible to generate a comprehensive picture of what genes are active at various stages of development.

For transcriptome analysis, the major programs are

- VELVET, SOAP denova - Transcriptome specific assembly programs.
- ESTScan software - Used to predict transcriptome coding regions and determine sequence direction.
- Blast2GO program - Used to obtain GO annotations for the unigenes, as well as for the KEGG and COG analyses.
- WEGO software - Used to perform GO functional classification of all unigenes.

Complete Chloroplast (cp) Genome Sequence Analysis of Coconut

Bioinformatic tools used for comparative analysis of the chloroplast genome of coconut (Huang *et al.*, 2013) are

- Gene annotation - DOGMA, Blast.
- To verify the exact gene and exon boundaries – MUSCLE.
- Confirming tRNA genes - tRNAscan-SE search server.
- Codon usage and base composition - Artemis.
- Prediction of Potential RNA editing sites– PREP suite.
- Phylogenetic analysis - Maximum Likelihood & Garli version 2.0.

Transcriptome Sequence Analysis of Coconut

Transcriptome sequence analysis of coconut provides a large quantity of novel genetic information for coconut (Haikuo *et al.*, 2013). This information will act as a valuable resource for further molecular genetic studies and breeding in coconut, as well as for isolation and characterization of functional genes involved in different biochemical pathways in this important tropical crop species.

- ESTScan software - To predict coding regions and to determine sequence direction.
- Blast2GO - GO annotation.
- WEGO - GO functional classification.
- Blastall - COG and KEGG pathway annotations.

The coconut transcriptome analysis (2013) showed total unigenes – 57,304.

Average length – 752 bp

Size range – 200bp - 3000bp

Predictive RNA editor for plants (PREP) suite predicted 83 potential RNA editing sites out of 27 genes. RT-PCR analysis confirmed editing at 64 of those sites. Of the genes investigated, NADH dehydrogenase (ndh) genes have the highest number of editing sites. The editing types in coconut were all non-silent and 100% C-to- U. Of these editing events, 82.67% occurred at the second base of the codon, 16% were at the first base of the codon and only 1.33% was at the third base of the codon.

Annotation and Classification of Coconut Transcriptome

The coconut transcriptome were compared against unigenes from various databases. Functional annotation showed that these unigenes covered every basic biological process, and 23,168 of these unigenes were also mapped into 215 KEGG pathways (Fan*et et al.*,2013)

GO Classification of Coconut Unigenes

Metabolic processes (5,538 unigenes, 9.66%), cell (10,874 unigenes, 18.98%) and catalytic activity (6,531 unigenes, 11.40%) were the most abundant GO slims in each of the biological processes, cellular component localization and molecular functionality categories respectively (Figure 13.1) (Fan *et al.*, 2013).

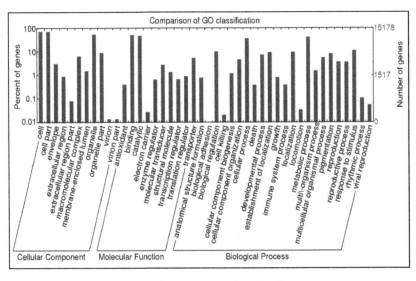

Fig. 13.1: Gene ontology classification

Complete cp Genome Sequence Analysis of Date Palm

Programs associated with the sequence analysis of date palm cp genomes are

- Genome annotation – DOGMA
- Sequence alignment – MUSCLE
- Identification of tRNA genes – DOGMA and tRNAscan-SE server
- Identification of repeat units – REPuter
- Detection of intravarietal SNP – Consed.

Other Palm Sequence Analysis

Microsatellites in palm sequences are reported (Manju et al., 2011).38, 083 microsatellites were localized in palm sequences. Dinucleotide repeats (49%) were found to be more abundant followed by mononucleotide (30%) and trinucleotide (19%).The SNPs were predicted in palms genomes. The SNPs were seen at a frequency of 2.84/100bp in the WRKY sequences of palms. About 143 candidate miRNAs were predicted. Of these, six potential miRNAs (miR160a, b, c, miR2936, miR2275, mi R1877, miR478e, miR395d) were detected by structure homology Manju *et al.*, 2011.

Bioinformatics for Improvement of Cocoa

Theobroma cacao is a diploid species (n = 10) with genome size of 380 Mbp (Argout, *et al.*, 2011). It is an important commercial crop of the tropics and approximately ten million people depend on cocoa plantations for their livelihood. Cocoa being the raw material for chocolate and confectionaries, it is one of the most traded commodities.

Expressed Sequence Tags (ESTs) Resource

Expressed Sequence Tags (ESTs) are single-pass, redundant, partial nucleotide sequences that represent the expressed portion of the genome. Advanced high-throughput sequencing technologies have generated vast number of ESTs and most of the sequences were deposited in public EST database. Expressed Sequence Tags provide researchers with a quick and inexpensive route for discovering new genes, for obtaining data on gene expression and regulation and for constructing genome maps. The primary step of genome sequencing project is to retrieve the EST sequences of respective organisms. Identification of functions of genomic sequences is a challenging task. It varies among organisms and depends upon genome size as well as the presence or absence of introns. We have analyzed the ESTs of cocoa and developed an annotated database that provides information on contigs, EST's, gene information and sequence similarity (Manimekalai *et al.*, 2009)

Presently NCBI-dbEST database contain 159996 (*Theobroma cacao*), 179119 (*Coffeaarabica*) and 49769 (*Camellia sinensis*) EST sequences. These EST sequences were extracted from different tissue types in different environmental condition. We have analyzed 4 libraries of ESTs derived from *Phytophthora megakarya* infected cocoa leaf and pod tissues. We identified 272 enzymes corresponding to 114 metabolic pathways. Functional annotation revealed that the most of the sequences are related to molecular function, stress response and biological process. The annotated enzymes are aldehyde dehydrogenase,

Application of Bioinformatics in Palm and Cocoa Research

catalase, acetyl-CoA C-acetyltransferase, threonine ammonia-lyase, acetolactate synthase, dihydroxy-acid dehydratase, O-methyltransferase and cinnamoyl-CoA reductase which play an important role in amino acid biosynthesis and phenyl propanoid biosynthesis (Naganeeswaran *et al.*, 2012).

Genome Resources

Genomes of two cocoa cultivars namely Matina 1-6 and Criollo B97-61/B2 (Argout, *et al.*, 2011; Motamayor *et al.*, 2013) were sequenced. More than 75% of the estimated genome of Criollowas assembled. Genome-wide shotgun strategy incorporating Roche/454, Illumina and Sanger sequencing technologies was used for cocoa genome sequencing. Whole-genome sequences are deposited under accession numbers CACC01000001–CACC01025912 (Argout *et al.*, 2011). Sequencing of cocoa cultivar Matina 1-6 genome generated 32,460,307 sequence reads. Total genome assembly comprises 99.2% (346.0 Mbp of 348.7 Mbp) of the cocoa genome. The genome assembly has been deposited in GenBank (accession number: ALXC01000000) (Motamayor *et al.*, 2013).

Cocoa Genome Database (http://www.cacaogenomedb.org/)

Annotated databases are the result of the applications of tools and software's for analysis of genomes and large amount of available scattered data. Annotated databases are designed to provide public access to ongoing research focused on the biotechnological aspects. Cocoa genome database contain cocoa genome sequence (Matina 1-6), predicted genes, predicted metabolic pathways information. CocoaGenDB is cocoa database provides information on phenotypic and genotypic data of cocoa developed at CIRAD (Centre de Coopération Internationale en Recherche Agronomique pour le Développement, France (http://cocoagendb.cirad.fr/).

Bioinformatics Tool for EST Analysis

We have developed an analysis pipeline "Standalone EST Microsatellite mining and Analysis Tool" (SEMAT) for high throughput discovery (mining and annotation) of SSR loci in EST sequence and automated design of primers for their PCR amplifications. A few other automated applications are available for EST functional analysis like EST-PAGE (Matukumalli *et al.*, 2004), ParPEST (D'Agostino *et al.*, 2005) and EST2uni (Javier *et al.*, 2008). But the SEMAT is specially designed for EST-SSR analysis. We compared SEMAT with similar pipeline, SSRPrimer (Robinson *et al.*, 2004). The SSRPrimer tool performs SSR search and primer design, but the process was not automated. In this way, SEMAT is a unique tool for automated EST processing and annotation. SEMAT pipeline was developed using Perl language and tested in Linux operating systems (Ubuntu and fedora). It is freely available at: http://semat. cpcribioinformatics.in/ (Naganeeswaran *et al.*, 2014).

Conclusion

The bioinformatics and genomics tools are well utilized in small genome crop plants. On contrary the current bioinformatics analysis tools for genomics and comparative genomics are not exploited/not available to complement the research and crop improvement in palms and other perennial crops. It is certain that cutting edge technologies in high-throughput sequencing and specialized bioinformatics tools will supplement the research and plant breeding in palms and cocoa. Hence, focused research on developing bioinformatics tools that can analyze huge genomics data of palms and cocoa is need of the hour.

References

1. Alberts., Bray., Hopkin., Johnson., Lewis., Raff., Roberts. and Walter. (2013) Essential Cell Biology. Interactive eBook preview. 4th edition published by Garland Science.
2. Altschul. S.F., Gish, W., Miller, W., Myers, E.W., Lipman, D. J. (1990) Basic local alignment search tool. J Mol Biol., 5;215- 403
3. Argout, X. *et al.* 2011. The genome of Theobroma cacao. Nat. Genet., 43:101-108.
4. Conte M.G., Gaillard, S., Lanau, N., Rouard, M. and Périn, C. (2008) GreenPhylDB: a database for plant comparative genomics. Nu. Acids. Res., 36: D991–D998.
5. D'Agostino N, Aversano M, Chiusano ML. (2005) ParPEST: a pipeline for EST data analysis based on parallel computing. BMC Bioinformatics, 6:S9
6. Dimitar, V., Asen, N., Atanas A., Dimov, G. and Lubomir, G. (2005) Application of bioinformatics in fruit plant breeding. Journal of Fruit and Ornamental Plant Research, 14: 1-18
7. Dwivedi, R. (2008) India 2008. TATA McGraw Hill's Series, 4-5
8. Fan, H., Xiao, Y., Yang, Y., Xia, W., Mason, A.S., Zhihui, X, Fei, Q., Songlin, Z. and Haoru, T. (2013) RNA-Seq Analysis of Cocosnucifera: Transcriptome Sequencing and De Novo Assembly for Subsequent Functional Genomics Approaches. PLoS One 8: e59997
9. Huang,Y.Y., Antonius, J.M. and Matzke Marjori M. 2013. Complete Sequence and Comparative Analysis of the Chloroplast Genome of Coconut Palm (*Cocos nucifera*). Plos One, 8:1-12.
10. Javier, F., Francisco, G. and Antonio, R. 2008. EST2uni: an open, parallel tool for automated EST analysis and database creation, with data mining web interface and microarray expression data integration. BMC Bioinformatics, 9:5
11. Manju, K.P., Manimekalai, R. and Arunachalam, V. 2011.Microsatellite in palm (*Arecaceae*) sequences. Bioinformation, 7: 347-351
12. Manimekalai, R., Anoop, Rakesh and Thomas, G.V. 2009. Putative gene database of cocoa and oilpalm. Journal of Plant Crops, 39(2): 311-318
13. Matukumalli, L.K., Grefenstette, J.J., Sonstegard, T.S. et al. (2004) EST-PAGE – managing and analyzing EST data. Bioinformatics, 20:286-288.
14. Meng, Y., Xiaowei, Z., Guiming, L., Yuxin Y., Kaifu, C., Quanzheng, Y., Duojun, Z., Ibrahim, S. and Jun, Y. (2010) The Complete Chloroplast Genome Sequence of Date Palm (*Phoenix dactylifera* L.). PlosOne, 5: e12762, 1-14.

Application of Bioinformatics in Palm and Cocoa Research 293

15. Mohod, A., Jain, S. and Powar, A.G. (2011). Cashew nut processing: sources of environmental pollution and standards. BIOINFO Environment and Pollution, 1:05-11.

16. Motamayor, U.C., Keithanne, M., Jeremy, S., Niina, H., Donald, L., Omar, C., Seth, D.F. et al. (2013). The genome sequence of the most widely cultivated cacao type and its use to identify candidate genes regulating pod color. Genome Biol., 14:R53. doi:10.1186/gb-2013-14-6-r53.

17. Naganeeswaran, S., Apshara, E. and Manimekalai, R. (2012). Analysis of expressed sequence tags (ESTs) from cocoa (*Theobroma cacao* L) upon infection with *Phytophthoramegakarya*. Bioinformation, 8 (2): 65-69.

18. Naganeeswaran, S., Manimekalai, R, Apshara, E, Manju KP., Malhotra, S.K. and Karun. A., (2014). Standalone EST Microsatellite mining and Analysis Tool (SEMAT): For automated EST-SSR analysis in plants. Tree Genetics and Genomes. DOI 10.1007/s11295-014-0785-2

19. Prapulla, M. and Indira, M. (2014). Impact of trade liberalization on Indian coffee exports. International Journal of Advanced Research in Management and Social Sciences, 3:85-99.

20. Richard, M. P. (2013). Power failure and its impact- a journey by the coconut farmers paving to less production. International Journal of Development Research, 3: 005-006.

21. Robinson, A.J., Love, C.G., Batley, J., Barker, G. and Edwards, D. (2004). Simple Sequence Repeat Marker Loci Discovery using SSR Primer. Bioinformatics, 20:1475–1476.

22. Sanchez, D.H., Pieckenstain, F.L., Szymanski, J., Erban, A. and Bromke, M. (2011). Comparative functional genomics of salt stress in related model and cultivated plants identifies and overcomes limitations to translational genomics. Plos One, 6: e17094.

23. Sankoff, D. and Blanchette, M. (1998). Multiple genome rearrangement and breakpoint phylogeny. J. Comput. Biol., (3):555-70.

24. Tang, J. and Moret, B.M.E. (2004). Phylogenetic Reconstruction from Arbitrary Gene-Order Data. Proceedings of the Fourth IEEE Symposium on Bioinformatics and Bioengineering, (BIBE'04) 0-7695-2173-8/04.

Chapter – 14

Application of Metagenomics in Agriculture

K. Hari Krishnan

Abstract

Microorganisms constituting the major fraction of the total biomass, are the main source of biodiversity in our planet and play an essential role in maintaining biogeochemical processes, which ultimately regulate the functioning of the Biosphere. Molecular ecological studies of microbial communities revealed that only a small fraction of total microbes in nature have been identified and characterized so far, since the majority of them are recalcitrant to cultivation. The emergence of 'metagenomics' have paved ways to the analysis and understanding of the genetic diversity, population structure, and ecology of complex microbial assemblages through culture - independent genomics-based approaches. The concept, metagenome, represents the total microbial genome in natural ecosystem consisting of genomes from both culturable microorganisms and viable but non-culturable bacteria. The construction and screening of metagenomic libraries in culturable bacteria constitute a valuable resource for obtaining novel microbial genes and products. Several novel enzymes and antibiotics have been identified through the metagenomic approach from many different microbial communities. This technology in the near future will provide breakthrough advances in the field of medicine, agriculture, energy production and bioremediation. This paper looks in to the various steps involved in metagenomic studies and how its potential can be utilized in the various aspects of agriculture to improve crop production and protection. The importance and role of bioinformatics in deciphering metagenomic data are also looked into.

Introduction

Microorganisms contribute significantly to the earth's biological diversity and they play a vital role in transforming the key elements of life - carbon, nitrogen, oxygen, and sulfur in to forms accessible to other living beings and facilitating the self-sustainable functioning of the Biosphere. They also make necessary nutrients, minerals, and vitamins available to plants and animals and also cleanup pollutants in the environment. All of these activities are carried out by complex microbial communities that have remarkable ability to adapt swiftly to environmental changes, rather than by individual microbes.

Though microbes are the most diverse life form in nature, relatively few of them have been cultured and characterized. Several studies on microbial biomass estimation has shown that prokaryotes are the most dominant organisms [1] and majority of them are not culturable [2,3,4]. Estimation on number of culturable bacteria in standard culture media is variable from 0.01% to 1% of total microorganisms, dependent on the different microbial communities [5]. Traditionally, microbiological studies has focused on single species in pure laboratory culture, and thus creating a lacuna in understanding of all individual members in any microbial community. Metagenomics offers a new way to understand the microbial world, most of which has never before been accessible to science, through the study of genomes recovered from environmental samples. Coined by Jo Handelsman and others in the University of Wisconsin, Department of Plant Pathology in 1998, metagenomics is the culture independent analysis of a mixture of microbial genomes [6,7,8].

The analysis of complex microbial communities in different environment will throw more insights on how these complex communities work which is not thoroughly understood till date. By applying the power of metagenomics, bypassing the need to isolate and culture individual community members - the total DNA extracted from an environmental sample representing the whole microorganism as a single genome (metagenome) and through its analysis the majority of microbes existing in nature can be characterized. It has become a very powerful tool to search for novel enzymes that are useful for biotechnological applications. A number of reviews have summarized the technology [9,10,11]. Since its first publication and the description of the basic technology [12], a remarkable number of reports were published, providing new

Application of Metagenomics in Agriculture

enzymes with a high potential for industrial applications [13,14,15]. Biocatalysts and bioactive compounds obtained from metagenomics libraries are shown in Table 14.1. The metagenomics process has already been used to identify novel antibiotics [16,17] and has revealed proteins involved in antibiotic resistance [18] and vitamin production [19].

Table 14.1: Biocatalysts and bioactive compounds from metagenomic libraries

Activity	Analyzed habit	Vector (insert size, kb)	Gene bank size (positive hits)	Reference
Agarase	Soil	Cosmid (38.1)	Not metioned	15
Alcohol oxidoreductase	Soil/sediment	Plasmid (4)	1,200,000 (16)	20
Amylase	Soil	BAC (27)	3,648 (8)	21
Biotin production	Soil	Cosmid (35)	50,000 (7)	22
Oxidation of polyois	Soil	Plasmid (7)	300,000 (15)	23
Cation transporter	Soil	Plasmid (7)	1,480,000 (2)	24
Cellulase	Anaerobic enrichment	Plasmid (8)	15,000 (23)	25
Chitinase	Seawater	Phage (6)	825,000(23)	26
Lipase/esterase	Soil	BAC(27)	3,648(2)	21
	Soil	Plasmid (10)	57,500(117)	27
Oxygenase	Soil	Plasmid(7)	3,600,000(5)	27
DNAse	Soil	BAC(27)	3,648(1)	21
Hemolysis	Soil	BAC(45)	24,576(29)	21
Metalloprotease	Soil	Plasmid (10)	117,000 (1)	27
Polyketide synthase	Marine sponges	Fosmid	Not mentioned	28
Indirubin	Forest Soil	BAC (35)	32,000	29
Turbomycin	Soil	BAC (44.5)	24,546 (3)	16
Violacein	Soil	Cosmid	Not Mentioned	30
Metabolite	Soil	BAC (63)	12,000(4)	31
formation	Soil	BAC(42)	3,648 (1)	21

Metagenomics Process

Metagenomics analyses are usually initiated by the isolation of environmental DNA and its purification to remove polyphenolic contaminants which will otherwise interfere with the further library generation steps (Fig 14.1). DNA isolation and purification is followed by the construction of DNA libraries in suitable cloning vectors and host strains. The classical approach includes the construction of small insert libraries (<10 kb) in a standard sequencing vector and in *Escherichia coli* as a host strain [32]. However, small insert libraries do not allow detection of large gene clusters or operons. Inorder to overcome this limitation researchers have been employing large insert libraries, such as cosmid

DNA libraries with insert sizes ranging from 25-35 kb or Bacterial Artificial Chromosome (BAC) libraries with insert up to 200 kb [33,34].

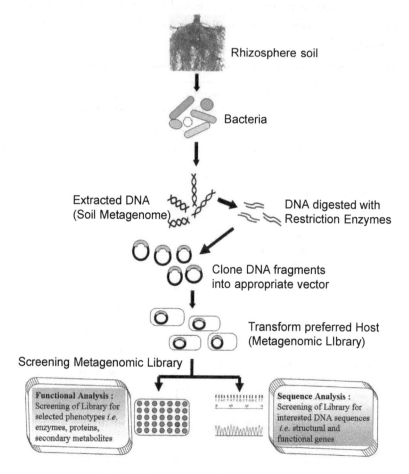

Fig. 14.1: Steps involved in metagenomics

Once libraries are generated, the clones have to be analysed. Generally two approaches are followed for the analysis of genetic material of a metagenomics library. They are Function-based and Sequence-based analyses. Function-based approach include either heterologous expression, in which clones that express the desired function are identified or selected, in which the clone expressing the desired function grows and others do not. An important limitation to heterologous expression is that the domesticated host bacterium must be able to express (transcribe and translate) the genes for the products to be detected. Selections provide the most powerful approach to finding rare clones with selectable characteristics such as antibiotic resistance, metal resistance, vitamin production etc. When a function of interest is detected, then the DNA coding

Application of Metagenomics in Agriculture

for that function is sequenced and compared with DNA from other organisms or communities. Function-based analysis enables identification of new enzymes and antibiotics in libraries from diverse environments. Phenotypic analysis of the introduced unknown genes in culturable bacteria could be an important way for predicting functional genomics of unculturable bacteria. However, estimation of the number of clones required to uncover the microbial diversity from various environments will be a challenging fete due to the enormous microbial diversity and various microbial population structure. This problem can be solved to a certain extent through large-scale construction of metagenomics libraries and development of high throughput screening technology.

Sequence-based approach typically involves either sequencing of random clones to accumulate vast stores of sequence information or identification of clones based on methods that detect a particular sequence. With both of these approaches, phylogenetic markers are sought on the clone of interest to link cloned sequences with the probable origin of the DNA. This analysis provides information on the distribution of functions in a community, linkage of traits, genomic organization and horizontal gene transfer.

Recently, a new DNA sequencing method, known as pyrosequencing, has been developed and this technique does not involve cloning and is therefore free of cloning biases. Pyrosequencing based on the principle "sequencing by synthesis" which is different from coventional Sanger sequencing, in that it relies on the detection of pyrophosphate release on nucleotide incorporation, rather than chain termination with dideoxynucleotides. The technology has many advantages in that vast amount of sequence information can be generated and is several times faster than Sanger-based sequencing and the sequencing costs per base pair are much lower. Pyrosequencing has imparted a great advancement in the analyses of the metabolic capabilities of microbial communities and the differences between microbial assemblages present in habitats with distinctly different biogeochemical parameters The advanced third generation pyrosequencing technology that is in use today can generate sequences of over 400 bp and have already proved to be valuable for the study of microbial diversity and metagenome analyses.

Metagenomics in Agriculture

Agriculture is one of the prime areas in which metagenomics will have many practical applications. Plants provide an excellent ecosystem for microorganisms that interact with plant cells and tissues with differing degrees of dependence. Bacteria present in the soil convert the atmospheric nitrogen into ammonia and increases the soil fertility. Microbes decompose the dead animal and plant materials and convert the elements like iron and manganese in to forms that can be easily used by plants as nutrients. Investigation on the relationship between roots and microbiota are essential to achieve innovations in agriculture and biotechnology. The group of microbes referred to as Plant Growth Promoting Bacteria (PGPB) are as effective as pure chemical on plant growth enhancement and disease control besides managing abiotic and other stresses in plants. Considering these facts it is very essential to understand the whole microbial diversity which will throw more light in to the basic concepts on plant–bacteria interaction, mineral–nutrient exchange, biofilm formation, role of bacteria in ethylene regulation via ACC deaminase, as well as the mechanism of action of PGPB-mediated antifungals to fight phytopathogens. In relation to plant health, the exploitation of such beneficial bacteria may improve agriculture system with economically sound production of human food and animal feed. Thus metagenomics enables us to explore how the microbial species are beneficial for plants and helps in producing more healthy crops.

Unraveling Rhizosphere Microbial Diversity

Plants uptake water and nutrients from the environment surrounding the root, known as rhizosphere. This biologically active zone of the soil harbour microbes including bacteria and fungi. The associations that occur between plant roots and soil microorganisms have been known for many decades. The rhizosphere contains an increased microbial biomass and activity compared with nonrhizosphere soil and the number of microorganisms in the rhizosphere is found to be 19–32 times larger than in root-free soil [35]. As the rhizosphere microbial community is intricately-intertwined with plant root, their interactions in the rhizosphere can be beneficial to the plant. Bacteria inhabiting the rhizosphere are referred to as plant growth-promoting rhizobacteria (PGPR) that exert beneficial effects on plant development which is often associated with increased rates of plant growth [36]. The well-known PGPR include members of the genera *Arthrobacter, Azoarcus, Azospirillum, Bacillus, Burkholderia, Enterobacter, Gluconacetobacter, Herbaspirillum, Klebsiella, Paenibacillus, Pseudomonas* and *Serratia*. These PGPR exert a direct effect on plant growth by inducing the production of phytohormones, supplying biologically fixed nitrogen, and increasing the phosphorous uptake by the solubilization of inorganic phosphates. They also promote plant growth by indirect mechanisms that involve suppression of bacterial, fungal, viral, and nematode pathogens [37,38,39].

The studies conducted on the rhizosphere soil microbial communities of white lupin [40] showed that white lupin have a greater influence on their microbial communities by modifying its rhizosphere, especially the space that surrounds its root cluster, by secreting organic acids and phenolic compounds, by changing pH, and so on. Further studies indicated that microorganisms living in the rhizosphere soil of cluster root improved the ability of the plant to absorb phosphorous from low phosphorous soil. The stimulation of microbial growth by roots is commonly known as the rhizosphere effect. However, the details of the rhizosphere effect on functional diversity of soil microorganisms are still unclear. The shift in the rhizosphere microbial community structure in response to the status of roots can be achieved through a comparative metagenomic analysis of the rhizosphere. In yet another study based on metagenomics the involvement of previously unknown microbial gene clusters in phytic acid utilization in the rhizosphere soil is being revealed. Though phytic acid is a major soil phosphorus compound and widely distributed in the world, it is known that plant do not exude phytase into the rhizosphere, which decompose phytic acid into inorganic phosphorus. Therefore it is strongly suggested that the rhizosphere soil microorganisms contribute this phenomenon and emphasize the importance to understand the role of functional genes related to phytic acid availability in the rhizoshpere soil, which can utilize phytic acid.

Recent developments in molecular biology methods are shedding some light on rhizospheric microbial diversity. One of the major difficulties that plant biologists and microbiologists face when studying these interactions is that many groups of microbes that inhabit this zone are not cultivable in the laboratory. Due to bias in studying those microorganisms that can be grown in the laboratory, there is limited knowledge on the abundance and activity of not-yet culturable PGPR. However, there are several examples of their existence and contribution to plant health, e.g., *Pasteuria penetrans*, not-yet-culturable bacterium parasitic to plant-pathogenic nematodes [41] and the nitrogen fixing activity by viable-but-not-culturable *Azoarcus* grass endophytes [42]. Bacteria belonging to the Acidobacteria and Verrucomicrobia are the most abundant group in many rhizospheres and their representatives are found to be the strains that mostly resist culturing [43]. However, it is yet unclear how their abundance is contributing towards plant health.

Several protocols have been developed for the isolation of metagenomic bacterial DNA from inside plant material and the subsequent generation of metagenomic library [44]. Although optimized for leaves and seeds, this method seems readily adaptable for use with root material, and thus of great use to the metagenomic exploration of microorganisms in the rhizosphere. For many of the traits or mechanisms known as plant growth-promoting [45], *in vitro* assays have been described and are exploitable for the functional screening of metagenomic library from rhizosphere DNA. Antibiotic activity towards plant pathogens can be assessed either directly by testing the clone or their extracts in confrontation assays using standard soil borne pathogen strains such as *Erwinia, Xanthomonas, Fusarium, Phytophthora etc.* [46,47]. Production of plant hormone indole 3 –acetic acid by metagenomic library clones can be measured using colorimetric assays [48]. Assays based on haloformatio are also available to identify PGPR-related phenotypes such as solubilization of mineral phosphate, siderophore formation and lytic enzyme production [49,50,51]. Activity screenings such as the ones described above have the potential to retrieve never-before-seen genes with PGPR activity from the metagenomic pool.

There is a clear potential for metagenomics to contribute to the study of microbial communities of the rhizosphere through the discovery of novel PGP genes and gene products, and the characterization of not-yet culturable PGPRs. An analysis of the rhizosphere by comparative metagenomics holds the promise to reveal several important questions regarding the unculturable fraction of the rhizosphere community. For one, it could expose what actually constitutes this fraction from a comparison of metagenomic DNA isolated directly from rhizosphere to DNA isolated from the entire culturable fraction from that same rhizosphere. The discovery of novel PGPR activities based on DNA sequence information from unculturables will add enormously to our understanding of the functional variation that exists in PGPR phenotypes. It will also benefit our ability to improve existing PGPR, by adding to the pool of exploitable PGPR genes and utilization of this pool to develop PGPRs with enhanced performance [52].

Suppressive Soil

Soil is the habitat on Earth that harbours the largest microbial diversity per unit mass or volume [53]. Traditional microbiological approaches already revealed that soils commonly harbour a broad array of antibiosis-related functions [54,55], some of which have been associated with the suppression of plant pathogens [56]. A soil is considered suppressive when, in spite of favorable conditions for disease to occur, a pathogen either cannot become established, es-tablishes but produces no disease, or establishes and produces disease for a short time and then declines. Generally suppressiveness is linked to the types and numbers of soil organisms, fertility level, and nature of the soil. The mechanisms by which disease organisms are suppressed in these soils include induced resistance, nutrient competition and direct inhibition through antibiotics secreted by beneficial organisms.

Suppressive soils often exert their antagonistic function by virtue of the activities of their microbiota. The level of disease suppressiveness is typically related to the level of total microbiological activity in a soil. The larger the active microbial biomass, the greater the soil's capacity to use carbon, nutrients and energy, thus lowering their availability to pathogens [57]. High competition coupled with secretion of antibiotics by some beneficial organisms and direct parasitism by others makes a tough environment for the pathogen. There is long-standing research into the organisms as well as the genes that underlie the suppressiveness of soil to phytopathogens and in many instances the suppressiveness could be related to the presence of phytopathogen antagonistic (antibiotic) functions in the soil microbiota [58]. The compounds involved in this antibiosis in soil are secondary metabolites produced by stationary-phase cells [59]. Although much is hitherto unknown, it is expected that a diversity of genes, gene clusters or operons is involved in the biosynthesis of antagonistic compounds, with sizes from roughly 8 to as much as 160 kb [60]. The microbiota of, in particular, disease-suppressive soils contains a wealth of antibiotic biosynthetic loci that are inaccessible by traditional cultivation-based techniques. For unlocking novel antibiotic genes from disease suppressive soil metagenomics-based approaches are well suited. Genetic screening of a metagenomic library generated from plant pathogenic fungus *Rhizoctonia solani* AG3 suppressive loamy soil based on hybridizations with generated probes for polyketide biosynthesis, non-ribosomal protein synthesis and gacA, revealed several inserts, of around 40-kb in size, with potential antibiotic production capacity [61]. The suppressing soil microbes' community is beneficial in agriculture and metagenomics offers bright perspectives, as it can provide quick access to the microbial genes which will aid in developing novel and much better microbial agents for management of the phytopathogens.

Biofilms on Plant Roots - Potential in Plant Pathology and Agriculture

Most microorganisms in the rhizosphere exist as biofilms rather than their planktonic mode and these plant root associated biofilms have been found to be beneficial for plant growth, yield, and crop quality [62]. Microbes in root-associated biofilms depend basically on root exudates for food and nutrition. By providing organic compounds as a nutrient source, these root exudates take a central role in being a major plant-derived factor and in triggering of root colonization and biofilm associations [63]. Though biofilm formation and plant growth promotion are governed by effective root colonization of the host plant, the exact mechanism of biofilm-mediated plant growth promotion have not been described adequately till date. The effect of PGPR biofilms in plant–microbe interactions and their possible mechanisms can be fully exposed only when complete diversity of these biofilms are brought to light.

Almost all species of bacteria use the quorum sensing (QS) based signaling pathway for the regulation of functions involved in relation to their environment. Quorum sensing relies upon production, accumulation, and perception of small diffusible molecules by the bacterial population, hence linking high gene expression with high cell population densities [64].This signaling mechanism modulates and coordinates bacterial interactions with plants, including antibiotic production and toxin release and is thought to be one of the main regulatory mechanisms in the formation of biofilms [65]. *N*-acyl *homoserine lactone* (*N*-AHSL) is an important class of signal molecules in bacteria amongst different classes of QS signal molecules. The AHLs-mediated cell-to-cell communication is mostly common among rhizospheric bacteria [66]. QS regulates pathogenicity or pathogenicity-related functions in bacteria of environmental importance such as the plant pathogens *Erwinia carotovora, Agrobacterium tumefaciens* and *Pseudomonas* [67] In pathogens such as *Erwinia* or *Pseudomonas*, *N*-AHSL-based QS is crucial to overcome the host defenses and ensure a successful infection.

Since QS is an important component of the adaptation strategy of bacteria to their environment, the competing microbes might surely have developed strategies to interfere with this communication system. Interference in QS was reported through the production of antagonists or the production of *N*-AHSL degradation enzymes (*N*-AHSLases) in various organisms [68]. They have been used to interfere efficiently with the expression of QS-regulated functions in bacteria. Thus, interfering with QS regulation, a strategy, termed as quorum quenching appears as one of the promising non-antibiotic based therapeutic strategies for the future [69]. Despite the large diversity of *N*-AHSL-degrading organisms identified to date, *N*-AHSL lactonase and *N*-AHSL amidohydrolase are the only families of *N*-AHSL-inactivating enzymes described so far [70].

It is evident that biofilm formation is a very common phenomenon in the rhizosphere and that quorum-sensing-based cell-to-cell communication could play a key role in green agricultural approaches. The importance of discovering novel and effective forms of quorum quenchers from rhizosphere biofilms may lead to discovery of novel pathogen defense strategies for their potential applications in futuristic agricultural systems. Metagenomic analysis of these biofilm communities will surely throw more light into this promising antibiotic-free antibacterial therapeutic strategy for crop protection.

Assessing Microbial Diversity Through 16S rRNA Gene Sequencing

Studying microorganisms recalcitrant to cultivation in the laboratory has been a major impediments to understanding natural microbial populations within the context of their environment. The high level of conservation of

16S ribosomal RNA gene (16S rRNA) commonly found in all life forms including prokaryotes makes this gene an ideal tool for the rapid identification and classification of microorganisms. A new era of microbial ecology was initiated when sequencing of ribosomal RNAs and the genes encoding them was introduced to describe uncultured bacteria in the environment. The molecular-based approaches have revealed an enormous phylogenetic diversity of microbial species found in natural environments that have not been cultured [71]. The number of major taxonomic divisions (phyla) within the bacterial domain of life alone has grown from 12 in the mid-1980s to greater than 80 today, of which less than 30 have cultured representatives [72,73]. The richness of the uncultured microbial world unveiled during the diversity studies employing 16S rRNA culture-independent methods is quite overwhelming. The public database, GenBank contained 21,466 16S rRNA genes from cultured prokaryotes and 54,655 from uncultured prokaryotes as on April 2004 and many of the uncultured organisms affiliate with phyla that contain no cultured members [74].

Here is an example for how the bacterial diversity associated with rice rhizosphere can be assessed through the generation of 16S rRNA clone library. The total DNA was isolated from the rhizosphere soil and amplified the 16S rRNA gene using eubacterial primers 27f and 1492r [75] (Fig 14.2). The amplicons were purified and ligated into a TA cloning vector (pGEM-T Easy, Promega) and transformed the host *Escherichia coli* JM 109 to generate the clone library. Positive clones were identified by colony PCR and subjected to Restriction Fragment Length polymorphism (RFLP) analysis (Fig 14.3).

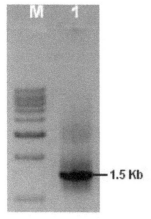
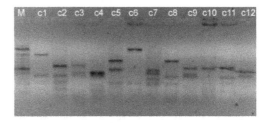

Fig. 14.2: Amplified 16SrRNA gene from rice rhizosphere metagenome. Lane M-1kb DNA ladder; Lane 1-16SrRNA amplication

Fig. 14.3: RFLaP pattern of rice rhizosphere associated bacterial 16SrRNA clones

Clones showing similar banding patterns were grouped together, one representative clone from each group were sequenced using ABI PRISM Big Dye terminator V3.1 cycle sequencing kit and ABI3730 DNA Sequencer. The sequences obtained, were compared with public data base (NCBI BLAST) to assign the sequence similarity of clone to the closest relative. The sequences are then compiled and aligned using BioEdit version 5.0.6 software [76] and generated the phylogenetic tree by neighbour-joining method with 1000 resampling bootstrap analysis using Mega v.4 software [77]. The 16S sequences were assigned to taxonomical hierarchy using DP classifier (online analysis of Ribosomal Database Project – 10). The phylogenetic tree generated (Fig. 14.4) based on the 16S rRNA sequences of the total 12 clones represented a

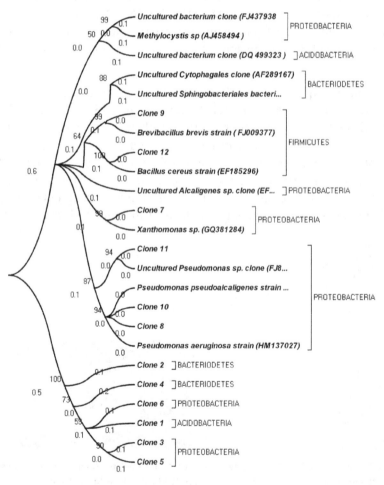

Fig. 14.4: Phylogentic tree based on 16SrRNA gene sequences from the rice rhizosphere metagenome

wide bacterial diversity with clones showing maximum identity with sequences retrieved from database as well as uncultured bacteria belonging to the taxa *Proteobacteria, Acidobacteria, Bacteriodetes* and *Firmicutes.*

In general, methods based on 16S rRNA gene analysis provide extensive information about the taxa and species present in an environment. Phylogenetic information is an essential requirement together with estimates of metabolic potential will link specific members of the community to biogeochemical processes. The metagenomic approach will help in exposing the hidden microbial diversity inhabiting environmental niches that resists conventional culture methods.

Bioinformatics Insights from Metagenomics

Metagenomics combines meta-analysis (the process of statistically combining separate analyses) and genomics (the comprehensive analysis of an organism's genetic material) [78]. Studies based on metagenomic approach generates huge amounts of sequences from environmental samples and bioinformatics tools are very essential for deciphering these sequences to understand gene function, evolution and organization and protein structure and function among the global prokaryotic diversity. Although processing metagenomic sequence data may seem to be straightforward, in fact it is just like as trying to reconstruct a puzzle with millions of pieces where most of them show similar colour and texture. Therefore, there should be a proper planning of experiments prior to implementing a metagenomic project, as these aspects can have major impacts on subsequent bioinformatics analyses.

The steps involved in a typical functional metagenomic study alternates between tasks performed in the wet lab and using bioinformatics tools (Fig 14.5). DNA extracted from environmental samples is digested with restriction enzymes and cloned in to appropriate vectors. The cloning strategy and primer sequences for the subsequent determination of the complete insert sequence are designed using bioinformatics tools. The reads obtained by sequencing are assembled using standard programs. Further bioinformatics analysis include finding Open Reading Frames (ORF), primers design for inverse PCR, BLAST searches for similar sequences, and annotation of the new sequences.

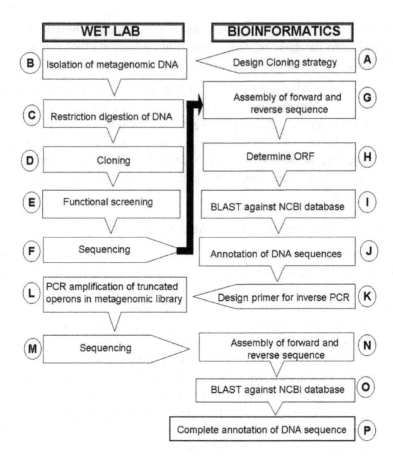

Fig. 14.5: A typical work flow in metagenomic analysis

The metagenomic data analysis follows the well established pattern of analysis for full genome sequencing projects [79]. After the raw reads have been obtained, either from large or short insert size clone libraries, assembly is usually the first step in data processing. Once longer contigs (continuous sequences) or scaffolds (contigs that still contain sequencing gaps) can be successfully established, gene calling and subsequent annotation is performed to gain insights into the phylogenetic and functional diversity of the sample. To get an overview of the functional and metabolic capacities represented in the sample, the protein coding genes are often mapped against the Kyoto Encyclopaedia of Gene and Genomes (KEGG) [80] and the Clusters of Orthologous Groups of proteins (COGs) [81]. Comparative metagenomics [82] and functional metagenomics approaches such as metatranscriptomics [83] and metaproteomics [84] are currently emerging techniques that aim at obtaining a more dynamic understanding of the differences and adaptations of the organisms to their environment.

Application of Metagenomics in Agriculture   309

Fragments obtained from large insert clones in the metagenomic library can be assembled into scaffolds and contigs of several kilobases in length following the shotgun sequencing approach [85]. About 400 sequencing reactions are required to sequence a DNA fragment (~ 40 kb) to eight-fold coverage. Since the resulting reads belong to the same large insert construct, assembly can be done easily with standard programs such as Phrap (www.phrap.com) or Arachne 2 [86]. But none of these currently used assembly programs has been specifically built for metagenomes. Aligners such as Phrap is primarily a single genome assembly program and will try to incorporate as many of the reads as possible, resulting in a large number of contigs. For metagenomes, programmes like Arachne assembler which utilizes a more conservative approach is preferable, as this significantly decreases the chance of misassemblies. An optimal balance between the number of contigs and the probability of chimeras can be maintained by performing repeated cycles of "binning" and assembly. Binning is the process in which sequence fragments of variable sizes are clustered into "bins", where each of the bins most probably resembles a single organism. The simplest binning approach is just to map occasionally occurring phylogenetic marker genes, or all genes on the contigs, to taxonomic groups based on best BLAST-hits [87].

Gene finding is one of the first and the most important step in understanding the genome once it has been sequenced. Classical gene predictors like ZCurve [88] and Glimmer [89] work reasonably well for high quality contigs from assembled shotgun sequences or large insert constructs. Gene predictors using extrinsic strategies such as similarity based searches against existing genomic and metagenomic databases to describe coding regions [90] require more time and tend to underpredict genes due to a lack of homologous information in the databases. Moreover, metagenomic gene predictors have to cope with (i) fragmented genes, (ii) low sequence quality leading to frameshifts and (iii) high phylogenetic diversity, which limits the performance of intrinsic gene finders. MetaGene is a genefinder programme for the analysis of metagenomic fragments which is capable to cope with most of these limitations. The tool utilises di-codon frequencies estimated by the GC content of a given sequence. This estimation, in combination with measures of the length distribution of ORFs, the distance from the leftmost start codon and the orientation and distance of neighbouring ORFs, extracted by statistical analysis of around 130 bacterial and archaeal genomes, are used for gene prediction[91]. GeneMark.hmm is another recently developed gene finder for metagenomic fragments, which utilizes around 357 bacterial and archaea genomes for gene prediction [92].

Functional annotation is the most important step in the process of analysing genomic fragments obtained from metagenomic studies, which gives a substantial insight into the abundance of the genetic potential available in the environmental sample being studied. Annotation should be carried out with care since poor annotations will lead to propagation of errors in the public databases due to the inconsistencies in functional assignments between and even within a single genome and by assigning potential functions to the genes found. Web-based bioinformatics supports for the annotation of metgenomes are provided by the Integrated Microbial Genomes (IMG/M) system[93] or CAMERA [94]. MG-RAST (MetaGenome Rapid Annotation using Subsystems Technology) (Fig 14.6) [95] is a recently released web-based platform providing support for gene prediction, annotation and also utilizes the annotations to reconstruct the metabolic networks that are functioning in the studied environment and then makes all this data available for download. One of the most advanced annotation system appropriate to meet emerging demands of metagenomics available for local installation is GenDB [96]. The standard tools and databases for providing functional assignments to the predicted genes are presented in Table 14.2.

Fig. 14.6: Screenshot of MG-RAST Home page(http://metagenomics.nmpdr.org)

Table 14.2: Standarad tools and databases providing functional assignments

BLASTn	GenBank	http://www.ncbi.nlm.nih.gov
	EMBL	http://www.ebi.ac.uk/
BLASTp or BLASTx	GenBank	http://www.ncbi.nlm/nih.gov
	Uniport	97
	Swiss-Port	98
HMMER	Pfam	99

Phylogenetic assignment of the ribosomal RNA genes is a rather straightforward approach to describe phylogenetic diversity. A commonly used method to show an overview of the taxonomic composition of the metagenome, is the analysis of the taxonomic affiliation of the best hit by searching for similarities of the contigs or genes to the UniProt or nr databases [100]. Phylogenetic assignment of the ribosomal RNA genes can be achieved by mapping it against the up to date ribosomal RNA (rRNA) sequence databases provided by SILVA [101] or the Ribosomal Database Project II (Fig. 14.7) [102]. The SILVA compatible ARB software suite can be used for detailed phylogenetic tree reconstruction, providing advanced alignment, tree reconstruction and visualization tools [103].

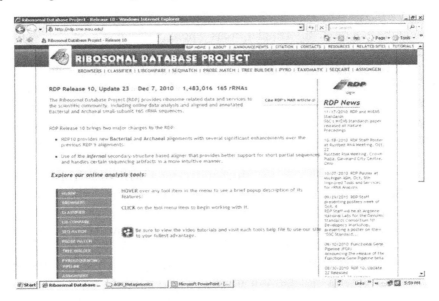

Fig. 14.7: Screenshot of RDB project home page (http://rdp.cme.msu.edu/)

Conclusion

The goals of the metagenomics are to study the microbial communities directly in their natural environments using the modern genomics techniques. It represents a combination of molecular and bioinformatics tools used to assess the genetic information of a community without prior cultivation of the individual species. Metagenomic data is usually generated from 40-150 kbp genomic fragments of uncultured microorganisms that are cloned directly from the environment. These fragments provide insight into the genetic composition of a microbial community and in some instances aid in the reconstruction of the complete genome of a microbe.

Sequencing the metagenome generates huge amounts of data from environmental samples, thus providing a glimpse of the global prokaryotic diversity of both species and genes in these sources. Metagenomic analysis generally follows the gene-centric approach, focusing on describing the environments by the study of the functional roles of the proteins encoded in the sequenced genes. In addition to functional assignment of proteins, accurate taxonomic assignments to the sequences would be essential, since it would help greatly in our understanding of the microbial community dynamics, to predict the effect of changes in their composition and to study key issues in the evolution of the community, such as the horizontal gene transfer in shaping the species.

Agriculture is one of the frontier areas where metagenomic studies can be applied with greater success. The rhizosphere represents one of the most complex ecosystems with almost every plant expected to have a chemically, physically, and biologically unique rhizosphere. Despite its intrinsic complexity, very little is known regarding the diversity of microbial communities which play an important role in the healthy sustenance of the plants. Recent advances in the analysis of environmental metagenomes suggested that metagenomics approach can be an extremely valuable tool for the characterization of complex microbial communities in the rhizosphere and will impart a thorough understanding to the processes like suppressive soil, root - biofilm and antibiosis etc. The insights achieved about rhizosphere biology through metagenomics will help in moulding strategies for an integrated management of soil microbial populations and to develop novel microbial formulations for plant growth promotion as well as effective management of the phytopathogens. This can lead to sustainable agriculture through cultivation of plant–microbial communities, which can reach a high productivity under minimal energy and chemical investments and with minimal pressures on the environment.

The global microbial diversity presents an enormous, largely untapped genetic and biological pool that could be exploited for the recovery of novel genes, biomolecules for metabolic pathways and various valuable products. As metagenomics approach progresses in addressing the global microbial diversity, new bioinformatics and data management tools also should be developed in parallel in order to effectively manage and analyze the enormous amounts of data generated through metagenomics studies.

Acknowledgements

The author is thankful to Prof. M. Radhakrishna Pillai, Director, Rajiv Gandhi Centre for Biotechnology for the facilities provided. I also express my gratitude to the Department of Biotechnology, Govt. of India for providing the funds.

References

1. Whitman, W. B., Coleman, D.C. and Wiebe, W. J. (1998) Prokaryotes: the unseen majority. Proc. Natl. Acad. Sci. USA, 95:6578- 6583.
2. Hugenholtz, P. and Pace, N.R. (1996) Identifying microbial diversity in the natural environment: a molecular phylogenetic approach. Trends Biotechnol., 14:190-197.
3. Rothschild, L. (2006) A microbiologist explodes the myth of the unculturables. Nature, 443:248–249.
4. Sekiguchi Y. (2006) Yet-to-be cultural microorganisms relevant to methane fermentation processes. Microbes. Environ., 21:1-15.
5. Amann, R.I., Ludwig, W and Schleifer, K.H. (1995) Phylogenetic identification and in situ detection of individual microbial cells without cultivation. Microbiol. Rev., 59: 143–169.
6. Handelsman, J. (2004) Metagenomics: application of genomics to uncultured microorganisms. Microbiol. Mol. Biol. Rev., 68: 669-685.
7. Riesenfeld, C.S., Schloss, P.D and Handelsman, J. (2004) Metagenomics: genomic analysis of microbial communities Annu. Rev. Genet., 38:525-52.
8. Susannah, G.T. and Edward, M.R. (2005) Metagenomics: DNA sequencing of environmental samples. Nature Rev. Gen., 6:805-814.
9. Streit, W.R. and Schmitz, R. (2004) Metagenomics – the key to the uncultured microbes. Curr Opin Microbiol., 7: 492–498.
10. Daniel, R. (2004) The soil metagenome – a rich resource for the discovery of novel natural products. Curr Opin Biotechnol., 15: 199–204.
11. Schmeisser, C., Steele, H. and Streit, W.R. (2007) Metagenomics, biotechnology with nonculturable microbes. Appl. Microbiol Biotechnol., 75: 955–962.
12. Schmidt, T.M., DeLong, E.F. and Pace, N.R. (1991) Analysis of a marine picoplankton community by 16S rRNA gene cloning and sequencing. J. Bacterio., 173: 4371–4378.
13. Ferrer, M., Golyshina, O.V., Chernikova, T.N., Khachane, A.N., Reyes-Duarte, D., Santos, V.A., Strompl, C., Elborough, K., Jarvis, G., Neef,.A., Yakimov, M.M., Timmis, K.N and Golyshin, P.N. (2005) Novel hydrolase diversity retrieved from a metagenome library of bovine rumen microflora. Environ. Microbiol., 7:1996–2010.
14. Ferrer, M., Golyshina, O.V., Plou, F.J., Timmis, K.N. and Golyshin, P.N. (2005) A novel alphaglucosidase from the acidophilic archaeon *Ferroplasma acidiphilum* strain Y with high transglycosylation activity and an unusual catalytic nucleophile. Biochem. J., 391:269–276.
15. Voget, S., Leggewie, C., Uesbeck, A., Raasch, C., Jaeger, K.E. and Streit, W.R. (2003) Prospecting for novel biocatalysts in a soil metagenome. Appl. Environ. Microbiol., 69: 6235–6242.
16. Gillespie, D.E., Brady, S.F., Bettermann, A.D., Cianciotto, N.P., Liles, M.R., Rondon, M.R., Clardy, J., Goodman, R.M. and Handelsman, J. (2002) Isolation of antibiotics turbomycin a and b from a metagenomics library of soil microbial DNA. Appl. Environ. Microbiol., 68: 4301–4306.
17. Brady, S. F., Chao, C. J. and Clardy, J. (2004) Longchain N-acyltyrosine synthases from environmental DNA. Appl. Environ. Microbiol., 70: 6865– 6870.
18. Riesenfeld, C.S., Goodman, R.M. and Handelsman J. (2004) Uncultured soil bacteria are a reservoir of new antibiotic resistance genes. Environ. Microbiol., 6:981-989.
19. Entcheva, P., Liebl, W., Johann, A., Hartsch, T. and Streit, W.R. (2001) Direct cloning from enrichment cultures, a reliable strategy for isolation of complete operons and genes from microbial consortia. Appl. Environ. Microbiol., 67:89-99.

20. Knietsch, A., Waschkowitz, T., Bowien, S., Henne, A. and Daniel, R. (2003) Construction and screening of metagenomic libraries derived from enrichment cultures: generation of a gene bank for genes conferring alcohol oxidoreductase activity on *Escherichia coli*. Appl. Environ. Microbiol., 69:1408-1416.

21. Rondon, M.R., August, P.R., Bettermann, A.D., Brady, S.F., Grossman, T.H., Liles, M. R., Loiacono, K.A, Lynch, B.A., MacNeil, I.A., Minor, C., Tiong, C.L., Gilman, M., Osburne, M.S., Clardy, J., Handelsman, J. and Goodman. R.M. (2000) Cloning the soil metagenome: a strategy for accessing in the genetic and functional diversity of uncultured microorganisms. Appl. Environ. Microbiol., 66:2541-2547.

22. Entcheva, P., Liebl, W., Johann, A., Hartsch, T. and Streit, W.R. (2001) Direct cloning from enrichment cultures, a reliable strategy for isolation of complete operons and genes from microbial consortia. Appl. Environ. Microbiol., 67: 89-99.

23. Knietsch, A. , Bowien, S., Whited, G., Gottschalk, G. and Daniel, R. (2003) Identification and characterization of coenzyme B12-dependent glycerol dehydratase- and diol dehydratase-encoding genes from metagenomic DNA libraries derived from enrichment cultures. Appl. Environ. Microbiol., 69: 3048-3060.

24. Majernik, A., Gottschalk, G. and Daniel, R. (2001) Screening of environmental DNA libraries for the presence of genes conferring Na_(Li_)/H_antiporter activity on *Escherichia coli*: characterization of the recovered genes and the corresponding gene products. J. Bacteriol., 183: 6645-6653.

25. Healy, F.G., Ray, R.M., Aldrich, H.C., Wilkie, A.C., Ingram, L.O. and Shanmugam, K.T. (1995) Direct isolation of functional genes encoding cellulases from the microbial consortia in a thermophilic, anaerobic digester maintained on lignocellulose. Appl. Microbiol. Biotechnol., 43: 667-674.

26. Cottrell, M.T., Moore, J.A. and Kirchman, D.L. (1999) Chitinases from uncultured marine microorganisms. Appl. Environ. Microbiol., 65:2552-2557.

27. Lorenz, P., Liebeton, K., Niehaus, F and Eck, J. (2002) Screening for novel enzymes for biocatalytic processes: accessing the metagenome as a resource of novel functional sequence space. Curr. Opion Biotechnology, 13: 572-577.

28. Schirmer, A., Gadkari, R., Reeves, C.D., Ibrahim, F., DeLong, E.F and Hutchinson, C.R. (2005) Metagenomic analysis reveals diverse polyketide synthase gene clusters in microorganisms associated with the marine sponge Discodermia dissoluta. Appl. Environ. Microbiol., 71:4840-4849.

29. Lim, H.K., Chung, E.J., Kim, J.C., Choi, G.J., Jang, K.S., Chung, Y.R., Cho, K.Y. and Lee, S.W. (2005) Characterization of a forest soil metagenome clone that confers indirubin and indigo production on *Escherichia coli*. Appl. Environ. Microbiol., 71:7768-7777.

30. Brady, S. F., Chao, C. J., Handelsman, J. and Clardy. J. (2001) Cloning and heterologous expression of a natural product biosynthetic gene cluster from eDNA. Org. Lett., 3:1981-1984.

31. MacNeil, I.A., Tiong, C.L., Minor, C., August, P.R., Grossman, T.H., Loiacono, K.A., Lynch, B.A., Phillips, T., Narula, S., Sundaramoorthi, R., Tyler, A., Aldredge, T., Long, H., Gilman, M., Holt, D. and Osburne, M.S. (2001). Expression and isolation of antimicrobial small molecules from soil DNA libraries. J. Mol. Microbiol. Biotechno., 13:301-308.

32. Henne, A., Daniel, R., Schmitz, R.A. and Gottschalk G. (1999) Construction of environmental DNA libraries in *Escherichia coli* and screening for the presence of genes conferring utilization of 4-hydroxybutyrate. Appl. Environ. Microbiol., 65: 3901-3907

33. Amjadi, M.., Garrigues, C., Jovanovich, S.B., Feldman, R.A. and DeLong, E.F. (2000) Construction and analysis of bacterial artificial chromosome libraries from a marine microbial assemblage. Environ. Microbiol., 2:516-529.

Application of Metagenomics in Agriculture

34. Beja, O., Suzuki, M.T., Koonin, E.V., Aravind, L., Hadd, A., Nguyen, L.P., Villacorta, R., Amjadi, M., Garrigues, C., Jovanovich, S.B., Feldman, R.A. and Delong, E.F. (2000) Construction and analysis of bacterial artificial chromosome libraries from a marine microbial assemblage. Environ. Microbiol., 2:516–29.

35. Bodelier, P.L.E., Wijlhuizen, A.G., Blom, C.W.P.M. and Laanbroek, H.J. (1997) Effects of photoperiod on growth of and denitrification by Pseudomonas chlororaphis in the root zone of Glyceria maxima, studied in a gnotobiotic microcosm. Plant Soil, 190:91–103.

36. Kloepper, J.W. and Schroth, M.N. (1978) Plant growth-promoting rhizobacteria on radishes. In: Station de pathologie vegetale et phyto-bacteriologie (ed). Proceedings of the 4th International Conference on Plant Pathogenic Bacteria, vol II. Gilbert-Clarey, Tours., 879-882.

37. Glick, B.R. (1995) The enhancement of plant growth by free-living bacteria. Can. J. Microbiol., 41:109–117.

38. Burdman, S., Jurkevitch, E. and Okon, Y. (2000) Recent advences in the use of plant growth promoting rhizobacteria (PGPR) in agriculture. In: Subba Rao NS and Dommergues YR (eds). Microbial interactions in agriculture and forestry, vol 2. Science Publishers Inc., Enfield, New Hampshire, 229–250.

39. Kirankumar, R., Jagadeesh, K.S., Krishnaraj, P.U. and Patil, M.S. (2008) Enhanced growth promotion of tomato and nutrient uptake by plant growth promoting rhizobacterial isolates in presence of tobacco mosaic virus pathogen. Karnataka J. Agric. Sci., 21:309–311.

40. Unno, Y., Ohkubo, K., Wasaki, J., Shinano, T. and Osaki, M. (2005) Plant growth promotion abilities and microscale bacterial dynamics in the rhizosphere of Lupin analysed by phytate utilization ability. Environmental Microbiology, 7:396-404.

41. Fould, S, Dieng, A.L., Davies, K.G., Normand, P. and Mateille, T. (2001) Immunological quantification of the nematode parasitic bacterium Pasteuria penetrans in soil. FEMS Microbiol Ecol., 37:187–195.

42. Hurek, T., Handley, L.L., Reinhold-Hurek, B., Piche, Y. (2002) Azoarcus grass endophytes contribute fixed nitrogen to the plant in an unculturable state. Mol. Plant Microbe. Interact., 15:233–242.

43. Buckley, D.H. and Schmidt, T.M. (2003) Diversity and dynamics of microbial communities in soils from agroecosystems. Environ. Microbiol., 5:441–452.

44. Jiao, J.Y., Wang, H.X., Zeng, Y. and Shen, Y.M. (2006) Enrichment for microbes living in association with plant tissues. J. Appl. Microbiol., 100:830–837.

45. Bloemberg, G. V. and Lugtenberg, B.J.J. (2001) Molecular basis of plant growth promotion and biocontrol by rhizobacteria. Current Opinion in Plant Biology, 4:343–350.

46. Rangarajan, S. Saleena, L.M., Vasudevan, P. and Nair, S. (2003) Biological suppression of rice diseases by Pseudomonas spp. Under saline conditions. Plant and Soil, 251:73- 82.

47. Kim, J., Kim, E., Kang, Y., Choi, O., Park, C.S. and Hwang, I. (2006) Molecular characterisation of biosynthetic genes of an antifungal compound produced by Pseudomonas fluorescens MC07. J. Microbiology and Biotechnology, 16:450–456.

48. Omer, Z.S., Tombolini, R., Broberg, A. and Gerhardson, B. (2004) Indole-3-acetic acid production by pink-pigmented facultative methylotrophic bacteria. Plant Groowth Regulation, 43:93-96.

49. Rodriguez, H., Gonzalez, T. and Selman, G. (2000) Expresion of mineral phosphate solubilizing gene from Erwinia herbicola in two rhizobacterial strains. J of Biotechnology, 84: 155-161.

50. Lee, E.T., Lim, S.K., Nam, D.H., Khang, Y.H. and Kim, S.D. (2003) Pyoveridin (2112) of Pseudomonas florescens 2112 inhibits Phytophthora capsici, a red–pepper blight-causing fungus. J Microbiology and Biotechnology, 13: 415-421.

51. Gobel, V., Megha, C., Vyas, P. and Chhatpar, H.S. (2004) Strain improvement of chitinolytic enzyme producing isolate *Pantoea dispersa* for enhancing its biocontrol potential against fungal plant pathogens. Annals of Microbiology. 54:503-515.

52. Timms-Wilson, T.M., Kilshaw, K. and Bailey, M.J. (2004) Risk assessment for engineered bacteria used in biocontrol of fungal disease in agricultural crops. Plant Soil, 266:57-67

53. Gans, J., Murray, W. and John, D. 2005. Computational improvements reveal great bacterial diversity and high metal toxicity in soil. Science, 309:1387-1390.

54. Garbeva, P., Postma, J., Van Veen, J. A. and Van Elsas, J. D. (2006) Effect of above-ground plant species on soil microbial community structure and its impact on suppression of *Rhizoctonia solani*. Environ. Microbiol., 8:233-246.

55. Adesina, M.F. Lembke, A., Costa, R., Speksnijder, A. and Smalla., K. (2007) Screening of bacterial isolates from various European soils for in vitro antagonistic activity towards *Rhizoctonia solani* and *Fusarium oxysporum*: site-dependent composition and diversity revealed. Soil Biol. Biochem., 39: 2818-2828.

56. Steinberg, C., Edel-Hermann, V., Alabouvette, C. and Lemanceau, P. (2006) Soil suppressiveness to plant diseases. In: van Elsas, J.D., Trevors, J.T., Jansson, J.K. (Eds.), Modern Soil Microbiology II. CRC Press, Boca Raton USA.

57. Andreas, T. (1992) Use of agricul-tural and municipal organic wastes to develop suppressiveness to plant pathogens- p. 35-42. *In:* E.C. Tjamos, G.C. Papavizas, and R.J. Cook (Ed.). Biological Control of Plant Diseases: Progress and Chal-lenges for the Future. NATO ASI Series No. 230. Plenum Press, New York, NY.

58. Garbeva, P., Voesenek, K. and Van Elsas, J.D. (2004) Quantitative detection and diversity of the pyrrolnitrin biosynthetic locus in soil under different treatments. Soil Biol Biochem. 36:1453-1463.

59. MacNeil, I.A., Tiong, C.L., Minor, C., August, P.R., Grossman, T.H., Loiacono, K.A., Lynch, B.A., Phillips, T., Narula, S., Sundaramoorthi, R., Tyler, A., Aldredge, T., Long, H., Gilman, M., Holt, D. and Osburne, M.S. (2001) Expression and isolation of antimicrobial small molecules from soil DNA libraries. J. Mol. Microbiol. Biotechnol., 31:301-308.

60. Ikeda, H., Nonomiya, T., Usami, M., Ohta, T and Omura, S. (1999) Organization of the biosynthetic gene cluster for the polyketide anthelmintic macrolide avermectin in *Streptomyces avermitilis*. Proc. Natl. Acad. Sci. USA, 96: 9509-9514.

61. van Elsas, J.D., Speksnijder, A.J. and van Overbeek, L.S. (2008) A procedure for the metagenomics exploration of disease-suppressive soils. J. of Microbio. Methods, 75:515-522.

62. Davey, M.E. and O'Toole, A.G. (2000) Microbial biofilms: from ecology to molecular genetics. Microbiol Mol. Biol. Rev., 64:847-867.

63. Bais, H.P., Weir, T.L., Perry, L.G., Gilroy, S. and Vivanco, J.M. (2006) The role of root exudates in rhizosphere interactions with plants and other organisms. Annu. Rev. Plant Biol., 57:233-266.

64. Reading, N.C. and Sperandio, V. (2006) Quorum sensing: the many languages of bacteria. FEMS Microbiol. Lett., 254: 1-11.

65. von Bodman, S.B., Bauer, W.D. and Coplin, D.L. (2003) Quorum sensing in plant-pathogenic bacteria. Annu. Rev. Phytopathol., 41:455-482.

66. Danhorn, T. and Fuqua, C. (2007) Biofilm formation by plant-associated bacteria. Annu. Rev. Microbiol., 61:401-422.

67. von Bodman, S.B. and Farrand, S.K. (1995) Capsular polysaccharide biosynthesis and pathogenicity in *Erwinia stewartii* require induction by an N-acylhomoserine lactone autoinducer. J. Bacteriol., 177:5000-5008.

68. Uroz, S., Dessaux, Y. and Oger, P. (2009) Quorum sensing and quorum quenching: the yin and yang of bacterial communication. Chem. BioChem., 10:205-216.

Application of Metagenomics in Agriculture

69. Finch, R.G., Pritchard, D.I., Bycroft, B.W., Williams, P. and Stewart, G.S. (1998) Quorum sensing: a novel target for anti-infective therapy. J. Antimicrob Chemother., 42:569–571.
70. Dong, Y.H., Xu, J.L., Li, X.Z. and Zhang, L.H. (2000) AiiA, an enzyme that inactivates the acylhomoserine lactone quorum-sensing signal and attenuates the virulence of *Erwinia carotovora*. Proc. Natl. Acad. Sci., USA, 97:3526–3531.
71. Pace, N. R. (1997) A molecular view of microbial diversity and the biosphere. Science 276: 734-740.
72. Hugenholtz, P., Goebel, B.M. and Pace, N.R. (1998) Impact of culture-independent studies on the emerging phylogenetic view of bacterial diversity. J. Bacteriol., 180: 4765-4774.
73. Fox, J.L. (2005) Ribosomal gene milestone met, already left in dust. ASM News, 71: 6-7.
74. Rappe, M.S and Giovannoni, S.J. (2003) The uncultured microbial majority. Annual Review of Microbiology, 57: 369–394.
75. Polz, M.F., Harbison, C. and Cavanaugh, C.M. (1999) Diversity and heterogeneity of epibiotic bacterial communities on the marine nematode *Eubostrichus dianase*. Appl. Environ. Microbiol., 65:231–240.
76. Hall, T. (2001) BioEdit version 5.0.6, North Carolina State University, Department of Microbiology.
77. Tamura, K, Dudley, J., Nei, M. and Kumar, S. (2007) MEGA4: Molecular Evolutionary Genetics Analysis (MEGA) software version 4.0. Molecular Biology and Evolution 10.1093/molbev/msm092.
78. Schloss, P.D. and Handelsman, J. (2003) Biotechnological prospects from metagenomics. Curr. Opin. Biotechnol., 14: 303-310.
79. Glockner, F.O. and Meyerdierks, A. (2006) Metagenome analysis. In: Stackebrandt E (ed) Molecular identification, systematics, and population structure of prokaryotes. Springer-Verlag, Heidelberg.
80. Kanehisa, M., Goto, S., Kawashima, S. and Okuno, Y. and Hattori, M. (2004) The KEGG resource for deciphering the genome. Nucleic Acids Res., 32:D277–D280.
81. Tatusov, R.L., Koonin, E.V. and Lipman, D.J. (1997) A genomic perspective on protein families. Science, 278:631–637.
82. DeLong, E.F., Preston, C.M., Mincer, T., Rich, V., Hallam, S.J., Frigaard, N.U., Matrinez, A., Sullivan, M.B., Edwards, R., Brito, B.R., Chisholm, S.W and Karl, D.M. (2006) Community genomics among stratified microbial assemblages in the ocean's interior. Science, 311:496–503.
83. Frias-Lopez, J., Shi, Y., Tyson, G.W., Coleman, M.L., Schuster, S.C., Chisholm, S.W and DeLong, E.F. (2008) Microbial community gene expression in ocean surface waters. Proc. Natl. Acad Sci. USA, 105:3805–3810.
84. Ram, R.J., Verberkmoes, N.C., Thelen, M.P., Tyson, G.W., Baker, B.J., Blake, R.C. II, Shah, M., Hettich, R.L. and Banfield, J.F. (2005) Community proteomics of a natural microbial biofilm. Science, 308:1915–1920.
85. Sambrook, J. and Russel, D.W. (2001) Molecular cloning: a laboratory manual, 3rd edn. Cold Spring Harbor Laboratory Press, Cold Spring Harbor, New York.
86. Jaffe, D.B., Butler, J., Gnerre, S., Mauceli, E., Lindblad-Toh, K., Mesirov, J.P., Zody, M.C. and Lander, E.S. (2003) Whole-genome sequence assembly for mammalian genomes: Arachne 2. Genome Res., 13:91–96.
87. Huson, D.H., Auch, A.F., Qi, J. and Schuster, S.C. (2007) MEGAN analysis of metagenomic data. Genome Res., 17:377–386.
88. Guo, F.B., Ou, H.Y. and Zhang, C.T. (2003) ZCURVE: a new system for recognizing protein-coding genes in bacterial and archaeal genomes. Nucleic Acids Res., 31:1780-1789.
89. Delcher, A.L., Bratke, K.A., Powers, E.C. and Salzberg, S.L. (2007) Identifying bacterial genes and endosymbiont DNA with Glimmer. Bioinformatics, 23:673–679.

90. Krause, L., Diaz, N.N., Bartels, D., Edwards, R.A., Puhler, A., Rohwer, F., Meyer, F. and Stoye, J. (2006) Finding novel genes in bacterial communities isolated from the environment. Bioinformatics, 22:E281–E289.

91. Noguchi, H., Park, J. and Takagi, T. (2006) MetaGene: prokaryotic gene finding from environmental genome shotgun sequences. Nucleic Acids Res., 34:5623–5630.

92. Zhu, W., Lomsadze, A. and Borodovsky, M. (2010) "*Ab initio* gene identification in metagenomic sequences". Nucleic Acids Res., 38:(2), e132, doi: 10.1093/nar/gkq 275.

93. Markowitz, V.M., Ivanova, N.N., Szeto, E., Palaniappan, K., Chu, K., Dalevi, D., Chen, I.M.A., Grechkin, Y., Dubchak, I., Anderson, I., Lykidis, A., Mavromatis, K., Hugenholtz, P. and Kyrpides, N.C. (2008) IMG/M: a data management and analysis system for metagenomes. Nucleic Acids Res., 36:D534–D538.

94. Seshadri., R., Kravitz, S.A., Smarr, L., Gilna, P. and Frazier, M. (2007) CAMERA: a community resource for metagenomics. PLoS Biol., 5(3):e75. doi:10.1371/journal.pbio.0050075.

95. Meyer, F., Paarmann, D., D'Souza, M., Olson, R., Glass, E.M., Kubal, M., Paczian, T., Rodriguez, A., Stevens, R., Wilke, A, Wilkening, J. and Edwards, R.A. (2008) The metagenomics RAST server - a public resource for the automatic phylogenetic and functional analysis of metagenomes. BMC Bioinformatics, 9: 386.

96. Meyer, F., Goesmann, A., McHardy, A. C., Bartels, D., Bekel, T., Clausen, J., Kalinowski, J., Linke, B., Rupp, O., Giegerich, R. and Puhler. A. (2003) GenDB – an open source genome annotation system for prokaryote genomes. Nucleic Acids Res., 31:2187–2195

97. Apweiler, R., Bairoch, A., Wu, C.H., Barker, W.C., Boeckmann, B., Ferro, S., Gasteiger, E., Huang, H., Lopez, R., Magrane, M. *et al.* (2004) UniProt: the Universal Protein knowledgebase. Nucleic Acids Res., 32:D115–D119.

98. Boeckmann, B., Bairoch, A., Apweiler, R., Blatter, M.C., Estreicher, A., Gasteiger, E., Martin, M.J., Michoud, K., O'Donovan, C., Phan, I., Pilbout, S. and Schneider, M. (2003) The SWISS-PROT protein knowledgebase and its supplement TrEMBL in 2003. Nucleic Acids Res., 31:365–370.

99. Bateman, A., Coin, L., Durbin, R., Finn, R.D., Hollich, V., Griffiths-Jones, S., Khanna, A., Marshall, M., Moxon, S., Sonnhammer, E.L.L., Studholme, D.J., Yeats, C. and Eddy, S.R. (2004) The PFAM protein families database. Nucleic Acids Res., 32:D138–D141.

100. Turnbaugh, P.J., Ley, R.E., Mahowald, M.A., Magrini, V., Mardis, E.R. and Gordon, J.I. (2006) An obesity-associated gut microbiome with increased capacity for energy harvest. Nature, 444:1027–1031.

101. Pruesse, E., Quast, C., Knittel, K., Fuch, B.M., Ludwig, W., Peplies, J and Glockner, F.O. (2007) SILVA: a comprehensive online resource for quality checked and aligned ribosomal RNA sequence data compatible with ARB. Nucleic Acids Res., 35:7188–7196

102. Cole, J.R., Chai, B., Farris, R. J., Wang, Q., Kulam-Syed-Mohideen, A.S., McGarrell, D. M., Bandela, A. M., Cardenas, E., Garrity, G. M. and Tiedje, J. M. (2007) The ribosomal database project (RDP-II): introducing myRDP space and quality controlled public data. Nucleic Acids Res., 35:D169–D172.

103. Ludwig, W., Strunk, O., Westram R., Richter, L., Meier, H., Yadhukumar., Buchner, A., Lai, T., Steppi, S., Jobb, G. *et al.* (2004) ARB: a software environment for sequence data. Nucleic Acids Res., 32:1363–1371.

Chapter – 15

Biodiversity Informatics

P.N. Krishnan, S. Sreekumar, C.K. Biju and M. Raveendran

Abstract

Our biological heritage, from genes to ecosystems, is threatened on a global scale. Biological species, the prime focus on biodiversity, provide all our basic requirements such as food, medicine, fuel and things for well being, make nature's sustainability through purifying air and water, pollinating crops, etc. and motivate us at a deep emotional level. During the course of human civilization and consequent urbanization process biological species and environment have been degraded. It is estimated that about 20% of all species are expected to be lost within 30 years and 50% or more by the end of the 21st century [36]. The current rates of species extinction are 1000–10,000 times higher than the background rate of 10^{-7} species per year inferred from fossil record. Today, we seem to be losing two to five species per hour from tropical forests alone. This amounts to a loss of 16 m population per year or 1800 populations per hour. This led to biodiversity crisis in the global level [23,26,35] and generated an outpouring of research on biodiversity for the last 25 years. As a result we could have a general understanding about the distribution of biodiversity, the proximate causes (habitat loss, habitat fragmentation, habitat degradation, invasive species, pollution, resource exploitation) contributing to its loss, sketches of the root causes of biodiversity loss, and the outlines of better *management policies. Since the establishment of the convention on biological diversity in 1992, conservation of biodiversity has become a priority concern on the international agenda.*

Introduction

The word '*Biodiversity*' was first coined by W.G. Rosen in 1985 while planning the *National Forum on Biological Diversity* organized by the National Research Council (NRC), USA which was held in 1986. But the term was first used in a publication by the Entomologist E. O. Wilson in 1988 as the title of the proceedings of *National Forum on Biological Diversity*. Generally, the word biodiversity is often used to describe all the species living in a particular area. However, scientists include not only living organisms and their complex interactions, but also interactions with the abiotic (non-living) aspects of their environment. There are so many definitions proposed by various authors to biodiversity. The globally accepted definition was coined by the Convention on Biological Diversity held in Rio de Janeiro in connection with United Nations Earth Summit, 1992 which defined 'Biodiversity' as "the variability among living organisms from all sources, including, 'inter alia', terrestrial, marine, and other aquatic ecosystems, and the ecological complexes of which they are part: this includes diversity within species, between species and of ecosystems". Based on this definition the elements of biodiversity can be categorized into three major groups. (1) Ecological diversity which consists of diversity in biomes, bioregions, landscapes, ecosystems, habitats, niches and population. (2) Organism diversity which include diversity among kingdoms, phyla, families, genera, species, population and individuals. (3) Genetic diversity which include diversity among population, individuals, chromosomes, genes and nucleotides. In this context, observed three levels of biological diversity were abserved; they are (i) molecular, (ii) organisms and (iii) ecological [68].

Biological diversity varies with latitude, altitude, precipitation, salinity, nutrient levels, etc. Assessments of diversity pose considerable problems. Generally, it is measured in terms of species composition and species abundance. Some of the frequently used diversity measures in ecological analyses are summarized [54]. They are (1) Alpha (a) diversity: The species diversity within a community or within a habitat, representing a balance between the actions of local biotic and abiotic elements, and immigration from other locations, comprises two components, i.e., species richness and evenness and can be measured by a variety of indices. (2) Beta (b) diversity: Inter community or inter habitat or differentiation diversity expressing the rate of species turnover per unit change in habitat, can be assessed by a variety of indices. (3) Gamma (g) diversity: Over all diversity at landscape level or that of the whole organisms or it can be

Biodiversity Informatics 321

expressed by number of species found in a landscape – It include both alpha and beta diversity. (4) Compositional diversity pattern: A measure of landscape complexity can be assessed as mosaic diversity (i.e., the variation in species richness among communities and variation in commonness or rarity among species) using affinity analysis. (5) Species-area relation: The number of species encountered is proportional to a power of the area sampled, i.e., $S\,a\,A^z$, where S is number of species encountered, A is area sampled and z is empirical constant.

Major role of biodiversity in ecosystem level are (1) generation of soils and maintenance of soil quality as well as nutrient level mainly by the activities of microbes, (2) purification of air quality through the process of photosynthesis, (3) maintenance of water quality through the absorption and recycling of nutrients and other wastes, (4) pest control through food chain, (5) detoxification and decomposition of wastes by the activities of decomposers /microbes, (6) pollination and crop production by the activities of various animal species, (7) climate stabilization by converting CO_2 into organic substance and moisture/ temperature control through the process of transpiration as well as luxurious vegetation of the forest cover, (8) prevention and mitigation of natural disasters by binding action of plant roots on soil and forest cover, (9) food security by providing wild genes which offer resistance to pests and diseases, (10) income generation by providing essential things and (11) aesthetic and spiritual satisfaction through the beauty of geographical and biological diversity. In addition to these, some more examples of biodiversity to human health are summarized [54] based on the report of Dobson which are listed below:

- One out of every 125 plant species studied at the Herb Research Foundation, Boulder, produced a major drug with a market value in the US of at least $ 200 million per year.

- Of the 118 (out of the top 150) prescription drugs in the US, 74% are based on plants, 18% on fungi, 5% on bacteria and 3% on vertebrates.

- Of the top 10 prescription drugs in the US, 9 are based on natural plant products.

- 80% of the world population relies on traditional plant medicine.

- Compounds from Gingko leaves are used by 80% Europeans older than 45 years to prevent senile dementia.

- Losing one tree species a day means losing 3-4 potentially valuable drugs every year, at a total of cost of $ 600 million.

Biodiversity has attracted world attention because of the growing awareness of its importance on the one hand and the anticipated massive depletion on the other [54]. Global biodiversity information is inevitable to a wide range of scientific, educational and governmental agencies to support well-informed decision making at regional, national and global level, yet information critical to such decision is not available readily. Part of the problem is associated with the complex nature of biodiversity data and information providers' uncertainties in terms of their existence and distribution. Moreover, currently most of the biodiversity related information is scattered and not organized properly for ready reference. Therefore, studies related to biodiversity and generation of new knowledge has been greatly hampered. The international conventions such as the Convention on Biological Diversity (CBD), Ramsar Convention, World Heritage Convention, etc. have called for extra efforts to generate better data/information. The baseline information on the status and distribution of each species on Earth is essential for the conservation and sustainable utilization of biodiversity. Management of information deluge and species identification and determining its' values before extinction are the major tasks faced by the biologists.

Bioinformatics

Bioinformatics is the science of management, analysis and invoking new information from the existing data in biology using computer and information technology. Computers are used to gather, store, analyse and integrate the information and information technology/internet is used to disseminate / access information at any time globally without any boundary. The ultimate goal of the field is to enable the discovery of new biological insights as well as to create a global perspective from which unifying principles in biology can be discerned. The term "Bioinformatics' was coined by Paulien Hogeweg in 1979 for the study of informatics process in biotic system and the discipline was well established in 1980s in couple with launching of several genome sequencing projects such as human, bacterial and mouse genome projects. However, remarkable contribution on this field was first laid by Margaret Oakley Dayhoff, who was a pioneer in the use of computers in chemistry and biology, beginning with her Ph.D. thesis in 1948. Her work was multi-disciplinary and used her knowledge of chemistry, mathematics, biology and computer science to develop an entirely new field. In 1965 she published the first computerized database, the "Atlas of Protein Sequences". The basic principles of bioinformatics are (1) database creation and management (2) analysis of biological data using computational tools and (3) development of computational tools/techniques for handling biological data. Although bioinformatics was evolved for handling genomic and proteomic data now it finds wide applications in all areas of biology.

Biodiversity Informatics

The term "Biodiversity informatics" may be first used in connection with the formation of Canadian Biodiversity Informatics Consortium in 1992 and eventually formed it into a corporate entry in 1993 (http://www.bgbm.org/BioDiviInf/The Term.htm). As much of information in biological domain (generally termed 'bioinformatics') now focuses on molecular aspects of life forms data, a new term may be needed to circumscribe the application of Information Technology Tools and Technologies at the organism level [52]. Biodiversity informatics is defined as 'the application of information technology to the management, algorithmic exploration, analysis and interpretation of primary data regarding life, particularly at the species level of organisation' [58]. It thus deals with information capture, storage provision, retrieval and analysis focused on individual organisms, populations and species and their interactions. It covers information generated by the fields of systematics, evolutionary biology, population biology and ecology as well as more applied fields such as conservation biology and ecological management. It is considered a part of biological informatics sandwiched between and strongly overlapping with environmental informatics and molecular bioinformatics [3].

Traditionally, biodiversity informatics has involved the discovery and compilation of specimen data, and application of these data to determine where organism lives. With development of computers and the internet, the capability now exists to aggregate and disseminate biodiversity data globally as well as to broaden the scope beyond that of establishing species ranges [8]. While 'biodiversity informatics' can be considered a new discipline, the use of automated techniques is not entirely new to the biodiversity domain. Indeed, the development of systematic techniques (often involving computers) has proven to be essential in the study and cataloguing of the speciation, distribution and evolution of life on Earth. However, to date, there has been limited integration and harmonization of biodiversity information within the context of molecular studies that are the focus of bioinformatics. Biodiversity informatics thus aims to identify linkages within and across all three levels of biological data relative to the organism. The fundamental tenet in biodiversity informatics is that biological information can be linked through the organism towards the development of new, testable hypotheses. The realms of biodiversity informatics are information systems, which uses taxa, specimens and (species) observation records as their references and index systems.

The major objectives of biodiversity informatics are (1) mobilizing existing information resources, (2) increasing research efficiency by timely provision of fundamental data for a steadily increasing number of problems, (3) providing the information base and the tools for diverse biodiversity modeling tasks. The field has got great potential with applications ranging from prediction of distribution of known and unknown species [50], prediction of geographical and ecological distribution of infectious disease vectors [2,11,48,45], prediction of species invasion [46,44] and assessment of impact of climate change on biodiversity [47,55,64]. In addition, the biodiversity information which exists today has economic value and represent an investment of billions of dollars worldwide [12]. Unfortunately, a comprehensive infrastructure that would allow this information to be easily accessed and effectively used so that society can reap the returns on its investments does not exist. The potential remains largely unexplored, as this field is now becoming a vibrant area where studies are being initiated. In the present scenario the major challenges in this field are (1) to provide consensus reference system to solve problems related to the complexity and variability of biodiversity data, (2) to encourage the data builders to follow subject wise standard data format and taxon based information system, (3) to organize the information system in such a way that it can be used easily by the non-specialist, (4) to interlink biological collections in various repositories and digital information about it, (5) to elucidate spatial reference system of biological collections, which are deposited in various museum, zoo, herbaria etc.

History of Biodiversity Informatics

Application of computers for the management of biodiversity data is well demonstrated. The historical review of computer based system in taxonomy and environmental science were published [56,24,66,40]. The first online identification system was developed [6] and the first batch working batch processed systems was developed [33], were the land mark in using computer to the field of systematics. In the past 10-15 years, advances in information technology, introduction of large capacity electronic storage media, the internet, world-wide-web, distributional database technology etc. and the policies of the owners of primary data sources (e.g., Large scale digitalization of data, creation of public access databases) are creating a revolution in the way that biodiversity information is created, maintained, distributed and used [4,37,32] with potential of much more to come [17].

On the global scenario Australia has been contributing much in biodiversity informatics. Since the mid-1970s, Australian herbaria have been digitizing their data cooperatively. The Environmental Resources Information Network (ERIN)

Biodiversity Informatics 325

was established in 1989 to provide geographically-related environmental information for planning and decision-making. Also in 1989, Herbarium Information Standards and Protocols for Interchange of Data (HISPID), a standard format for interchange of electronic herbarium specimen information, developed by a committee of representatives from all Australian herbaria, was first published. ERIN's experience set an example for several other initiatives, such as Mexico's Comisión Nacional para el Conocimiento y Uso de la Biodiversidad (CONABIO), Costa Rica's Instituto Nacional de Biodiversidad (INBio) and Brazil's Base de Dados Tropical (BDT). INBio and CONABIO both became fully engaged in biodiversity informatics after the exchange of experience with ERIN experts in 1993. In the early 1990s, researchers from diverse fields of expertise held meetings, and began the Biodiversity Information Network – Agenda 21 (BIN21) initiative. This group established what was called a Special Interest Network with a Virtual Library. BIN21 was set up as an informal, collaborative, distributed network consisting of a series of participating "nodes" aiming at complementing existing or planned actions. BIN21 was actively involved in discussions of the Clearing House Mechanism (CHM) to the Convention on Biological Diversity (CBD), and produced a document proposing the structure being used today, composed of focal points and thematic network (http://:www.bdt.org.br/bin21/wks95/chm_doc.html). At the time, owing to technological limitations, what was envisaged was creation of directories of people, institutions, and data sources. In 1998, a research project was launched at the University of Kansas Natural History Museum and Biodiversity Research Center, The Species Analyst (TSA). TSA's main objective was to develop standards and software tools for access to world natural history collection and observation databases. This project was one of the first networks to draw on distributed data sources from biological collections worldwide, setting an example for other initiatives, which were attracted by examples based on an associated modeling tool, the Genetic Algorithm for Rule-set Prediction (GARP), originally developed [60]. Since then there are several global and regional efforts aiming at organizing data stakeholders and making data available for conservation and sustainable development and research in this area.

Taxonomic Problems in Biodiversity Information

There are an estimated two million organisms that are associated with taxonomic treatments [68,67]. A taxonomic treatment generally includes an organism name, morphological description, distributional information and other related (e.g., phylogenetic) information. These taxonomic treatments can be used to supplement contemporary molecular data, especially in the context of identifying 'genotype– phenotype' correlations (e.g., molecular patterns can be associated with morphological descriptions between taxonomic groups). However, using organism names as identifiers to link information can be problematic, especially in a historical context [52]. Organism names change over time, for example before 1919, data associated with *Escherichia coli* were labeled with *Bacillus coli* or *Bacterium coli*. Issues remain even in light of an array of regulatory bodies that strive to develop systematic rules to stabilize names and minimize ambiguity [18,51,56]. Reconciliation techniques are thus needed to interconnect multiple names, either objectively (e.g., *Doryteuthis pealeii* and *Loligo pealeii* are names that refer to the common squid) or subjectively (e.g., *Brucella abortus* and *Brucella suis* are names that refer to the causative agent for Brucellosis, which affects a range of hosts). Disambiguation methods are also needed to distinguish different organism concepts associated with the same names that refer to more than one species (*e.g.*, *Peranema* refers to a genus of both a fern and a euglena). The successful development of comprehensive scientific name indices can be used to identify relevant data across a wide range of resources. A centralized index might also foster the development of applications that can be used to infer linkages between organisms across heterogeneous data sources [52].

The cataloguing of scientific names into a single, publicly accessible resource is a paramount first step to develop a framework for organizing biological knowledge [49,68]. The Unified Medical Language System (UMLS) began development in the mid-1980s as a means to create a standard language for biomedicine that could be used by computer-based clinical information systems (House of Commons 99th Congress). The UMLS includes terms from over 100 biomedical terminologies and ontologies organized into over one million concepts [5]. The UMLS does contain some scientific name terminologies, most notably National Center for Biotechnology Information (NCBI) Taxonomy. However, in addition to NCBI Taxonomy, there are a number of other resources that maintain lists of scientific names. These include Species2000 (http://www.sp2000.org/) and the Integrated Taxonomic Information System (ITIS; http://www.itis.usda. gov/), both of which are associated with the Catalogue of Life Project [16]. Organism names are also maintained by groups of researchers focused on a particular taxonomic group, e.g., IndexFungorum (Fungi; http://www.bioone.org/), AlgaeBase [53], CephBase Cephalopods; [10], Deutsche Sammlung von Mikroorganismen und Zellkulturen (DSMZ; Microorganisms derived from Euzeby's List [15] and FishBase [38]. To organize these different lists from multiple sources, the Universal Biological Indexer and Organizer (uBio; http://www.ubio.org) project has been working towards the integration of scientific and vernacular names. The uBio databases are designed to function much in the same way as the UMLS, as an aggregator of lists of concepts and hierarchies into a single resource. Currently, uBio contains over 10 million organism name strings, which have been collected from a range of existing scientific name resources, including all of the previously mentioned.

Biodiversity Data

Biodiversity information is complex, voluminous and can be categorized into different levels such as molecular sequences, gene diversity, individuals, species, genera, populations, habitats, ecosystems, biomes, etc. The United Nations Environment Programme (UNEP) outlines eight major categories of biodiversity data for country studies [68]. These datasets will serve three main objectives of Convention on Biological Diversity (CBD) *viz.* the conservation of biodiversity, the sustainable use of biological resources and the equitable sharing of the benefits from using those resources. The categories are as follows

Biological: Information on ecosystem, species and genetic resources.

Physical: Information on physical factors such as climate, topography and hydrology that allows biological data to be placed within a physical context.

Socio-economic: Information on socio-economic attributes such as population, population distribution and transport routes.

Cost and Benefits: A value of biodiversity that takes into account the cost and benefits of management options.

Pressure and Threats: Information on both potential and actual threats to biological diversity.

Sustainable management: Information on current and past management activities particularly the use of biological resources.

Sources and Contacts: Information models, standards and technologies and appropriate agencies or experts who can be contacted.

Interrelationships: Information on the interrelationship between and among species and ecosystems so as to forecast the effects of proposed actions.

Thus biodiversity informatics mainly focuses on species and specimen data as the primary information component of a global comprehensive data net work on biodiversity. It also uses environmental and ecological data for modeling distribution pattern of species and populations.

Nature of Biological Data

The most striking feature of data in life science is not its volume but its diversity and variability. In fact, biological data sets are intrinsically complex and are organized in loose hierarchies that reflect our understanding of the complex living systems, ranging from genes and proteins, to protein-protein interactions, biochemical pathways and regulatory networks, to cell and tissues, organisms and populations and finally the ecosystems on earth. The variability in biological data is quite natural, because individuals and species vary tremendously. For example, structure and function of organs vary across age and gender, in normal and different disease states, and across species. Essentially, all features of biology exhibit some degree of variability [8].

Biological Data Types (data heterogeneity)

Sequences: Sequence data, such as those associated with the DNA of various species, rapidly growing with the advent of automated sequencing technology.

Graphs: Biological data indicating relationships can be demonstrated as graphs, e.g., Pathways data (metabolic pathways, signaling pathways, gene regulatory networks, etc.) genetic maps and structural taxonomies, workflow chart, etc.

High-dimensional data: System biology is highly dependent on comparing the behavior of various biological units, data points that might be associated with the behavior of an individual unit must be collected for thousands or tens of thousands of comparable units, e.g., Gene expression studies.

Geometric information: Biological function depends on the 3D structure of the biological molecules. For example, the 'docking' behavior of molecules at a potential binding site depends on the three-dimensional configuration of the molecules and the site. In this instance, molecular structure data are very important. Graphs are one way of representing three-dimensional structure (e.g., proteins), but ball-and-stick models of protein backbones provides a more intuitive representation.

Scalar and vector fields: These data are relevant to natural phenomena that vary continuously in space and time. Scalar and vector field properties are associated with chemical concentration and electric charge across the volume of a cell, current fluxes across the surface of a cell or through its volume, and chemical fluxes across the cell membranes, as well as data regarding charge, hydrophobicity, and other chemical properties that can be specified over the surface or within the volume of a molecule or a complex.

Biodiversity Informatics

Patterns: Within the genome are patterns that characterize biologically interesting entities. For example, the genome contains patterns associated with genes (i.e., sequences of particular genes) and with regulatory sequences (that determine the extent of a particular gene's expression). Proteins are characterized by particular genomic sequences. Patterns of sequence data can be represented as regular expressions, Hidden Markov Models (HMMS), stochastic context-free grammars (for RNA sequences), or other types of grammars. Patterns are also interesting in the exploration of protein structure data, microarray data, pathway data, proteomics data and metabolic data.

Constraints: Consistency within a database is critical if the data are to be trustworthy, and biological databases are no exception. For example, individual chemical reactions in a biological pathway must locally satisfy the conservation of mass for each element involved. Reaction cycles in thermodynamic databases must satisfy global energy conservation constraints.

Spatial information: Real biological entities, from cells to ecosystems, are not spatially homogeneous, and a great deal of interesting science can be found in understanding how one spatial region is different from another. Thus, spatial relationships must be captured in machine-readable form, and other biologically significant data must be overlaid on top of these relationships.

Biodiversity Informatics – on Action and Searches

About 1.8 million species has been described, of which information about 10% of the world wide specimens are available on the electronic domain [31]. There are about 1,600 botanic gardens and arboreta world wide as member of Botanic Garden Conservation International, of which majority are located in Europe and North America [25]. In addition, most of the developing countries have established several botanic gardens for conserving a large number of accessions. In all these gardens a total of 3.2 million accessions consisting of 80,000 species are maintained. This represents about 30% of known species of flowering plants and ferns. The global database maintained by Botanic Garden Conservation International (BGCI) has already documented 250,000 accessions that include records from 350 institutions, representing 30,000 species [34], which might be increased to a large extent now. Similarly, there are billions of specimen records and observational data that exist in natural history collections worldwide, which are increasing day-to-day [7,62]. The Global Biodiversity Information Facility (GBIF; http:// www.gbif.org/ [14] and the Taxonomic Database Working Group (TDWG; http://www.tdwg.org) are organizations that strive to develop structured formats to represent and share biodiversity data. An overview of the emerging formats for biodiversity data have been reviewed [27]. These structured data can be used to complement existing stores of genomic and biomedical knowledge (e.g., as stored in GenBank and Medline, respectively), leading towards the integration of knowledge across a range of biological resources. The topic of knowledge integration in Biodiversity Informatics is rather timely—the Encyclopedia of Life (EOL) project, which is inspired by Wilson [57], will depend on the development of the requisite informatics infrastructure to identify, validate and manage information such that they can be presented through a single portal. As EOL strives to create a web site for all species known to be present on earth, the scope of issues associated with organizing and linking data across the plethora of current and future repositories is immense. The ultimate goal of the EOL is to build a consumer-driven product that provides the most authoritative information on all species and the means to add, mine and analyze the information. The challenge is particularly acute, since biodiversity knowledge predominantly exists in collection institutions, especially natural history museums and herbaria. This knowledge includes studies on the evolution, speciation and distribution of life from around the globe.

Biodiversity Database Integration

Integrating diverse sources of digital information on biodiversity is a major challenge. Not only there are numerous disparate data providers each with their own specific user communities but also the information interested among the users is diverse, which includes taxonomic names and concepts, specimens in museum collections, scientific publications, genomic and phenotypic data and images [39]. Of course, the problem posed by integration is not unique to biodiversity informatics — the wider bioinformatics community is keenly aware of this challenge [59]. However, most bioinformatics integration efforts link together relatively few databases built upon similar data (e.g., macromolecular sequences and their annotations). At the time of writing the Global Biodiversity Information Facility (GBIF: http://www.gbif.org) lists some 217 different biodiversity data providers, serving a total of 145 660 886 records, mostly (but not limited to) museum specimens. The Catalogue of Life (http://www.catalogueoflife.org) contains over a million names contributed by 47 sources. If we add the contents of the 'traditional' bioinformatics databases GenBank and PubMed, along with the taxonomic literature accumulated since 1758 (much of it yet to be digitized), then the magnitude of the challenge facing biodiversity informatics becomes readily apparent.

Biodiversity database integration requires globally unique shared identifiers, in other words a way to determine whether two items of data refers to the same entity or not, this was reviewed [9]. The obvious candidate for shared identifier is the taxonomic name of an organism [52]. It is a natural link between different

Biodiversity Informatics

databases that store information about that organism [41] and the basis of current tools that aggregate information from multiple sources. However, taxonomic names have serious limitations as identifiers in databases [29] as mentioned earlier they are not completely stable and globally unique. Names may change due to taxonomic revision, there may be multiple names (Synonyms) for the same taxon, and the same name may refer to different taxa (homonyms) [39]. As reviewed and discussed [9], the key prerequisite for integrating biological information from diverse sources is the use of globally unique identifiers to consistently identify objects. To solve this problem the bioinformatics community has focused primarily on three alternatives, HTTP URIs (Uniform Resource Identifiers), DOIs (Digital Object Identifiers), and LSIds (Life Science Identifiers) (http://wiki.tdwg.org/GUID.

Biodiversity Information on the Web

Today a large number of web URLs are providing Biodiversity information, which can be categorized into five major groups as follows:-

- Global Databases, e.g., SPECIES 2000 (www.sp 2000.org)

- Geographical, Regional and National databases, e.g., Integrated Taxonomic Information System (IT IS) (www.itis.usda.gov)

- Database for specific taxonomic group, e.g., ICTVdB (Virus Database) (www.ncbi.nlm.nih.gov/ictvdb)

- Government, Scientific and Advocacy organization promoting biodiversity study, e.g., DIVERSITAS (www.diversitas-international.org/index.html)

- Other resources, e.g., Biodiversity and Biological Collection Web Server (http://biodiversity.uno.edu)

A comprehensive detail of Biodiversity information resources and related web links are available on the URL www.sciencemag.org.

Applications of Biodiversity Informatics

- The complex and voluminous data of biodiversity can be digitized for easy accession, analysis and interpretation.

- It makes easy survey, documentation and measurement of biodiversity data.

- Based on the available data, future biodiversity of a particular area c a n be predicted and model can be formulated by computational methods, thereby appropriate measures can be taken for its conservation and sustainable utilisation.

- It helps to predict species invasions using ecological niche modeling.
- The electronic information may serve as the raw material for augmenting future developments in all areas of biology.
- The digital databases can easily provide the current status of the biodiversity of a particular area.
- The biodiversity extinction rate can be easily documented and theoretical studies and modeling can be formulated for its conservation on priority basis.
- The computational analysis makes easy understanding of the phylogenetic relationship among the species/individuals, causes of range limitation or species, reaction to changing environment.
- Through internet biodiversity databases can be linked together and the information can be shared.
- The researchers can easily identify the priority materials for their studies.
- The potential indigenous material can be easily identified for biotechnological intervention.
- Assessment of the pest damage to crops and evaluation of the possible routes for invasive species or diseases, can be done.

DNA Barcoding

Another recent development in molecular biology and taxonomy in conjunction with bioinformatics is the DNA taxonomy or DNA barcoding [22,63,42], which has made tremendous impacts on Biodiversity exploration and data documentation. It is a well acknowledged fact that traditional taxonomic practices are insufficient to cope up with the growing demand for accurate and accessible taxonomic information. Generally, for species identification taxonomist depends on morphological characteristic features. This method has four significant limitations. First, both phenotypic plasticity and genetic variability in the characters employed for species recognition can lead to incorrect identification. Secondly, this approach overlooks morphologically cryptic taxa, which are common in many groups [25,30]. Third, since morphological keys are often effective only for a particular life stage or gender, many individuals cannot be identified. Fourth, although modern interactive versions represent a major advance, the use of taxonomy keys often demands

Biodiversity Informatics

high level of expertise for diagnosis and misdiagnoses are common. The limitations inherent in morphology-based identification system and dwindling pool of taxonomists signal the need for a new approach for taxon recognition. Not only the inherent characters of many species which flower once in their life time like bamboos, and some other organisms which flower for a day or hours make it difficult for identification but also guide the scientists to opt for a new and advanced approach.

It is also striking to note that so far approximately 1.8 million species has been described by taxonomists, since the introduction of binomial nomenclature system by Linnaeus in 1753, but total number of species on earth is unknown and expected to range between 10 – 100 million. In fact, since few taxonomists can critically identify more than 0.01% of the estimated 10-15 million species [21], a community of 15,000 taxonomists will be required, in perpetuity, to identify life if our reliance on morphological diagnosis is to be sustained.

Several techniques that promised to separate species using simple molecular biology have come and gone over the past 30 years. One such approach focused on allozymes, enzymes that vary only slightly among species. However, researchers eventually found that the allozyme technique could not distinguish between closely related species. Because allozyme data are subject to interpretation, individual scientists could read the same results in different ways, effectively blurring species boundaries. In 2003, Paul D. N. Hebert, a population geneticist at Guelph University, Ontario, Canada proposed the DNA barcoding as a system to aid species recognition and identification through the characterisation of a standard gene region across all organisms. As the Universal Product Code (UPC) that identifies items sold in most countries today, a sliver of DNA could identify every living species on Earth [23]. They proposed that the mitochondrial DNA is the potential source of genome for DNA barcoding. This is based on the simple concept that most eukaryotic cells contain mitochondria. Mitochondrial DNA (mtDNA) has a relatively fast mutation rate, this lead mtDNA to have a significant variance between species and a comparatively small variance within species. A 684 bp region of the mitochondrial gene, known as cytochrome c oxidase I (COI) has been proposed as potential barcode for animal life forms. But the CO1 gene has proved to be of no use in plants because of the intrinsically lower rates of sequence evolution in plant genomes. In plants the Internal Transcribed Spacer (ITS) region of nuclear ribosomal cistron is the most commonly sequenced locus [1]. This region has shown broad utility across photosynthetic eukaryotes (with the exception of ferns and fungi) and has been suggested as a possible plant barcode locus [61]. For phylogenetic investigation, the plastid genome has been more readily exploited than the nuclear genome, and may offer for plant barcoding what the mitochondrial genome does for animals. It is uni parently inherited, non-recombining and is found structurally stable. To identify a locus that is universal and readily sequenced and has sufficiently high sequence divergence at the species level in plants, The Consortium for the Barcode of Life - Plant working group - take the responsibility to find out uniform barcode region among plant species under the leadership of Kew Gardens with the participation of more than ten international organizations. Consequently The Consortium for the Barcode of Life (CBOL) received two well-documented proposals for the barcode regions for land plants. The first proposal *rbcL* and *matK* and the second proposal a three-locus barcode consisting of *rbcL*, *matK*, and *trnH-psbA*. After considering various factors such as *Universality* (Is it easy to obtain sequences?), *Sequence quality* (Are the sequence traces unambiguous and are bidirectional reads obtained?) and *Discriminatory power* (Is it good at telling species apart?) the Executive Committee of CBOL in 2009 recommended *rbcL* and *matK* as the barcode regions for land plants based on the condition that it will be reviewed (*rbcL+matK* barcode) after 18 months.

The important steps of DNA barcoding protocol is depicted below:

Specimen collection
↓
Genomic DNA isolation and purification
↓
Isolation of barcode region from genomic DNA using primer
↓
PCR amplification of the barcode region
↓
PCR product cleanup and sequence analysis using a DNA sequencer
↓
Sequence editing using software package and deposit the sequence data and details of species information on the web enabled database for further reference

This technique has already led to the discovery of several new species of bird, butterflies and fishes. Thousands of species have already been coded using this simple and novel technique. Considering the success and the potential the Biodiversity Institute of Ontario at the University of Guelph, Canada has been established (2007) at a total cost of $ 4.2 million. It's the first of its kind in the world and houses the Canadian Centre for Barcoding. This Institute when becomes fully operational can identify ca. 50000 samples annually as against 1000 species using conventional methods. This technique of barcoding could allow scientists to identify an estimated 10 million species in the next 20 years. In India, Department of Biotechnology has recognized its importance and already initiated preliminary steps for barcoding of selected taxa. Kerala State Council for Science Technology and Environment (KSCSTE) has established

Biodiversity Informatics

a DNA barcoding centre at JNTBGRI Puthenthope campus and initiated R & D on this line.

The appeal and utility of the barcoding system is through the development of a comprehensive and rigorous database that is widely accessible. The goal of this method is that anyone, anywhere, anytime be able to identify quickly and accurately the species of a specimen whatever its condition.

Conclusion

Biodiversity informatics is starting to gain momentum as a scientific discipline. Biodiversity informatics is not meant to replace existing biological disciplines any more than bioinformatics is intended to replace 'wet-bench' work. Instead, biodiversity informatics aims to bring together relevant information into a form that can be used by biodiversity researchers. Furthermore, it strives to develop resources and services that may further the initiatives that can benefit from the use of biological data—from basic biology to biomedical science to general knowledge. The range of available data types and formats for biodiversity knowledge is humbling—e.g., climate, geographic and disease knowledge. Recent advanced technology of DNA barcoding fastens the identification of species through DNA tags, will help in documenting the biodiversity using bioinformatics tools.

References

1. Alvarez, I. and Wendel, J.F. (2003) Ribosomal ITS sequences and plant phylogenetic inference. Mol. Phylogenet. Evol., 29:417-434.
2. Beard, C.B., Pye, G., Steurer, F.J., Salinas, Y., Campman, R., Peterson, A.T., Ramsey, J.M., Wirtz, R.A. and Robinson, L.E. (2002) Chagas disease in a domestic transmission cycle in southern Texas, USA. Emerging Inf. Dis., 9:103-105.
3. Berendsohn, W.G. (2001) Biodiversity Informatics, http://www.bgbm.org/BioDivInf/def-e.htm
4. Bisby, F.A. (2000) The quiet revolution: biodiversity informatics and the Internet. Science, 289: 2309–2312.
5. Bodenreider, O. (2004) The unified medical language system (UMLS): integrating biomedical terminology. Nucleic Acids Res., 32: D267–270.
6. Boughey, A.S., Bridges, K.W. and Lkeda, A.G. (1968) An automated biological identification key. Museum of Systematic Biology. University of California. Irvine Research Series No 2. pp 1-36.
7. Causey, D., Janzen, D.H., Peterson, A.T., Vieglais, D., Krishtalka, L., Beach, J.H. and Wiley, E.O. (2004) Museum collections and taxonomy. Science, 305:1106–1107.
8. Chung, S.Y. and Wooley, J.C. (2003) Challenges Faced in the Integration of Biological Information. In: Lacroix Z, Critchlow T, (eds). Bioinformatics: Managing Scientific Data. San Francisco: Morgan Kaufmann, pp 11–34.
9. Clark, T., Martin, S. and Liefeld, T. (2004) Globally distributed object identification for biological knowledge bases. Brief Bioinfor, 50:59–70.

336   Agriculture Bioinformatics

10. Conservation News 2, (2): 61-64.
11. Costa, J., Peterson, A.T. and Beard, C.B. (2002) Ecological niche modeling and differentiation of populations of *Triatoma brasiliensis* Neiva, 1911, the most important Chagas disease vector in northeastern Brazil (Hemiptera, Reduviidae, Triatominae). Am. J. Trop. Med. Hyg., 67:516-520.
12. Cotter, G.A. and Bauldock, B.T. (2000) Biodiversity informaics infrastructure: an information common for biodiversity community. Proceedings of 26th International Conference on very Large Data bases, Cairo, Egypt., pp 201-204.
13. Dobson, A.P. (1995) Conservation and Biodiversity. Scientific American Library. New York.
14. Edwards, J.L., Lane, M.A. and Nielsen, E.S. (2000) Interoperability of biodiversity databases: biodiversity information on every desktop. Science, 289: 2312-2314.
15. Euzeby, J.P. (2006) List of Prokaryotic Names (http://www. bacterio.cict.fr/).
16. Gewin, V. (2002) All living things online. Nature, 418:362–363.
17. Godfray, C. (2002) Challenges for taxonomy. Nature, 417: 17– 19.
18. Greuter, W., McNeill, J., Barrie, F.R., Burdet, H.M., Demoulin, V., Filgueiras, T.S., Nicolson, D.H., Silva, P.C., Skog, J.E., Trehane, P., Turland, N.J. and Hawksworth, D.L. (2000) International Code of Botanical Nomenclature (St Louis Code). Ko"nigstein: Koeltz Scientific Books.
19. Guralnick, R.P., Wieczorek, J., Beaman, R., Hijmans, R.J. and the BioGeomancer Working Group (2006) BioGeomancer: Automated Georeferencing to Map the World's Biodiversity Data. PLoS Biol 4(11): e381. doi:10.1371/journal.pbio.0040381.
20. Hawksworth, D.L. (1995) Botanic Gardens Conservation International: A Handbook for Botanic Gardens. www.cbd.int/doc/case-studies/abs/cs-abs-rio-bot-gar.pdf.
21. Hawksworth, D.L. and Kalin-Arroyo, M.T. (1995) Magnitude and distribution of biodiversity. In *Global biodiversity assessment* (ed. V H. Heywood), Cambridge University Press. pp.107–191.
22. Hebert, P.D.N., Cywinska, A., Ball, S.L. and deWaard, J.R. (2003) Biological identifications through DNA barcodes. Proc. R. Soc. Lond. Biol. Sci., 270:313–322.
23. Heywood, V.H. and Watson, R.T. (1995) Global biodiversity assessment. Cambridge University Press, Cambridge. pp. 1–1152.
24. Hull, D.L. (1988) Science as a process. University of Chicago Press, Chicago, pp 586.
25. Jarman, S.N. and Elliott, N.G. (2000) DNA evidence for more DNA-based identifications PDN Hebert and others 02PB0653.9phological and cryptic Cenozoic speciations in the Anaspididae, 'living fossils' from the Triassic. J. Evol. Biol., 13:624–633.
26. Jenkins, M. (2003) Prospects for biodiversity. Science, 302: 1175–1177.
27. Johnson, N.F.(2007) Biodiversity informatics. Annu. Rev. Entomol., 52:421–38.
28. Kellings, S. (2008) Significance of organism observations: data discovery and access in biodiversity research. Report for the Global Biodiversity Information Facility, Copenhagen.
29. Kennedy, J., Hyam, R., Kukla, R. and Paterson, T. (2006) Standard data model representation for taxonomic information. OMICS, 10: 220-230.
30. Knowlton, N. (1993) Sibling species in the sea. A. Rev. Ecol. Syst., 24:189–216.
31. Krishtalka, L. and Humphrey, P.S. (2000) Can natural history museums capture the future? Bioscience, 50: 611–617.
32. Krishtalka, L., Peterson, A.T., Vieglais, D.A., Beach, J.H. and Wiley, E.O. (2002) The Green Internet: a tool for conservation science. In: Conservation in the Internet age: strategic threats and opportunities (Ed. JN Levitt), Washington, DC: Island Press, pp. 143–164.
33. Lapage, S.P., Bascomb, S., Willcox, W.R. and Curtis, M.A. (1970) Computer identification of bacteria. In: Baillie A and Gilbert RJ (Eds.) Automation, Machanization and data handling in Microbiology. Academic Press, London. pp 1-22.

Biodiversity Informatics

34. Leadley, E., Wyse Jackson, D., and Wyse Jackson, P. (1993) Developing the BGCI database of botanic gardens and their collections worldwide. Botanic Gardens. Conservation News, 2 (2): 61-64.

35. Loreau, M., Montreal, Q., Arroyo, M.T.K., Babin, D., Barbault, R., Donoghue, M., Gadgil, M., Häuser, C., Heip, C., Larigauderie, A., Ma, K., Mace, G., Mooney, H.A., Perrings, C., Raven, P., Sarukhan, Schei P., Scholes, R.J. and Watson, R.T. (2006) Diversity without representation. Nature, 442: 245–246.

36. Myers, N. (1993) Tropical forests: the main deforestation fronts. Environmental Conservation, 20:9–16.

37. Oliver, I., Pik, A., Britton, D., Dangerfield, J.M., Colwell, R.K. and Beattie, A.J. (2000) Virtual biodiversity assessment systems. Bioscience, 50: 441–449.

38. Page, R.D.M. (2005) A taxonomic search engine: federating taxonomic databases using web services. BMC Bioinformatics, 6:48.

39. Page, R.D.M. (2008) Biodiversity Informatics: the challenge of linking data and the role of shared identifiers. Brief. Bioinformatics, 9: 345-354.

40. Pankhurst, R.J. (1991) Practical Taxonomy computing. Cambridge University Press, Cambridge. pp. 202.

41. Patterson, D.J., Remsen, D., Marino, W.A. and Norton, C. (2006) Taxonomic indexing - extending the role of taxonomy. Syst. Biol., 55:367-73.

42. Pennisi, E. (2003) Modernizing the tree of life. Science., 300:1692–1697.

43. Peterson A.T. (2003) Predicting the geography of species' invasions via ecological niche modeling. Q. Rev. Biol., 78:419-433.

44. Peterson, A.T. (2003) Projected climate change effects on Rocky Mountain and Great Plains birds: generalities of biodiversity consequences. Global Change Biol., 9:647–655.

45. Peterson, A.T. and Shaw, J.J. (2003) *Lutzomyia* vectors for cutaneous leishmaniasis in southern Brazil: Ecological niche models, predicted geographic distributions, and climate change effects. Int. J. Parasitol., 33:919-931.

46. Peterson, A.T. and Vieglais, D.A. (2001) Predicting species invasions using ecological niche modeling: new approaches from bioinformatics attack a pressing problem. BioSci, 51:363-371.

47. Peterson, A.T., Ortega-Huerta, M.A., Bartley, J., Sanchez-Cordero, V., Sobero´n, J., Buddemeier, R.H. and Stockwell, D.R.B. (2002a) Future projections for Mexican faunas under global climate change scenarios. Nature, 416: 626–629.

48. Peterson, A.T., Stockwell, D.R.B. and Kluza, D.A. (2002b) Distributional prediction based on ecological niche modeling of primary occurrence data. In: Predicting species occurrences: issues of accuracy and scale (ed. Scott JM, Heglund P and Morrison ML),Washington, DC: Island Press. pp. 617–623.

49. Polaszek, A. (2005) A universal register for animal names. Nature, 437: 477.

50. Raxworthy, C.J., EMartínez-Meyer, N., Horning, Nussbaum, R.A., Schneider, G.E., Ortega-Huerta, M.A. and Peterson, A.T. (2003) Predicting distributions of known and unknown reptile species in Madagascar. Nature, 426:837-841.

51. Ride, W.D.L., Cogger, H.G., Dupuis, C., Kraus, O., Minelli, A., Thompson, F.C. and Tubbs, P.K. (1999) International Code of Zoological Nomenclature. International Trust for Zoological Nomenclature. London, UK.

52. Sarkar, I.N. (2007) Biodiversity informatics: organizing and linking information across the spectrum of life. Brief. in Bioinformatics, 8(5):347–357.

53. Schatz, B., Mischo, W., Cole, T., Bishop, A., Harum, S., Johnson, E., Neumann, L., Chen H. and Dorbin, N.G. (1999) "Federated Search of Scientific Literature," Computer, 32(2): 51-59. doi:10.1109/2.745720.

338   Agriculture Bioinformatics

54. Singh, J.S. (2002) The biodiversity crisis: A multifaceted review. Current Science, 82(6):638-647.
55. Siqueira, M.F. de and Peterson, A.T. (2003) Consequences of global climate change for geographic distributions of cerrado tree species. *Biota Neotropica* 3(2). Electronic journal, http://www.biotaneotropica.org.br/.
56. Sneath, P.H.A. and Sokal, R.R. (1973) Numerical Taxonomy, WH Freeman San Francisco, pp. 359.
57. Sneath, P.H.A. (1992) 'International Code of Nomenclature for Bacteria'. Washington, DC: American Society for Microbiology.
58. Soberon, J. and Peterson, A.T. (2004) Biodiversity informatics: managing and applying primary biodiversity data. Philos. Trans. R. Soc. London, B 359: 689-98.
59. Stein, L. (2003). Integrating biological databases. Nat. Rev. Genet., 4:337–45.
60. Stockwell, D.R.B. and Peters, D.P. (1999) The GARP modelling system: Problems and solutions to automated spatial prediction. International Journal of Geographical Information Science, 13:2 143-158.
61. Stoeckle, M. (2003) Taxonomy, DNA, and the bar code of life. BioScience 53:796-797.
62. Suarez, A.V., Tsutsui, N. (2004) The value of museum collections for research and society. BioScience, 54:66–74.
63. Tautz, D., Arctander, P., Inelli, A., Thomas, R.H. and Vogler, A.P. (2003) A plea for DNA taxonomy. Trends Ecol. Evol., 18: 70–74.
64. Thomas, C.D., Cameron, A., Green, R.E., Bakkenes, M., Beaumont, L.J., Collingham, Y.C., Erasmus, B.F.N., de Siqueira, M.F., Grainger, A., Hannah, L., Hughes, L., Huntley, B., van Jaarsveld, A.S., Midgley, G.F., Miles, L., Ortega-Huerta, M.A., Peterson, A.T., Phillips, O.L. and Williams, S.E. (2004) Extinction risk from climate change. Nature, 427:145-148.
65. UNEP (1993) Guidelines for country studies on Biological Diversity, United Nations Environment Programme, Nairobi, Kenya.
66. Vernon, K. (1988) The founding of numerical taxonomy. British J. History of Sci., 21: 143-159.
67. Wilson, E.O. (2005) Systematics and the future of biology. Proc. Natl. Acad Sci., USA, 102 (Suppl 1):6520–1.
68. Wilson, E.O. (2003) The encyclopedia of life. Trends Ecol. Evol., 18:77–80.

Chapter – 16

Omics: What Next?

Prashanth Suravajhala and Rajib Bandopadhyay

Abstract

There has been consistent narrowing down of gap in our knowledge about various 'omics' as the omics' based disciplines are filled with valuable information emerging from research. During the last two decades, the practice of genetics has not only changed the broad horizons of medicine but also found its niche through "high-dimensional biology" there by allowing geneticists to reframe the good-old definition of gene from a locatable region of genome sequence involved in heritability to a locatable region of genome sequence associated with a functional region or a functor. Various other scientists viewed genetics "as the study of single genes and their effects" and genomics as "the study not just of single genes, but of the functions and interactions of all the genes in the genome." We will cover the qualitative difference that bridges various disciplines casing Omics.

Introduction

Omics is a general term for broad discipline of science of interactions and expressions taking place in living organisms. The central dogma of molecular biology has given a clear dimension for defining genes: a DNA transcript to give mRNA and mRNA translating to give a bigger macromolecule in the form of amino acids. Summation of amino acids is protein. Genes, the bits of DNA and the genetic instructions for making proteins are located on chromosomes. Chromosomes in turn are present in the nucleus of every cell in a eukaryote. Every gene is composed of two inter-linked strands of DNA. Genomics usually involves sequencing all the bases in an organism's DNA, each A, T, C and G (adenine, thymine, cytosine and guanine respectively). Once a genome is sequenced it is used for studies of function of numerous genes (functional genomics), or comparison of the genes in one organism with those of another (comparative genomics). Genomics can be seen as an entry point for looking at the other 'omics' sciences. The information in the genes of an organism, its genotype, is largely responsible for the final physical makeup of the organism-scientifically referred to as the phenotype. The environment also has some influence on the phenotype – so organisms are often more than the sum of their parts. The DNA in the genome is only one aspect of the complex mechanism that keeps an organism running – so decoding the DNA is one step towards understanding, but by itself it does not specify everything that happens within the organism. The building blocks that make up each strand of DNA are called nucleotides. Nucleotides are made up of sugar, phosphate groups and nitrogen-containing rings better known as bases. The four different types of nucleotides found in DNA differ from each other by the bases that are attached to their sugar and phosphate groups: guanine, adenine, thymine, and cytosine as discussed. Different genes vary in the sequence or pattern of these nucleotides along the DNA strands. During the process of gene making up a protein, several metabolites, enzymes, RNA molecules are processed. The study of many such entities could be attributed to genomics, transcriptomics and proteomics. The word 'Omy' means many or "money". For example, gen'ome' could be defined as study of number of genes, proteomes-the study of many number of proteins in a cell while their respective disciplines are called as genomics and proteomics. However, many people have recently argued that, the concept of the gene resulting in omics has begun to outlive its usefulness. People tried to reach consensus in defining gene proposing an alternative intending to define gene based on a richer explanation: Genetic functor, or genitor, is a sweeping extension of the classical genotype/phenotype paradigm that describes the 'functional' gene [8].

Origin of Omics

We are familiar with the word 'chromosome' and then also 'genome'. As genome refers to the complete genetic makeup of an organism, some people have made the inference that there exists some root, *"-ome-", of Greek origin referring to wholeness or to completion, but such root is unknown to most or all scholars [6]. The complete sequencing of the human genome and later on other genome have shown us a new era of systems biology referred to as omics. Bioinformaticians and molecular biologists started to apply the "-ome" suffix widely. The term omics refers to the comprehensive analysis of biological systems. The three most common omics are 'Gen-omics', 'Prote-omics' and 'Transcript-omics'. Except these three omics, bioinformatics and biology researchers coined and used various omics such as phenomics, physiomics, metabolomics, lipidomics, glycomics, interactomics, cellomics and so on. Researchers are taking up the omes and omics very rapidly as shown in the use of the terms in Pubmed in the last decade *(Refer 'Omics, Bioinformatics and Systems biology' paragraph)*. A variety of omics subdisciplines have begun to emerge, each with their own set of instruments, techniques, reagents and software. The omics technology has driven in new areas of research consists of DNA and protein microarrays, mass spectrometry and a number of other instruments that enable high-throughput analyses.

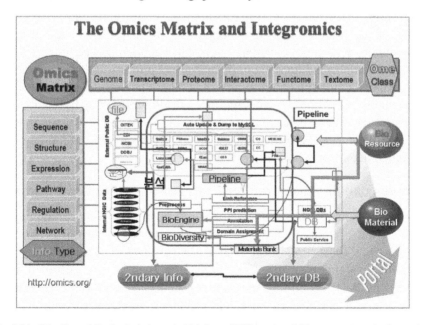

Fig. 16.1: Pipeline of Omics in integrated biology: Different entities are networked together in present day integrated biology ~ integromics (adapted from http://www.omics.org, Bio license)

Reductionist Approach vs. Dynamic Approach in defining 'Omics'

The simple definition is reductionistic approach while a bit elaborate hypothesis to produce knowledge en masse is defined through dynamism. Several biologists and multi- disciplinarians felt that the approach of defining these terms with suffix ome is relevant. This is what they mean by reductionism approach. Defining classically, some biologists try to understand how several genes make up the proteins and how these metabolites are processed in a living system, how they interact with each other [11]. Certainly, this view could be compared to localists and globalists or generalists view [1]. The figure 1 gives a noticeable idea of quantitative nature of important omics' networked together. Admittedly, the fields of omics have revolutionized biomedicine and by far means there needs to be a focus on change in defining these omics'

While genomics forms a main hierarchy of classification, there are many other Omics' which fall under a clad of primary (gen)omics'. For example, functional genomics, comparative genomics, phylogenomics etc as we discussed earlier is a part of genomics. With over 874 microbial genomes sequenced *till date* and still the number increasing, metagenomics, aims to access the genomic potential of an environmental sample. It would answer to some of the questions posed like: how to investigate the normal flora of mammals, analysis of ancient genomes, and exploration of the distribution of novel pathways. In addition, the development of new bioinformatics approaches and tools allow innovative mining of these data. This environmental 'omics' bridges integration of metagenomics with complementary approaches in microbial ecology. The use of genomics, metagenomics and high throughput proteomics have emerged in the last years facilitating above approaches. These powerful approaches offer the possibility of exciting new findings that will allow analyzing the community as a microbial system, determining the extent to which each of the individual participants contributes to the process, how they evolve in time to keep the conglomerate and environment healthy [36].

Genomes can be Compared for Proteomes and for Other 'Omes'

Bioinformatics has enabled *All-against-All* comparison distinguishing unique proteins from proteins that are members of families made up of paralogs resulting from gene duplication events. The last two decades have seen an avalanche in databases and algorithms like BLAST that allowed such comparisons. Post genomic era, the genomes sequenced so far would essentially cover the future of omics in them. The genomes enable predicting not only evolutionary relationships but also make use of different approaches used in identifying function of genes. This functional genomics is the cause of understanding how proteins interact with each other, how they network in the living organism. The gene or protein function could be ascertained based on physiological characterization or if the two proteins are known to be physically interacting with each other or virtually interacting with each other. The Systems Biology approaches in present day bioinformatics has brought in a special emphasis on associated based networks in the form of virtual interactions thereby making up the possibility of phenome-genome networks grow bigger [11]. Ultimately it makes sense when such interactions bring out a function and finding a candidate for a disease.

Omics, Bioinformatics and Systems Biology

The increase of GenBank accessions resulted not only in number of genes identified but also the number of citations these accessions refer to. There are more than 23064 articles available in PubMed when a keyword 'Genomics' is made. While various databases and terms have been defined, several omics' are being reported at http://www.omics.org

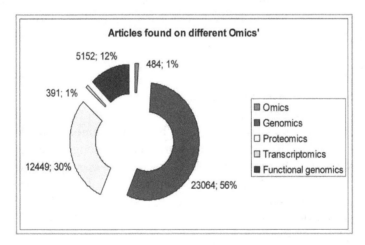

Fig. 16.2: Articles found on different Omics' in PubMed as on December 13, 2007.

The last 10 years have not only seen the rise of bioinformatics producing an unprecedented amount of genome-scale data from many organisms but also the wet-lab research community has been successful in exploring these data using bioinformatics, although many challenges still persist. One of them is the effective integration of such data sets directly into approaches based on mathematical modelling of biological systems. This is where Systems Biology has bud resulting in top-down and bottom-up approaches. The advent of functional genomics has enabled the molecular biosciences to come a long way towards characterizing the molecular constituents of life. Yet, the challenge for biology overall is to understand how organisms function. By discovering how function arises in dynamic interactions, systems biology is everywhere addressing the missing links between molecules and physiology. Top-down systems biology identifies molecular interaction networks on the basis of correlated molecular behaviour observed in genome-wide "omics" studies. On the other hand, bottom-up systems biology examines the mechanisms through which functional properties arise in the interactions of known components [4]. Applications in cancer are a good example to counteract these two major types of complementary strategies [32]. Several web-based repositories have been established to store protein and peptide identifications derived from MS data, and a similar number of peptide identification software pipelines have emerged to deliver identifications to these repositories. Integrated data analysis is introduced as the intermediate level of a systems biology approach and as a supplementary to bioinformatics to analyse different 'omics' datasets, i.e., genome-wide measurements of transcripts, protein levels or protein-protein interactions, and metabolite levels aiming at generating a coherent understanding of biological function[31]. Furthermore, existing and potential problems/solutions like de facto experimental and bioinformatics challenges would hold prospective in the near future.

Omics and Wet-lab

Fundamental biological processes can now be studied by applying the full range of omics technologies (genomics, transcriptomics, proteomics, metabolomics, and beyond) to the same biological sample. Clearly, it would be desirable if the concept of sample were shared among these technologies, especially as up until the time a biological sample is prepared for use in a specific omics assay, its description is inherently technology independent. Wide array of assays including high-throughput methods like MS/MS, Yeast two hybrids and pull down assays are preferentially used to navigate them. However, the compulsion for accurate analyses of all these high through put methods is to remove redundant and false-positive data. Redundancy of data has been a biggest threat for causing errors in data usage. Sharing a common information representation would encourage data sharing leading to decrease of redundant data and the potential for error. This would result in a significant degree of harmonization across different omics data standardization activities, a task that is critical if we are to integrate data from these different data sources [19]. The bioinformatics applied to –omics' are varied and particularly noteworthy or characteristic for proteomics research, for example in 2DE analysis or mass spectrometry. Another important task of bioinformatics is the prediction of functional properties. Several ontology based functional network has been built recently through vast amount of databases.

Metabolomics and Metabonomics: What's the Difference?

Metabolomics has come into sight as one of the newest "omics" science with dynamic portrait of the metabolic status of living systems. The analysis of the metabolome is particularly challenging due to the diverse chemical nature of metabolites. Metabolites are the product of the interaction of the system's genome with its environment and are not merely the end product of gene expression but may also form part of the regulatory system in an integrated manner. With rise of integrated disciplines, metabolomics also has arisen to its distinct status. Metabolomics has its roots in early metabolite profiling studies but is now a rapidly expanding area of scientific research in its own right. Metabolomics (or metabonomics) has been labelled one of the new "omics", joining genomics, transcriptomics, and proteomics as a science employed towards the understanding of global systems biology [27]. The metabolomic tools aim to fill the gap between genotype and phenotype permitting simultaneous monitoring molecules in a living system. The smartness of using metabolic information could be applied in translating into diagnostic tests where in they might have the potential to impact on clinical practice, and might lead to the supplementation of traditional biomarkers of cellular integrity, cell and tissue homeostasis, and morphological alterations that result from cell damage or death [5]. Some of the applications of metabolomics include facilitating metabolic engineering to optimise mircoorganisms for biotechnology even as it spreads to the investigation of biotransformations and cell culture. Metabolomics serves not only as a source of qualitative but also quantitative data of intra-cellular metabolites essential for the model-based description of the metabolic network operating under *in vivo* conditions. To collect reliable metabolome data sets, culture and sampling conditions, as well as the cells' metabolic state are crucial. Hence, application of biochemical engineering principles and method standardisation efforts become important. Together with the other more established omics technologies, metabolomics will strengthen its claim to contribute to the detailed understanding of the *in vivo* function of gene products, biochemical and regulatory networks and, even more ambitious, the mathematical description and simulation of the whole cell in the systems biology approach. This concurrent knowledge will allow the construction of designer organisms for process application using biotransformation and fermentative approaches making effective use of single enzymes, whole microbial and even higher cells [20]. While Metabonomics is considered measuring plural or multiparametric response of living systems to genetic modification, there is a consistent debate of synonymising it with metabolomics. Admittedly, there is a consensus of the former being associated with NMR while the latter with mass spectroscopy. This part of microbial transformation has lead to study of metagenomics. Several standards have been recently debated while setting standards for these two Meta-omics' [21].

Mitochondriomics

Mitochondria are semiautonomous organelles, presumed to be the evolutionary product of a symbiosis between a eukaryote and a prokaryote. The organelle is present in almost all eukaryotic cells in an extent from 10^3-10^4 copies. The main function of mitochondria is production of ATP by oxidative phosphorylation and its involvement in apoptosis. The organelles contain almost exclusively maternally inherited mtDNA, and they have specific systems for transcription, translation and replication of mtDNA. Mitochondrial dysfunction has been correlated with mitochondrial diseases where the clinical pathologies are believed to include infertility, diabetes, blindness, deafness, stroke, migraine, heart, kidney and liver diseases [25].

Recently cancer was added to this list when investigations into human cancer cells from breast, bladder, neck, and lung, revealed a high occurrence of mutations in mtDNA. With the emerging understanding of the role of mitochondria in a vast array of pathologies, research of mitochondria and mitochondrial dysfunction have in the last decade yielded a huge amount of data in form of publications and databases. Nevertheless, the field of mitochondrial research is still far from exhausted with many unknown factors yet to be discovered. The recent identification of number of proteins targeting mitochondria has enabled immense interest to understand the function of some genes unnoticed in mitochondrion. With only 13 proteins sitting inside mitochondria through oxidative phosphorylation, and over 1500 estimated proteins targeting this tiny organelle, identifying complete protein repertoire in this machinery could decipher the biology behind mitochondria or what makes us breathe.

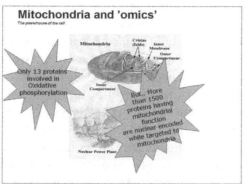

Fig. 16.3: Mitochondria and omics
[Picture taken from http://www.puc.edu/Faculty/Gilbert_Muth/art0072.jpg]

Omics: What Next? 347

As thousands of new genes are identified in genomics efforts, the rush is on to learn something about the functional roles of the proteins encoded by those genes. Clues to protein functions, activation states and protein-protein interactions have been revealed in focused studies of protein localization. Adding to the various high-throughput methods, technical breakthroughs such as GFP protein tagging and recombinase cloning systems, large-scale screens of protein localization are now being undertaken to understand the function of the proteins [23].

The application of high-throughput technologies to aging-related research has the potential to dramatically enhance our understanding of how long life is determined at a molecular level. The high-throughput technologies have not only enabled to understand how longevity could be determined but also put forth several clues in understanding neurological disorders like Schizophrenia, Alzheimer's etc. Genome-scale studies are being carried out in every major model system used for aging-related research, and new technologies are being developed to rapidly identify mutations or small-molecules that increase life span. While recently dog has also been known to play a role as a model system for cancer in drug discovery development research because of its similarities to human anatomy and physiology, it may prove invaluable in research and development on cancer drugs, because dogs naturally develop cancers that share many characteristics with human malignancies. This probably would extra polate genome-wide, cross comparison of organisms for finding function of more genes [16]. A meta-analysis of data derived from genome-wide studies of aging in simple eukaryotes will allow the identification of conserved determinants of longevity that can be tested in other mammals [14].

Interactomics

The mapping of protein-protein interactions is a key to understanding biological processes which led to a new facet of omics called interactomics [17]. Many technologies have been reported to map interactions widely applied in yeast. To date, the number of reported yeast protein interactions that have been truly validated by at least one other approach is low and the amount of throughput takes process. This is because of the false discovery rate of proteins interacting with their partners. With the advent of virtual interactions, the growth of false positives also increased thereby allowing the researchers to keep a track of finding these false positives through statistics. Any data set of interaction map is complex while tools to decipher true positives are being developed in the form of mark-up languages[16]. The mapping of human protein interaction networks is even more complicated. Thus, it has been suggested that it is unreasonable to try to map the human interactome; instead, interaction mapping in human cell lines should be focused along the lines of diseases or changes that can be associated with specific cells. [7]. This "-omics revolution" has forced us to re-evaluate our ability to acquire, measure, and handle large data sets. Omics platforms such as expression arrays, mass spectrometry and other high-throughput methods have enabled quantitation of proteins and metabolites derived from complex tissues. Applying systems biology, the integrated analysis of genetic, genomic, protein, metabolite, cellular, and pathway events are in flux and interdependent. With onset of various datasets, it has necessitated that the use of a variety of analytic platforms as well as biostatistics, bioinformatics, data integration, computational biology, modelling, and knowledge assembly protocols. Such sophisticated analyses have provided new insight into the understanding of disease processes through phenome-genome networks and interactomics studies [15]. In this regard, systems biology clubbed with interactomics, more appropriately considered as a process containing a series of modules, aims to provide tools and capabilities to carry out such tasks [18]. While protein analysis is a field of research with a long history, a recent development of a series of proteomics approaches including Mass spectrometry opened the door for a synergistic combination with genomic sequence analysis. There need a focus on aspects of genome-wide transcription control, regulomics in analogy to all the other -omics, and how a combination of MS-based proteomics with *in silico* regulomics analyses can produce synergistic effects in the quest to understand how cells function [38] Carrying this further, it has been suggested that the term "translatome" could be used to describe the members of the proteome weighted by their abundance, and the "functome" to describe all the functions carried out by these. However, there are still many difficulties resulting from the noisiness and complexity of the information. As discussed earlier, removing false positives could be enhanced using various tools to some degree. However, these can also be overcome through averaging with broad proteomic categories such as those implicit in functional and structural classifications [10].

What is Plant Omics?

Plant-omics is related to function and metabolism of plant traits while human genome sequencing project showed us the new path in plant genome sequencing. Recent EST analysis of the bean root nodule transcriptome yielded thousands of individual genes, many of which were involved in carbon and nitrogen metabolism within nodules [24]. Based on EST analysis using macroarrays, various metabolic routes were defined with higher expression in nodules than in other tissues, and key enzymes in nodule carbon and nitrogen metabolism were determined. Expression analysis of some of the transcripts for selected key enzymes in different plant tissues revealed that genes of N and C metabolism were expressed at variable levels in tissues other than nodules particularly pods. The data suggested that selected genes in metabolism may be co-regulated by similar genetic elements leading to coordinated expression. There are some progresses with metabolic flux analysis in plant omics. Single gene changes may alter metabolic flux, but flux analysis can be used to study multigene traits. Pioneering studies using stable and radio-labelled isotopes in conjunction with mass balance equations for modelling flux have contributed to rapid progress in metabolite flow and control analysis. Although not easily usable on complex systems, flux through multiple metabolic pathways in various cellular compartments can be described using metabolic flux analysis (MFA) based on metabolic map based models. This method and approach is essential to the integration of gene action and biochemical output. When applied to organs and tissues, MFA allows one to model the genetic changes that contribute to the chemical composition important to crop quality. The MFA does not need a precise knowledge of the rate constants of the various reactions, but helps in predicting metabolic flow according to environmental conditions. For example, MFA was used successfully on heterotrophic tissues such as maize root tips, fruits or germinating lettuce seeds [29].

Transgenic seed models have also been created to understand seed metabolism. The approaches range from biochemical and histological methods to high-throughput genomics. Seed development is genetically programmed but seed metabolites, beneath their nutritive role, also act as signals to regulate assimilate import and influence seed development [37]. Recently, unique research tools have been developed to investigate seed metabolism at spatial resolution by imaging metabolites directly in tissue sections. Transgenic seed models have been created with defined changes within the metabolism based on the current understanding of the mechanisms that regulate seed storage activity. Such models can improve our knowledge of pathway regulation and can achieve desirable changes in seed composition and yield. In biological systems a range of genetic and biological networks extends from cells to organs and plants. Understanding how these networks function and are regulated within a given environment to give rise to crop components remains crucial to agriculture. While integration of genomic science into crop physiology, biochemistry, agronomy and plant breeding is a priority for crop improvement, inter-disciplinary collaborations are required to maximise the integration and translation of genomic information into legume crops. The integration of knowledge arising from the different approaches towards applied uses (human or animal nutrition, plant breeding, agriculture, food technology, etc.) will result in a more comprehensive understanding of processes which limit plant productivity and will lead to better strategies for improvement of crops.

Arabidopsis as the Model Plant: Omics Perspective

Plant as a bioreactor gives us food, chemicals, pharmaceuticals and renewable sources of materials and energy. The growth processes are still poorly understood although some of the key factors for plant organ growth have already been identified. The circuitry that links the different levels of organisation (whole plant, organ, cell, molecular module and molecule) remains to be uncovered. To understand the above problem one attempt has recently been taken to characterise a multi-cellular system exhaustively at all relevant levels in Arabidopsis. One five year project entitled "Arabidopsis GROwth Network integrating OMICS technologies" (AGRON-OMICS) had been started in November 2006 with a total budget of • 12 million. The main research goals of the project are:

- To survey systematically what are the components controlling growth processes in plant cells (genome sequences, proteins, metabolites)

- To understand how they coordinate their action and

- To explain quantitative growth phenotypes at the molecular level.

Plant-Omics will not only have an impact on our understanding of biological processes, but the prospect of more accurately diagnosing and treating disease will soon become a reality and thus the study of plant omics will help to understand in better way about the plant-pathogen interactions. The study of plant omics will help us to protect plant diseases and increase production. Much more knowledge of gene function and its interactions with other is needed. Large number of data has been generated after genome sequencing of Arabidopsis, Rice using high-throughput technologies. Scientists are not only interested to study a single gene and its function but also are interested to study which genes, proteins and metabolites are being involved in plant growth and how they interact to result in morphogenesis and function of plant organs. There is still a long way for researchers to understand and utilize the molecular networks underlining the plant's response to heavy metals and plant pathogen. Tools such as *Omics Viewer*, a user data visualization and analysis tool, allows a list of genes, enzymes, or metabolites with experimental values to be printed as a full pathway map of the gene/protein of interest.

Omics: What Next?   351

Physical Mapping and Sequencing

Genome mapping

The ultimate goal of mapping is to identify the gene(s) responsible for a given phenotype (traits) or the mutation responsible for a specific variant or to assign the gene(s) to a particular chromosome. The basic principles of linkage mapping have been used by scientists for many years, to establish whether a particular disease is associated with a specific chromosomal region. The first loci were mapped in Duffy blood group in 1969 [26]. When a genome of animal or plant was first investigated, this map was non existent. The map improves with the scientific progress and is perfect when the genomic DNA sequencing of the species has been completed. During this process, and for the investigation of differences in strain, the fragments are identified by small tags. With the development of molecular makers (RFLP, RAPD, AFLP, SSR, STS, EST and SNP), high efficiency cloning vector and deletion lines (as in wheat) mapping become easy and accurate technique. There are two major types of genome mapping while here we will discus only physical mapping because it is considered to be more accurate than genetic maps.

- Genetic maps depict relative positions of loci based on the degree of recombination. This approach studies the inheritance/assortment of traits by genetic analysis. The unit of genetic map is centimorgan (cM).

- Physical maps show the actual (physical) distance between loci (in nucleotides) and the map unit is kilobases (Kb).

Physical mapping

Physical Mapping is the process of determining how DNA contained in a group of clones overlap without having to sequence the entire DNA in the clones. Once the map is determined, we can use the clones as a resource to efficiently contain stretches of genome in large quantity. Physical mapping of the genome recovers different levels. A broad definition would say that it consists of placing nucleotide sequences with respect to a DNA matrix. For instance, placing a gene responsible for a disease on the chromosome in which it is contained. More generally, physical mapping concerns to any nucleotide sequence (the probe) which positions onto a longer nucleotide sequence for which the target has to be known. Cutting DNA is performed by restriction enzymes. The resulting fragments are usually inserted into bacteria or other micro-organisms (or clones). This allows for their conservation and mass production of DNA. How are all these cloned fragments reorganized in the corresponding order on the chromosomes they come from? - is the role of physical mapping techniques to give the most precise answer to this question.

Different types of Physical Mapping of Genomes

- **Top down (restriction mapping) approach:** This approach locates the positions and distances between endonuclease recognition sites on a DNA molecule. In this method, restriction map of large DNA fragments or high molecular DNA is digested with a restriction enzyme having a low number of restriction sites (macrorestriction). With this approach, long-range restriction maps locate the positions of rare-cutting endonuclease recognition sites on a DNA molecule by pulse field gel electrophoresis (PFGE). It is a reductionist or top down approach because here we start with isolation of individual chromosomes or their large segments using hybrid cell lines of known chromosome composition. Then from the existing genetic maps of these isolated individual chromosomes (or segments), we use probes that serve as anchor points on the physical map. The gaps between these anchor points on the physical map are then filled by extensive southern analysis. Initially it was applied for yeast.

- **Bottom up (clone or contig maps) approach:** It consists of libraries of overlapping clones where the relationship of each clone to other clones is resolved.

- **Fluorescent In Situ Hybridization (FISH):** locates the position of a marker by hybridizing a labelled probe to intact chromosomes.

- **Optical maps:** Visually inspects and measures the positions of endonuclease recognition sites on a DNA molecule.

- **EST (Expressed Sequence Tags) maps:** Plot the location of transcribed sequences.

- **STS-content maps:** Consists primarily of random genomic Sequenced Tagged Sites (STS).

The importance of Physical Maps with respect to localization and isolation of genes leads to better understanding of genome organization and evolution. Of all the above techniques discussed, Fiber-FISH is an efficient technique for determining the physical size of gaps on molecular contig maps [30].

Advances and Future Perspectives

Personalized medicine in genomics: person-'omics'

Personalized medicine is the treatment best suited for an individual patient, based on genomics. With several 'omics' technologies being increasingly used for this purpose, personalized drug-discovery efforts are in progress in major therapeutic areas. As discussed, molecular biomarkers prove to be an important link between drug discovery efforts and diagnostics, driving multi-faceted fields like RNA interference and nano-biotechnology to be used for applications related to drug discovery. The limitations of various approaches to personalized medicine needs attention which could ultimately pave way for development of drugs and drug candidates [12]. Personal sequence of genomes would play a pivotal role in understanding what makes us human.

Oncoproteomics

Of late, another omics called oncoproteomics has emerged to provide insight into the disease's pathological mechanisms, discovery of biomarkers and diagnosis. Oncoproteomics is the application of proteomic technologies in cancer. With various high-throughput methods used to identify proteins, there are even methods through which proteins can be identified from the blood or directly from the tumor tissue by laser capture microdissection (LCM) and tissue microarrays. Nanobiotechnology has refined the use of proteomics resulting in nanoproteomics. However, oncoproteomics alone is not enough to provide a complete picture of cancer while it would definitely play an important role in diagnosis of cancer and personalized medicine. Examples of applications of oncoproteomics are given for cancers of various organs such as the brain, breast, colon and rectum, prostate, and leukaemia [13]. Based on its capacity to separate and identify proteins, including those with post-translational modifications, proteomics provides new ways to understand post-genomic events that contribute to transformation and to identify new therapeutic targets. Another challenge and a major perspective in the so-called omics technology are the tools in drug discovery. Currently, many tools are available to profile drug response. The novel discipline of systems biology has offered interpretative cues to the amount of data generated by these omics technologies [9]. While metabolome analysis has become very popular recently, and novel techniques for acquiring and analyzing metabolomics data continue to emerge that are useful for a variety of biological studies, there is much need to understand how metabonomics as a tool would play a role in drug-discovery processes. This is applied much in pharmaceutical industry and pediatric hospitals where it would detect and help for the treatment of new born babies with inborn errors of metabolism. In this process, it was also proposed that the metabonomic biomarkers could be used in clinical drug development [35].

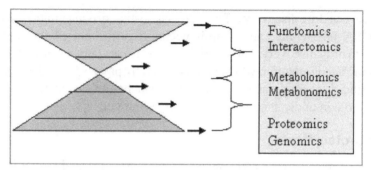

Fig. 16.4: Quantitative picture of various Omics'.

The lower base represents Genomics that pile up number of proteins in the form of proteomics. While Metabolomics and metabonomics make up the two small bases, they interact with each other making up a very larger base for interactomics and functomics. In conclusion, larger the genome, larger may be the number of proteins that make up from the genome and greater are the interactions among the metabolites. With many terms being framed, there has been consistent narrowing down of gap in our knowledge about various 'omics' as these disciplines are filled with valuable information emerging from research. The integration of this knowledge in the medical education curriculum and in continued professional education programs is urgently required to ensure applications of genomics in the provision of healthcare. During the last two decades, the practice of genetics has not only changed the broad horizons of medicine but also found its niche within this discipline there by allowing geneticists to reframe the good-old definition of gene from a locatable region of genome sequence involved in heritability to a locatable region of genome sequence associated with a functional region or a functor [8]. Various other scientists viewed genetics "as the study of single genes and their effects" and genomics as "the study not just of single genes, but of the functions and interactions of all the genes in the genome". In conclusion, there are many terms, quantitative differences among the *similar* fields—the study of multiple genes as opposed to one gene. Although scientists argue that classical disciplines like genetics to be a part of genomics, the distinction is not fully under acceptable terms while setting up one critical goal for all these terms. That being said, there is a qualitative difference bridging among these disciplines in medicine making up one goal from the concept of disease in genetics to the concept of information in genomics, thus yielding another omics called regulatory genomics.

Challenges in High-dimensional Biology (HDB)

Recently, the term "high-dimensional biology" (HDB) has been proposed for investigations involving high-throughput data [32]. The HDB information includes whole-genome sequences, expression levels of genes, protein abundance measurements, and other permutations. The identification of biomarkers, effects of mutations and effects of drug treatments and the investigation of diseases as multifactorial phenomena can now be accomplished on an unprecedented scale.

Finding the Function of Hypothetical Proteins

Another trendsetting feature of protein-protein interaction maps is to find the function of unknown proteins. Protein-Protein Interaction or PPI has become a very common step in annotation of a protein. Various tools like iHOP, STRING and other association networks aid the researcher to find if there are interacting partners of protein of interest. The data could be visualized through tools like Cytoscape and Osprey for further analyses. The nearest partners would essentially mean that the hypothetical or uncharacterized protein could play a function similar to its interactor(s). In the context of PPI networks, we could consider if a model is to be developed from the network or a network is to be generated with an already established model. Precisely, the putative function of a protein could be better known from a PPI network to develop a model from it. Information on "known" or "unknown" protein-protein interactions is still mostly limited but integrating tools such as these could generalize a way to find the function of hypothetical proteins [33]. Thus the field 'hypomics' could bring out a complete set of hypothetical proteins in human adding to set of another omics.

References

1. Bilello, J.A., (2007) SEB 'Omics' in translational medicine: are they lost in translation? Exp. Biol. Ser., 58:133-43.
2. http://bioinfo.mbb.yale.edu/what-is-it
3. http://biowiki.net/hypome/index.php/Main_Page
4. Bruggeman, F.J. and Westerhoff, H.V. (2007) The nature of systems biology. Trends Microbiol., 15(1):45-50.
5. Claudino, W.M., Quattrone, A., Biganzoli, L., Pestrin, M., Bertini, I. and Di Leo, A. (2007) Metabolomics: available results, current research projects in breast cancer, and future applications. J. Clin. Oncol., 25(19): 2840-6.
6. Dell, H.G., Scott, R., et al. (1996) A Greek-English Lexicon.
7. Figeys, D. (2004) Brief Funct Genomic Proteomic. combining different 'omics' technologies to map and validate protein-protein interactions in humans, 2(4):357-65
8. Fox Keller, E. and Harel, D. (2007) Beyond the Gene. PLoS One, 2(11).
9. Giorgetti, L., Zanardi, A., Venturini, S. and Carbone, R. (2007) ImmunoCell-Array: a novel technology for pathway discovery and cell profiling. Expert, Rev. Proteomics, 4(5): 609-16.
10. Greenbaum, D., Luscombe, N.M., Jansen, R., Qian, J. and Gerstein, M. (2001) Interrelating different types of genomic data, from proteome to secretome: 'oming in on function. Genome Res., 11(9):1463-8.

11. Huang, S. (2004) Brief Funct Genomic Proteomic. Back to the biology in systems biology: what can we learn from biomolecular networks? 2(4):279-97.
12. Jain, K.K. (2006) Challenges of drug discovery for personalized medicine. Curr. Opin. Mol. Ther., 8(6):487-92.
13. Jain, K.K. (2007) Recent advances in clinical oncoproteomics. J. BUON. Sep;12 Suppl., 1:S31-8.
14. Kaeberlein, M. (2004) Aging-related research in the "-omics" age. Sci. Aging. Knowledge Environ., (42): 39.
15. Kasper Lage, E. Olof Karlberg, Zenia, M. Størling, Páll Í Ólason, Anders G Pedersen1, Olga Rigina, Anders M Hinsby, Zeynep Tümer, Flemming Pociot, Niels Tommerup, Yves Moreau & Søren Brunak. (2007) A human phenome-interactome network of protein complexes implicated in genetic disorders. Nature Biotechnology, 25: 309 - 316.
16. Khanna, C. The dog as a cancer model. Nature Biotechnology, 24(9): 1065-1066.
17. LidBenning, C. and Stitt, M. (2004) Current Opinion in Plant Biology, 7: 231–234.
18. Morel, N.M., Holland, J.M., van der Greef, J., Marple, E.W., Clish, C., Loscalzo, J., Naylor, S. and Mayo (2004) Primer on medical genomics. Part XIV: Introduction to systems biology—a new approach to understanding disease and treatment. Clin. Proc., 79(5): 651-8.
19. Morrison, N., Cochrane, G., Faruque, N., Tatusova, T., Tateno, Y., Hancock, D. and Field, D. (2006) Concept of sample in OMICS technology. OMICS. Summer,10(2):127-37.
20. Oldiges, M., Lütz, S., Pflug, S., Schroer, K., Stein, N. and Wiendahl, C. (2007) Metabolomics: current state and evolving methodologies and tools. Appl. Microbiol Biotechnol., 76(3):495-511.
21. Oliver Fiehn, Bruce Kristal, Ben Van Ommen, Lloyd, W. Sumner, Susanna-Assunta Sansone, Chris Taylor, Nigel Hardy, And Rima Kaddurah-Daouk, (2006) Establishing Reporting Standards for Metabolomic and Metabonomic Studies: A Call for Participation. Omics, 10(2):158-163.
22. http://www.omics.org
23. O'Rourke, N.A., Meyer, T. and Chandy, G. (2005) Protein localization studies in the age of 'Omics'. Curr. Opin. Chem. Biol., 9(1):82-7.
24. Ramírez, M., Graham, M.A., Blanco-López, L., Silvente, S., Medrano-Soto, A., Blair, M.W., Hernández, G., Vance, C.P. and Lara, M. (2005) Sequencing and analysis of common bean ESTs. Building a foundation for functional genomics. Plant Physiol., 137(4):1211-27.
25. Reichert, A.S. and Neupert, W. (2004) Mitochondriomics or what makes us breathe. Trends Genet., 20: 555-62.
26. Renwick, J.H. (1969) Progress in mapping human autosomes. Br. Med. Bull., 25(6).
27. Rochfort, S. (2005) Metabolomics reviewed: a new "omics" platform technology for systems biology and implications for natural products research. J. Nat. Prod., 68(12):1813-20.
28. Sarah Calvo, Mohit Jain, Xiaohui Xie, Sunil, A., Sheth, Betty Chang, Olga, A. Goldberger, Antonella Spinazzola, Massimo Zeviani, Steven, A. Carr and Vamsi, Mootha, K. (2006) Systematic identification of human mitochondrial disease genes through integrative genomics. Nature Genetics, 38: 576-582.
29. Schwender, J., Ohlrogge, J. and Shachar-Hill, Y. (2004) Understanding flux in plant metabolic networks. Curr. Opin. Plant Biol., 7(3):309-317.
30. Scott, A., Jackson, Ming, Li Wang, Howard, Goodman, M. and Jiming Jiang (1998) Application of fiber-FISH in physical mapping of Arabidopsis thaliana. Genome, 41: 566–572.

31. Steinfath, M., Repsilber, D., Scholz, M., Walther, D. and Selbig, J. (2007) Integrated data analysis for genome-wide research. EXS., 97:309-29.
32. Stransky, B., Barrera, J., Ohno-Machado, L. and De Souza, S.J. (2007) Modeling cancer: integration of "omics" information in dynamic systems. J. Bioinform. Comput. Biol., 5(4):977-86.
33. Suravajhala, P., hypo, hype and 'hyp' human proteins. Bioinformation, 2(1):31-33.
34. Tapan, S., Mehta, Stanislav, O., Zakharkin1, Gary, L., Gadbury, David, B., Allison (2006) Epistemological Issues in Omics and High-Dimensional Biology: Give the People What They Want. Physiol. Genomics., 28(1): 24-32.
35. Vangala, S. and Tonelli, A. (2007) Biomarkers, metabonomics, and drug development: can inborn errors of metabolism help in understanding drug toxicity? AAPS J., 9(3): E284-97.
36. Ward, N. (2006) New directions and interactions in metagenomics research. FEMS Microbiol. Ecol., 55(3):331-8.
37. Weber, H. (2004) In: Proc. 5th Eur. Conf. on Grain Legumes and 2nd ICLGG, 7–11 June 2004, Dijon, France, 139–140 (Ed. AEP). AEP, Paris, France.
38. Werner, T. (2004) Proteomics and regulomics: the yin and yang of functional genomics. Mass Spectrom Rev., 23(1):25-33.

Chapter – 17

Computational Methods in Plant Genome Sequence Analysis

Archana Pan and Ipsita Chanda

Abstract

Advances in genome sequencing technologies have made it possible to uncover the whole genome sequences of numerous organisms from microbes to higher eukaryotes. Haemophilus influenza, a free living bacterium, was the first complete genome to be sequenced in 1995 by a team headed by J. Craig Venter and Hamilton Smith at The Institute for Genomic Research (TIGR). Immediately after this, Mycoplasma genitalium, a bacterium responsible for reproductive-tract infections and well-known for having the shortest genome of all free-living organisms was sequenced. Subsequently, TIGR unearthed the genome sequences of many microbes like Methanococcus jannaschii, Archaeoglobus fulgidus, Helicobacter pylori, Borrelia burgdorferi etc. During the years 1996-2000, the complete genome sequences of several model organisms such as bacterium Escherichia coli K-12, baker's yeast Saccharomyces cerevisiae, nematode Caenorhabditis elegans, fruit fly Drosophila melanogaster etc were published. Since then the complete genome sequences of many other eukaryotes have been released by the genome sequencing community.

Introduction

A landmark achievement in the biological research is the publication of the first draft of the human genome in 2001 by two scientific communities separately - the International Human Genome Sequencing Consortium, headed by Francis Collins and Celera Genomics Corporation, led by J. Craig Venter[1&2]. Other than animals, different plant genomes have been sequenced. The genomes of *Arabidopsis thaliana*, a model flowering plant, rice (*Oryza sativa*), the most important food plant of the world and black cottonwood or poplar tree were three initially sequenced plant genomes[3,4&5]. Since then sequencing of many plant genomes have been completed and many are in progress.

The rapid explosion of genome sequence data has been the key factor in the foundation of the field of bioinformatics that focuses on the acquisition, storage, manipulation, and distribution of this enormous amount of data in a systematic manner. This results in the development of different databases and methods/tools. Plants are the important resources for food, oxygen and different useful commodities like fiber, fuel, timber, paper, medicines etc. Computational analysis of the plant genome is far behind to that of animals. The research in this aspect helps plant biotechnologists to improve the products of the plants for betterment of mankind. In order to understand the biology of plant genome in depth, a plant genomics group focuses on the analysis of plant genomes using bioinformatics techniques. In this chapter, we will discuss different plant genome resources and computational methods, mainly sequence alignment methods, gene identification methods, identification of regulatory regions and phylogenetic analysis.

Plant Genome Databases

A database is a collection of data that are stored in a systematic order with definite format so that it can easily be accessed, retrieved, managed, and updated. Biological databases store biological data that are collected from a variety of resources, such as scientific experiments, published literature, high-throughput experimental tools, computational analyses *etc.* Biological data include nucleotide or protein sequences, protein sequence patterns or motifs, macromolecular 3D structures, gene localization (both cellular and chromosomal), gene functional annotation, clinical effects of mutations, gene expression, metabolic pathway data etc. Researchers deposit biological data directly to a database. The curators check the type and degree of errors, redundancy, consistency or conflict of submitted data and then update it. Based on the contents, biological databases can be roughly divided into three categories: primary databases, secondary databases and specialized databases[6].

Primary databases contain raw sequence or structural data submitted by the worldwide scientific community. GenBank (National Centre for Biotechnology Information), EMBL (Europen Molecular Biology Laboratory), DDBJ (DNA Databank of Japan) and PDB (Protein Data Bank) are the examples of primary sequence and structural databases. Secondary databases include information that is obtained by analysis of raw biological data deposited in the primary databases. The examples of secondary databases are SWISS-PROT (database of sequence annotation derived from EMBL), TrEMBL (Database of Translated Nucleic Acid Sequences) PROSITE, Pfam, PRINTS (Database of Protein family, Domain, Motif etc), DALI (Database of Protein Secondary Structure). Specialized databases serve a specific research community and focus on a particular organism. These databases may include primary and secondary data of specific organisms. For example, Flybase, Oryzabase, AceDB, MATDB etc are the example of specialized databases. Biological databases are enriched with a large collection of data obtained from plants. A list of plant genome databases/resources[7&8] are shown in Table 17.1.

Sequence Alignment

Sequence alignment is a process of comparing biological sequences by searching for a series of individual residues or residue patterns which are in the same order in the sequences. Sequences are written row wise in such a way that identical or similar residues are placed in the same columns, whereas non-identical or non-similar residues can either be placed in the same column as a mismatch or opposite to a gap in one of the other sequences. In an optimal alignment, non-identical residues and gaps are so placed to bring as many identical or similar residues as possible into vertical register. The aligned sequences, which are significantly similar, are biologically meaningful. The sequence similarity may be produced as a consequence of functional, structural, or evolutionary relationships between the sequences. From the evolutionary point of view, the similar sequences should not be mistaken as homologous sequences. Homologs are similar sequences that have been derived from a common ancestor. Homologs can be described as either orthologs or paralogs. Orthologs are similar sequences in two different organisms that have arisen due to a speciation event. Orthologs retain their functionality throughout evolution. Paralogs are similar sequences within a single organism that have arisen due to a gene duplication event. There is another type of sequences, called xenologs, which are similar sequences but do not share the same evolutionary origin, rather arise out of horizontal transfer events through symbiosis, viruses, etc.

Table 17.1: List of plant genome databases/resources

Databases	Description	Website
PlantGDB	Plant Genome Database provides tools and resources for plant genomics	http://www.plantgdb.org
PlantsDB	PlantsDB aims to provide a data and information recourse for individual plant species	http://mips.helmholtz-muenchen.de/ plant/genomes.jsp
PPMdb	Plant Plasma Membrane Database is a proteome database dedicated to proteins from plant plasma membranes	http://sphinx.rug.ac.be:8080/
PlantCARE	A database of plant *cis-acting* regulatory elements and a portal to tools for *in silico* analysis of promoter sequences	http://sphinx.rug.ac.be:8080/ PlantCARE/
PlantProm DB	A plant promoter database, is an annotated, non-redundant collection of proximal promoter sequences for RNA polymerase II with experimentally determined transcription start site (s), TSS, from various plant species	http://mendel.cs.rhul.ac.uk/ and http:// www.softberry.com
PLEXdb	Gene expression resources for plants and plant pathogens	http://www.plexdb.org
PlantsP	A functional genomics database for plant phosphorylation	http://PlantsP.sdsc.edu
ChromDB	Plant chromatin database	http://www.chromdb.org
ATGC	The Arabidopsis Genome Centre provides Physical map and genetic map information	http://signal.salk.edu/genome/
TAIR	The Arabidopsis information resource includes the complete genome sequence along with gene structure, gene product information, metabolism, gene expression, DNA and seed stocks, genome maps, genetic and physical markers, publications, and information about the Arabidopsis research community	http://www.arabidopsis.org
AtGDB	Arabidopsis thaliana genome database provides a convenient sequence-centered genome view for Arabidopsis thaliana with a narrow focus on gene structure annotation	http://www.plantgdb.org/AtGDB/
MaizeGDB	The community database for biological information about the crop plant *Zea mays* with genetic, genomic, sequence, gene product, functional characterization, literature reference	http://www.maizegdb.org
Maiztedb	Maize transposable element (TE) database stores information about TEs and make the information available in a variety of different formats	http://maizetedb.org/~maize/

(Contd.)

Computational Methods in Plant Genome Sequence Analysis

Databases	Description	Website
Oryzabase	A comprehensive rice science database maintained by National Institute of Genetics, Japan, which contains genetic resource stock information, gene dictionary, chromosome maps, mutant images and fundamental knowledge of rice science	http://www.shigen.nig.ac.jp/rice/oryzabase/top/top.jsp
RGAP	The Rice Genome Annotation Project Database and Resource is a NSF project, which provides sequence and annotation data for the rice genome	http://rice.plantbiology.msu.edu
RGP	Rice genome research program	http://rgp.dna.affrc.go.jp/E/index.html
SGN	A collection of data resource of the Solanaceae species including tomato, potato, pepper, eggplant, petunia, nicotiana	http://solgenomics.net
Gramene	A curated open-source data resource for comparative genome analysis in the grasses including rice, maize, wheat, barley, sorghum etc, as well as other plants including Arabidopsis, poplar and grape	http://www.gramene.org
SoyBase	Integrating Genetics and Molecular Biology for Soybean Researchers	http://soybase.org
CottonDB	A database that contains genomic, genetic and taxonomic information for cotton (*Gossypium* spp)	http://www.cottondb.org/wwwroot/cdbhome.php
IGGP	International Grape Genomics Program is a collaborative genome project dedicated to determining the genome sequence of the grapevine *Vitis vinifera*	http://www.vitaceae.org/index.php/nternational_Grape_Genome_Program
IWGSC	International Wheat Genome Sequencing Consortium	http://www.wheatgenome.org
IPGC	The international populous genome consortium	http://www.ornl.gov/sci/ipgc/
ICuGI	The web portal for the International Cucurbit Genomics Initiative including melon, cucumber, watermelen, pumpkin, etc.	http://www.icugi.org
BrGP	Brassica rapa genome project	http://www.brassica-rapa.org/BRGP/index.jsp
Barley SNP Database	A database of SNPs of barley	http://bioinf.scri.ac.uk/barley_snpdb/

There are two main types of sequence alignment: Pairwise sequence alignment and Multiple sequence alignment. In this section, we will discuss both types of sequence alignment.

Pairwise Sequence Alignment

A pairwise sequence alignment is a process of aligning two sequences of DNA, RNA, or protein in order to locate the regions of similarity that may result from the structural, functional and/or evolutionary relationships between the sequences. Pairwise sequence alignment is of two types, global and local alignment [9&10]. In global sequence alignment, the alignment is carried out across the entire length between two sequences to get the best possible alignment. The global alignment is more applicable for aligning two sequences which are closely related with almost equal length. On the other hand, local alignment finds local regions having highest level of similarity between two sequences without regard for the alignment of rest of the sequence regions. This alignment is useful for aligning divergent biological sequences containing functionally conserved regions and aligned sequences may be of different lengths. Fig. 17.1 illustrates global and local alignment of two sequences.

```
A C C T T A G T T T T C A C            — — — —T T G C— — — —
  |      | | |            | |                   | | |
T C G G T A G AC A CG A C            — — — —T T G A— — — — —
```

Global alignment   Local alignment

Figure 17.1: Global and local alignment of two DNA sequences

There are different methods for pairwise sequence alignment, namely Dot matrix method, Dynamic programming, k-tuple (word) method.

Dot matrix method: Dot matrix method is the graphical representation of similarity between two sequences in a two dimensional matrix [11&12]. In this method, the rows in the matrix correspond to the residues of one sequence (X) whereas the columns in the matrix correspond to the residues of the other sequence (Y) (Fig.17.2). At the beginning, the first residue in X is compared with all residues of Y and dot is placed in the respective column of the first row if there is a match. This process continues for rest of the residues in X until the last one. A diagonal row of dots represents region of similarity between two sequences (Fig. 17.2). Isolated dots that are not on the diagonal indicate random matches. Anti-diagonal rows of dots signify inverted repeats.

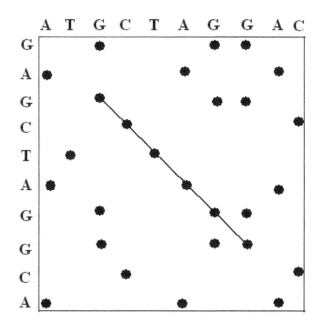

Fig. 17.2: A dot matrix plot of two DNA sequences, GAGCTAGGCA (X) and ATGCTAGGAC (Y). The diagonal of dots represent internal repeat, GCTAGG of either sequence.

Detection of matching regions can be improved by using a sliding window while comparing the two sequences. Rather than comparing single residue in the sequences, a window of adjacent residues in the two sequences is compared and a dot is plotted in the matrix if a certain minimum number of matches occur. For DNA sequences a typical window size is 15 and minimum number of matches required is 10. On the other hand, for protein sequences window size is 2 or 3 and minimum match requirement is 2.

Dot matrix method can also be used to find out direct and inverted repeats in a single sequence (DNA/protein)[13]. If a sequence is plotted against itself the regions that share significant similarities appear as parallel lines (direct repeat) off the main diagonal or as anti-diagonal lines (inverted repeat). Regions containing insertions/deletions can be readily determined by this method. The method is useful for predicting self-complementary regions in RNA which have the potential of forming secondary structure. The limitation of dot matrix method is that it only provides qualitative measurement of sequence similarity between two sequences. A list of programs based on Dot matrix method is given in Table 17.2.

366   Agriculture Bioinformatics

Table 17.2: List of programs based on Dot matrix method

Program	Website
Vector NTI software package (under AlignX)	
Dotlet Java applet	http://www.isrec.isb-sib.ch/java/dotlet/Dotlet.html
Dotter	http://www.cgr.ki.se/cgr/groups/sonnhammer/Dotter.html
GCG software package	
Compare	http://www.hku.hk/bruhk/gcgdoc/compare.html
DotPlot+	http://www.hku.hk/bruhk/gcgdoc/dotplot.html
Emboss software package	
Dotmatcher	bioweb.pasteur.fr/seqanal/interfaces/dotmatcher.html
Dotpath	bioweb.pasteur.fr/seqanal/interfaces/dotpath.html
Dotup	bioweb.pasteur.fr/seqanal/interfaces/dottup.html
DNA strider	
Pipmaker	http://bio.cse.psu.edu/pipmaker/
dotmatcher	http://www.hku.hk/bruhk/emboss/dotmatcher.html
Dothelix	http://www.genebee.msu.su/services/dhm/advanced.html
MatrixPlot	http://www.cbs.dtu.dk/services/MatrixPlot/

Dynamic Programming

In order to get a measure of sequence similarity quantitatively between sequences, Dynamic programming[14] can be used. Dynamic programming aligns two sequences by comparing all possible pairs of residues, one from each sequence, using a scoring scheme for matches, mismatches, and gaps to determine the best (optimal) alignment between two sequences. This approach is mathematically guaranteed to provide the optimal alignment for a particular scoring scheme. Matches, mismatches, and gaps can be included into the score calculated by the algorithm; this maximizes the number of matched characters. The first step to align two sequences by dynamic programming is to create a two dimensional matrix (Fig.17.3). The matrix size would be (m+1 x n+1), where n and m are number of residues of two sequences, respectively. In this matrix, row 0 and column 0 will represent gaps and the following rows and columns are labeled with the residues of two sequences. A suitable scoring scheme (substitution matrix) is to be selected to compute three parameters: match, mismatch and gap. After the selection of the scoring scheme, there are three steps involved in calculating the optimal scoring alignment: matrix initialization, matrix filling (scoring), traceback (alignment). Dynamic programming uses two separate algorithms for global and local alignments of

Computational Methods in Plant Genome Sequence Analysis

the sequences. These are Needleman-Wunsch and Smith-Waterman algorithm, respectively [9&10]. In the initialization step of global alignment, each row of 0th column $(S_{i,0})$ is set to w*i and each column of 0th row $(S_{j,0})$ is set to w*j, where i, j and w represent row, column and gap penalty, respectively. The matrix fill step starts filling of the matrix from the upper left hand corner and moves to the lower right corner filling one row and one cell at a time. This step finds the maximum global alignment score, $S_{i,j}$ for each position in the matrix. In order to find out the maximum score, $S_{i,j}$ for any i,j, it is necessary to know the scores for the matrix positions to the left, above and diagonal to i,j $i.e$, $S_{i-1,j}$, $S_{i,j-1}$ and $S_{i-1,j-1}$.

So, for each position, $S_{i,j}$ is defined to be the maximum score at position i,j, i.e.,

$S_{i,j}$ = MAXIMUM

$[S_{i-1, j-1} + s (a_i, b_j)$ (match/mismatch in the diagonal),

$S_{i,j-1} + w$ (gap in sequence #1),

$S_{i-1,j} + w$ (gap in sequence #2)]

Here, $s(a_i, b_j)$ is the score of the pair of residues, a and b in ith row and jth column.

When all the cells are filled with scores, a best alignment is determined through a trace-back procedure that search for the path with the best total score.

In order to obtain a local sequence alignment two slight modifications are required in global alignment algorithm. Temple Smith and Mike Waterman did the modifications to the Needleman Wunsch algorithm in 1981. The first modification is to apply negative scores for mismatches. The second modification is that when the dynamic programming scoring matrix value becomes negative, the value is set to zero, which terminates any alignment up to that point. This results in the matrix score to :

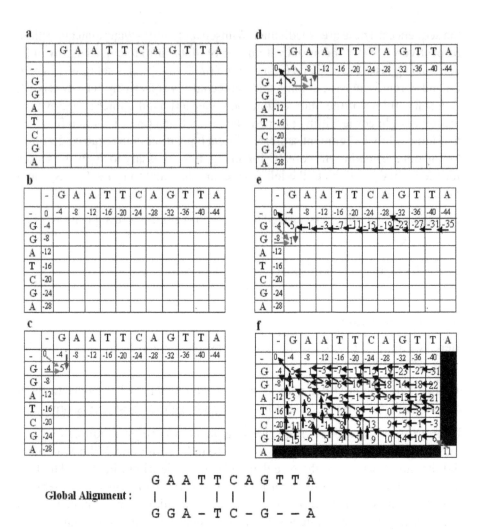

```
                    G A A T T C A G T T A
Global Alignment:   |   | |   |     |
                    G G A - T C - G - - A
```

Fig. 17.3: Example of dynamic programming to align two sequences GAATTCAGTTA and GGATCGA. Two dimensional matrix creation by writing two sequences across the top and down the left side of the matrix (a), matrix initialization (b), matrix filling (scoring) (c,d,e), trace back (f). Green, blue, red and black arrows indicate the scores for the matrix positions to the left, above, diagonal to a cell and trace back path, respectively.

$S_{i,j}$ = MAXIMUM

[$S_{i-1, j-1}$ + s (a_i,b_j) (match/mismatch in the diagonal),

$S_{i,j-1}$ + w (gap in sequence #1),

$S_{i-1,j}$ + w (gap in sequence #2),

0]

Computational Methods in Plant Genome Sequence Analysis

The local alignments are produced by starting at the highest-scoring positions in the matrix and reaching to a position that scores zero following a trace path. The limitation of the dynamic programming is that it may become slow and memory intensive for aligning large sequences. A list of programs for pairwise sequence alignment is given in Table 17.3.

Table 17.3: Pairwise sequence alignment programs

Program	Alignment type	Website
FASTA suite of programs		
GAP	global	http://bioinformatics.iastate.edu/aat/align/align.html
SIM	local	http://bioinformatics.iastate.edu/aat/align/align.html
SSEARCH	local	http://pir.georgetown.edu/pirwww/search/pairwise.html
LALIGN	local	http://www.ch.embnet.org/software/LALIGN form.html
EMBOSS software package		http://bioweb.pasteur.fr/seqanal/alignment/intro-uk.html#EMBOSS
Needle	global	
Stretcher	global	
Water	local	
Matcher	local	
SWIFT suit	local	http://bibiserv.techfak.uni-bielefeld.de/swift/
GapMis	SemiGlobal	http://www.inf.kcl.ac.uk/pg/gapmis/

k tuple (word) method : k-tuple means consecutive occurrence of k number of symbols in a string forming a word. The method is much faster than dynamic programming and ideal for database searches although it does not guarantee to find optimal alignment. This is a heuristic algorithm that tries to find out identical short stretches of sequences (words or k-tuples) between two aligning sequences. Once the perfect match is found, the algorithm tries to extend the alignment in both directions by the dynamic programming method until either one of the sequences ends or the score drops below some threshold. The word or k-tuple methods are used by the FASTA and BLAST algorithms [15,16&17]. There are different subprograms of Blast and Fasta (Table 17.4) available at websites, http://www.ncbi.nlm.nih.gov/BLAST and http://www.ebi.ac.uk/Tools/sss/fasta, respectively.

370   Agriculture Bioinformatics

Table 16.4: BLAST and FASTA programs

Program	Description
BLASTN	Search nucleotide database using a nucleotide query
BLASTP	Search protein database using a protein query
BLASTX	Search protein database using a translated nucleotide query
TBLASTN	Search translated nucleotide database using a protein query
TBLASTX	Search translated nucleotide database using a translated nucleotide query
FASTA	Search nucleotide or protein database using a nucleotide or a protein query
FASTX, FASTY	Search protein database using a translated nucleotide query
TFASTX, TFASTY	Search translated nucleotide database using a protein query

BLAST and FASTA perform almost equally well in database searching. FASTA provides more sensitive results than BLAST with a better coverage rate for homologs, though the procedure is slower than BLAST. On the other hand, BLAST has higher specificity than FASTA, so the chances of occurrence of potential false positives are reduced in BLAST. BLAST sometimes gives many best-scoring alignments from the same sequence whereas FASTA returns only one final alignment.

Database searching is one of the most effective ways to assign putative functions to newly determined sequences. The structure of a query protein can be determined if it shows similarity with a protein of known structure present in the database. Database similarity search also helps to determine evolutionary relatedness between sequences.

Substitution Matrices

Different scoring matrices are used for sequences alignment. Percent Accepted Mutation (PAM) [18&19] and Blocks Amino Acid Substitution Matrices (BLOSUM) [20,21&22] are two important substitution matrices. PAM matrix lists the likelihood of change of one amino acid to another in homologous protein sequences during evolution. The first PAM matrix, PAM1, was initially calculated by looking at the differences between sequences that were 85% similar. It was assumed that each amino acid change at a site is independent of the previous changes at the sites. The increasing PAM numbers correlate with the increasing evolutionary distances of protein sequences. For example, PAM250 is applied to the sequences that have 20% similarity. It is obtained by multiplying PAM1 by itself 250 times. This number of evolutionary changes approximately corresponds to an evolutionary span of 2,500 million years. In PAM matrices, the lower serial numbers are more suitable for aligning more closely related sequences. Thus, PAM120, PAM80, and PAM60 matrices are more suitable to align sequences that are 40%, 50%, and 60% similar, respectively. BLOSUM is based on the amino acid substitution rate in highly

Computational Methods in Plant Genome Sequence Analysis

conserved blocks. The BLOSUM matrices are not explicitly based on an evolutionary model, as the PAM matrices are, but are based instead on related families of proteins. For example, BLOSUM62 indicates that the sequences selected for constructing the matrix share an average identity value of 62%. Unlike PAM, the lower the BLOSUM number, the more divergent sequences they represent.

In case of nucleotide sequence alignment, in addition to using a match/mismatch scoring scheme, nucleotide mutation matrices can be constructed. They are based upon two different models of nucleotide evolution. One is Jukes-cantor model [23&24] and other is Kimura Model [25]. Jukes-Cantor model assumes that mutation rates are uniform among nucleotides, whereas Kimura model assumes that mutation rates are different for transitions and transversions.

Multiple Sequence Alignment

A Multiple Sequence Alignment (MSA) is a process of aligning three or more sequences of DNA, RNA, or protein to achieve optimal matching of the sequences. It allows the identification of conserved sequence patterns or motifs in the related sequences, which may not be detected by pairwise sequence alignment. By protein multiple sequence alignment many conserved and functionally important critical residues can be detected. MSA also helps to identify the hydrophobic, hydrophilic and "gappy" regions (loop/variable regions) in the related protein sequences which facilitate the prediction of secondary structure of proteins. MSA may create a profile or a pattern from a family of protein sequences that can be used to match against a sequence in the database and identify new family members. MSA is also essential prerequisite to carry out phylogenetic analysis of sequence families and prediction of protein structures. Consensus patterns which are derived from MSA can be used to design PCR primers.

The purpose of the most multiple sequence alignment algorithms is to achieve maximum SP scores (sum of pair scores). SP scores is the sum of the scores of all possible pairs of sequences in a multiple alignment, based on a particular scoring matrix. MSA exploits either exhaustive or heuristic programs. Exhaustive programs include dynamic programming algorithm. The limitation of dynamic programming is that the computing time and memory requirement increases exponentially with the increase of the number of sequences. This results in the prohibition of computation of large sequence data. Dynamic programming cannot be used for aligning more than 10 sequences. The maximally used algorithms for MSA are heuristic. The heuristic algorithms are of three types: progressive alignment, iterative alignment and block-based alignment.

The progressive alignment method [26&27] involves stepwise alignment of multiple sequences. At first, it performs pairwise alignments for all possible pair of sequences using global alignment algorithm [28]. The similarity scores from the pairwise comparisons are then used to generate a distance matrix. This matrix is used to build a phylogenetic tree using the neighbor-joining method, referred to as a 'guide tree'[29-30]. As referred by the guide tree, two most closely related sequences are first globally realigned. These two aligned sequences are converted to a single consensus sequence which is then aligned with the next closely related sequence in a similar fashion and converted to a single consensus sequence. This process is repeated until all sequences involved are not aligned. The most well-known progressive alignment program is Clustal (www.ebi.ac.uk/clustalw/). The two varieties of Cluster [31&32] are Clustal W and Clustal X. The W version provides a simple text-based interface and the X version provides a more user-friendly graphical interface. The advantage of Clustal is that it can apply different scoring matrices while aligning sequences, depending on degrees of similarity. The other benefits of Clustal are that it introduces less insertion or deletion gaps in the conserved regions and fast in speed. The limitation of the method is that it only uses the global alignment method and thus, is generally applicable to the sequences of almost equal length. Moreover, the method is "greedy" in nature i.e. any error made in the initial step will be fixed and cannot be corrected. To alleviate some of the limitations, a new generation of algorithms has been developed. For example, T-Coffee (Tree-based Consistency Objective Function for alignment Evaluation; (www.ch.embnet.org/software/TCoffee.html) and DbClustal (http://igbmc.u-strasbg.fr:8080/DbClustal/dbclustal.html) which use both global and local pairwise alignment for all possible pairs involved. Moreover, POA (Partial order alignments, www.bioinformatics.ucla.edu/poa) instead of depending on guide tree, build a graph profile that retains the information of the original sequences and eliminates the problem of error fixation as found in Clustal alignment. PRALINE (http://ibivu.cs.vu.nl/programs/pralinewww/) has the capacity to perform profile-based alignment, which is based on protein structure information and is thus much more accurate than Clustal[33].

The iterative approach starts with a low-quality alignment and improves it by iterative realignment to get an optimal solution. This method reduces the greedy problem of progressive alignment method and thus provides better result. However, the method gives no guarantee to find out the optimal alignment. The other limitations are that as this method uses only global alignment, it fails to recognize conserved domain and motif and not applicable for sequences of divergent lengths. An example of iterative alignment is PRRN (http://prrn.ims.u-tokyo.ac.jp/)[34].

Computational Methods in Plant Genome Sequence Analysis

The block-based method is based on the local sequence alignment program which identifies the regions of similarity in the form of highly ungapped block. This type of program is useful for carrying out alignment of sequences with divergent lengths and finding out the regions conserved among the sequences. The program DIALIGN2 (http://bioweb.pasteur.fr/seqanal/interfaces/dialign2.html), [35] is based on block-based method. Match-Box (www. sciences. fundp.ac.be/biologie/bms/matchbox submit.shtml) is a web-based server that identifies conserved blocks among sequences.

Gene Identification Methods

The fundamental and critical step in genome annotation process is to locate genes along a genome. In this section, we will briefly discuss different important gene identification methods, particularly the protein-coding genes for eukaryotic sequences. Before that we will discuss the general gene structure of eukaryotic genes.

Eukaryotic gene structure

In eukaryotes, a typical protein coding gene consists of two types of regions: coding regions, known as exons and non coding regions, known as introns (Fig. 16.4). Thus in eukaryotes, a coding sequence is not a continuous sequence segment, it is interrupted by non-coding intron sequences. Introns are removed through splicing mechanism that leads to the formation of mature mRNA. A mature mRNA is flanked by the untranslated terminal regions (5´-UTR and 3´-UTR) which are the non-coding transcribed regions. 5´-UTR is located at upstream of the translation initiation site, whereas 3´-UTR is positioned downstream of the translation stop codon (Fig. 17.4). These UTRs play a critical role in the post-transcriptional regulation of gene expression, such as regulation of translation, control of mRNA decay *etc*. There is a considerable variation in overall gene size and intron size. The number of exons/introns in a gene also varies significantly. Thus, in most eukaryotes, gene organization is so complex that gene identification poses a great challenge to the computational biologists.

In silico gene prediction strategies exploit different types of information that are embedded in DNA sequence. Mainly two types of information are regularly used to locate genes in the genomic sequence: *content sensor* and *signal sensor* [36, 37 & 38].

Content sensors are measures that classify a DNA region into coding region and non-coding region. The coding regions have specific statistical properties which can be used to distinguish them from non-coding regions. Several statistical measures such as nucleotide composition, especially GC-content (intron being more AT-rich than exon, particularly in plant), codon composition, hexamer frequency, base occurrence periodicity etc. are usually used to differentiate coding region from non-coding one [39]. Among these coding measures, hexamer ie., six nucleotide usage was found to be the most discriminative variable between coding and non-coding sequences. Signal sensors are measures that detect the presence of functional sites within and around a gene. A key component of the most successful gene-finding algorithms is the ability to recognize the DNA sequence patterns that are critical for transcription, splicing, and translation. These signals are usually characterized as short sequence patterns in the genomic DNA that correspond directly to regions on mRNA or pre-mRNA that have a key role in splicing or translation. The signals most commonly used by computational methods are translational start codon, stop codon and the splice junctions (GT-AG) surrounding introns [40]. In addition to splice sites and start/stop codons, signal recognition algorithms address transcription promoters (TATA box), terminators, branch points (CU[A/G]A[C/U], located 20-50 bp upstream of the AG acceptor), polyadenylation sites (a consensus AATAAA hexamer, located 20 to 30 bp downstream of the coding region), ribosomal binding sites, and various transcription factor binding sites.

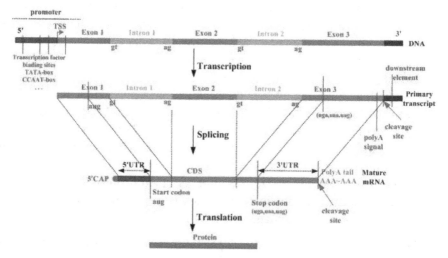

Fig. 17.4: Typical eukaryotic protein-coding gene structure
[Image resource: website (http://www.nslij-genetics.org/gene/) maintained by Wentian Li]

Gene Predictor Programs for Eukaryotic Sequences

In silico gene prediction methods can be broadly classified into two main categories: extrinsic approach and intrinsic/*ab initio* approach [37].

Computational Methods in Plant Genome Sequence Analysis

Extrinsic Gene-finders

Extrinsic gene-finders, also known as 'sequence similarity-based gene-finders', rely on similarity searches between an uncharacterized genomic DNA region (query sequence) and a known database (protein, cDNA, dbEST, or genomic DNA). Thus, sequence similarity information can be obtained by different type of sequence comparisons: genomic DNA/protein, genomic DNA/cDNA, genomic DNA/EST or genomic DNA/genomic DNA. It has been estimated that approximately 50% genes can be identified by comparison of a genomic DNA (that is translated into all six reading frames) with protein sequence databases. However, even when a good hit is obtained, a complete exact identification of gene structure can not be always predicted as homologous proteins may not share all of their domains. Comparison of a genomic DNA with cDNA is quite reliable for the identification of exons in query genomic sequence, particularly if the genomic sequence is aligned against a cDNA from the same organism, or a closely related organism. EST-based sequence similarity generally has a drawback as they only correspond to small part of the gene sequence and therefore, it is often difficult to predict the complete gene structure of a given genomic DNA. Similarity search with genomic DNA can be done by two approaches: intra-genomic comparisons and inter-genomic (cross-species) comparisons. Intra-genomic comparisons can provide data for multigenic families, apparently representing a large percentage of the existing genes (e.g. 80% for *Arabidopsis*). Inter-genomic comparisons can help the identification of orthologous genes, without any prior knowledge of them. The important strength of similarity-based approach is that gene predictions rely on known biological databases. They should give biologically relevant predictions. However, the weakness of this approach is that no hit will be obtained if the database does not contain a sufficiently similar sequence. Most of the sequence similarity-based programs employ the combination of similarity information with signal information obtained from signal sensors.

Programs which Rely on Genomic DNA/Protein Comparison

The program Procruster, pioneered "by Gelfand et al. in 1996"[40] involves 'spliced alignment' of a genomic sequence with a homologous protein sequence. Homologous protein can be identified by BLASTX search. Procruster then explores all possible exon assembles in query DNA sequence by translating the exons and aligning them with homologous protein. Other programs such as GeneWise, PredictsGenes, ORFgene, ALN have been developed for similar task [37]. The program like INFO (INterruption Finder and Organizer) has been used to find intron-exon splice junctions in the query DNA sequence by comparing the six conceptual translations of the query sequence with sequences present in protein databanks using a similarity matrix and windowing algorithm. ICE program utilizes the dictionary-based approach for gene annotation and exon prediction [37]. The program first constructs dictionaries of k-tuples (k>4) from nonredundant protein OWL database and dbEST databases. Then by using a look-up procedure it finds those k-tuples in the query sequence that have match in the dictionary.

Programs which Rely on Genomic DNA/cDNA Comparison

The programs like AAT, GeneSeqer, SIM4, Spidey *etc* are based on alignment of the query DNA sequence against a cDNA database. AAT, GeneSeqer, Spidey, GeneWise, ALN, etc have been used for annotation of various genome like human (International Human Genome Sequencing Consortium, 2004), *Caenorhabiditis elegans* (The *C. elegans* Sequencing Consortium, 1998), *Arabidopsis thaliana* (The Arabidopsis Genome Initiative, 2000) and *Aspergillus oryzae.*

Programs which Rely on Genomic DNA/dbEST Comparison

The programs such as EbEST, Est2genome, TAP, PAGAN predict gene structure from EST matches.

Programs which Rely on Genomic DNA/genomic DNA Comparison

Many of the programs in this category rely on the fact that closely related organisms harbor more conserved regions than distantly related organisms. ROSETTA is the first program for gene annotation in human based on comparison of two closely related organisms (human-mouse). The program CEM (Conserved Exon Method) also perform similar task like ROSETTA. Both the programs first perform alignment of two related sequences and identify conserved exon pairs from the pairwise alignment, then find the best chain of exons representing the genes in both sequences, using dynamic programming. Later on in 2001, SGP-1 (Syntenic Gene Prediction) has been developed, which is also based on comparison of homologous genome sequences, but unlike the previous two programs it is not species-specific. It has been applied to several sets of homologous sequence pairs from vertebrates and plants. Program AGenDA finds the exon candidates on both sequences under study by comparing them using Dialign, a program that assembly pairwise alignment from gap-free local segment alignments. All the above mentioned programs assume conservation of exon-intron structure in related genomic sequences. Unlike those programs Pro-Gen and Utopia do not consider the conservation of exon-intron structure and thereby they can be applied to analyze relatively distant homologs. The program SLAM introduces a probabilistic cross-species gene finding algorithm. The novel characteristic of this program is that it can predict conserved non-coding sequence (CNS) which may be useful for the annotation of UTRs, regulatory elements and other non-coding features. There are some other programs, such as MUMmer, WABA, PipMaker *etc* that collect information on conservation or synteny between organisms from genomic alignments [37].

Ab initio Gene-finders

Current *ab initio* gene-finders (intrinsic approach) utilize content sensors as well as signal sensors. Several programs have been developed based on the assumption that exons and introns exhibit distinct nucleotide distribution i.e. the mono nucleotide usage, codon preference, or hexamer usage of exon is significantly different from that of intron sequence. The simplest program CODONUSAGE utilizes open reading frames, start codons and codon preference to find out protein coding regions in long DNA sequences [39]. Programs like SORFIND, GenView2 exploit hexamer composition, coupled with different types of statistical models. In fact, a variety of algorithms such as Markov models (MM), Dynamic Programming (DP), Neural Networks (NN), Linguist method, Discriminant Analysis etc, have been applied to predict gene structure [36].

Markov Models

Markov chain model is the most frequently used model in gene prediction programs. A Markov model (MM), being a stochastic model, considers that probability of a particular nucleotide appearing at a given position depends on the k previous nucleotides where k represents the order of MM. Such a model is defined by the conditional probabilities P(X/k previous nucleotide), where X=A, T, G or C. To build a Markov model, a training set of DNA sequences (known coding sequences) is required. The simplest Markov model is homogeneous zero order Markov mode which assumes that each nucleotide occurs independently with a given frequency. These types of models are frequently used for non-coding regions. It is worth mentioning at this point that intrinsic methods use two types of content sensors: one for coding sequences and another for non-coding sequences. Many *ab initio* gene finding programs including GeneMark (for prokaryotic sequences), Genscan usually rely on more complex three-periodic Markov models of order five (use hexamer composition) in order to characterize coding sequences. In fact, fifth order markov model is widely used for gene prediction as it is known that hexamer usage is a good discriminator between exon and intron. However, there is a limitation of using fifth order Markov model in gene finding. The method's efficacy may be reduced if certain hexamers are absent (which often happens in short gene sequences). To cope with this limitation, interpolated Markov models (IMMs) have been introduced to different gene finding programs. Here, for each conditional probability, an IMM combines statistics from different order Markov models (1st to 8th order), according to the information available. The program like GLIMMER exploits IMM to predict genes in prokaryotic genomes (initial version). Later on, modified version of this program, GlimmerM and GlimmerHMM utilize these IIMs coupled with splice sites prediction algorithm to identify genes in eukaryotic genomes including *Arabidopsis thaliana,* rice.

Many eukaryotic gene prediction programs use Hidden Markov model and Generalized hidden markov model (GHMM, an extension of HMM). First hidden markov model based program was developed by Krogh, named ECOPARSE for gene prediction in *Escherechia coli.* Other HMM-based programs are Genie, GeneMark. hmm, FGenesh, GRPL. The model utilized in Genscan is similar in its overall architecture to GHMM approach adopted in the program Genie. Genscan and GeneMark.hmm and GlimmerM have been used to predict *Arabidopsis thaliana* genes. HMMgene and VEIL utilize a slightly different method called CHMM (class HMM).

Dynamic Programming and Neural Network

Many intrinsic gene finders use Dynamic programming algorithm in order to determine the most likely gene structures among the exponential number of possible gene structures defined by potential signals. Program GeneParser uses splice sites information and intron- and exon-specific content measures (codon usage, local compositional complexity, hexamer frequency, length distribution, periodic asymmetry) to calculate scores for all subintervals in the test sequence. First a neural network is utilized to combine the various measures into the log-likelihood ratio for each subinterval in order to accurately typify an exon or intron. Then DP is applied to find the optimal combinations of exons and introns. Geneview program that predicts ORF's using splice sites information and hexamer frequency also employs DP to construct the best gene structure from the numerous possible exon assemblies. Grail (Gene Recognition and Assembly Internet Link), based on neural network algorithm, only identifies the positions of candidate exons on a sequence but does not attempt to produce assembled genes. The gene assembly program GAP III exploits DP to assemble candidate exons predicted by GRAIL III for constructing optimal gene model. Programs like GeneID, Glimmer, VEIL, *etc* also use DP with other models. GeneID, which efficiently handles very large genomic sequences, has been designed with a hierarchical structure (signal to exon to gene). Program exploits position weight matrices (PWMs), Markov model of order 5 and Dynamic programming algorithm.

Linguistic Methods

Program GenLang explains the usual coding measures and signal strengths in a linguistic context as 'leaf rules' associated to a cost. A formal grammar, optimized on a training set, is then utilized to construct a gene structure as the parse that minimizes the total cost [41].

Discriminant Analysis

Discriminant analyses such as linear discriminant analysis (LDA), quadratic discriminant analysis (QDA) are the techniques in multivariate analysis that can be used to combine different measures (content sensor and signal sensor) to discriminate exons and introns. FGENES (Find Genes) exploits LDA whereas program MZEF (Michel Zhang's Exon Finder) uses QDA for exon prediction [37].

None of the gene finding programs under extrinsic and intrinsic approaches can predict genes in a genomic sequence with 100% accuracy. Both the approaches have their merits and demerits. The limitation of extrinsic approach is that it relies on the presence of homologous sequences in the database. If homologous sequences of a query are absent in the database the method cannot be used. Combination of these two approaches may improve the quality of gene prediction significantly.

Combined Methods

In this approach, two different types of gene prediction programs or two or more similar types of programs can be combined to construct gene structure [36]. A pioneer in this area was the GSA program (Gene Structure Assembly), born from the fusion between AAT and Genscan, which provides better results than those obtained with the two programs separately. GenomeScan is an extension of Genscan that incorporates similarity with a protein retrieved by BLASTX or BLASTP. GenomeScan is thus able to accurately predict coding regions missed by both Genscan and BLASTX used alone. FGENESH+ and FGENESH_C are the extension of existing algorithm, FGENESH, that use similarity to a protein or a cDNA sequence, respectively to improve gene prediction. FGENES-H2 and Twinscan use the FGENESH and Genscan programs respectively. A highly integrative approach is used in the EuGeÁne program: it combines NetGene2 and SplicePredictor for splice site prediction, NetStart for translation initiation prediction, IMM-based content sensors and similarity information from protein, EST and cDNA matches. GAZE allows integration of arbitrary prediction information from multiple sources supplied by the user. DIGIT integrates three programs namely FGENESH, Genscan and

HMMgene. AUGUSTUS+ is based on a novel GHMM for gene prediction that integrates intrinsic and extrinsic information in one single probabilistic model. The tracking of exons shared in common by two or more gene-finders carries an advantage in that it significantly reduces the number of over-predictions, but may also exhibit poor sensitivity and possible inconsistencies at the gene level. Platforms such as Genotator, MagPie and Ensembl gather evidence acquired from ab initio or homology-based prediction programs, and are considered to be relatively useful tools, which facilitate both human driven and automated annotations.

A list of gene prediction programs used in plant genome annotation is given in Table 17.5.

Table 17.5: Gene prediction programs used in plant (Possibly with Homology Integration)

Program	Algorithm*	Homology	Website
Genscan	GHMM		http://genes.mit.edu/GENSCAN. html
GlimmerM	DP, IMM		http://www.tigr.org/tdb/glimmerm/ glmr_form.html
GlimmerHMM	IIM, GHMM		http://www.cbcb.umd.edu/software /GlimmerHMM/
GeneMark.hmm	GHMM		http://opal.biology.gatech.edu/ GeneMark/eukhmm.cgi
Grail	DP, NN	EST/cDNA	http://compbio.ornl.gov/Grail-1.3
GeneID	DP, MM		http://www1.imim.es/geneid.html
GenLang	Grammar rule		http://www.cbil.upenn.edu/~sdong/ genlang_home.html
MZEF	Quadratic discriminant analysis		http://rulai.cshl.org/tools geneûnder/
SIM4		cDNA/DNA	http://pbil.univ-lyon1.fr/sim4.php
GeneSeqer		EST/protein	http://www.maizegdb.org/ geneseqer. php
FGENESH	HMM		http://www.softberry.com/ berry.phtml?topic=index &group = programs&subgroup=gfind
AUGUSTUS	IMM,WWAM		http://augustus.gobics.de/

* DP, Dynamic programming; MM, Markov model; IIM, Interpolated Markov model; HMM, Hidden Markov model; GHMM, Generalized HMM; NN, Neural network; WWAM, Windowed weight array model

Prediction of Regulatory Elements

Regulatory elements are the regions around the genes that control the gene expression by regulating the process of gene transcription. There are different kinds of regulatory elements that assist in the initiation of gene transcription and the number is more in eukaryotes than in prokaryotes. In eukaryotes, regulatory elements comprise of promoter, initiator sequence (Inr), different types of transcription factors binding sites etc (Fig.17.5). Promoter is the RNA polymerase binding site and in eukaryotes RNA polymerase binding to the promoter site depends on a number of transcription factors. The core element of the promoter is TATA box which consists of a consensus motif TATA(A/T)A(A/T) located at 30 bps upstream from translation initiation site. Many vertebrate genes contain high density CG dinucleotides in promoter regions known as CpG islands that coincide to transcription initiation site. Most of the transcription factor binding sites are located about 500 bps upstream of the transcription start site. Inr is the unique sequence that is made up of the consensus pyrimidine rich element, (C/T)(C/T)CA(C/T)(C/T) and overlaps the transcription initiation site. There are many regulatory sites which lie downstream of transcription start regions. *In silico* predictions of the regulatory elements try to detect the consensus sequence patterns of the regulatory regions of genes. The main problem of computational detection of these regions is that these consensus elements are very short with a length of about 6-8 nucleotides and this property increases the chances of obtaining false positives in any sequence. To overcome this problem, different sophisticated algorithms are developed considering various features of regulatory regions which reduce the random chance of their occurrence in any sequence [42, 43 & 44]. The currently used computational methods are *ab initio* method, phylogenetic foot-printing based method and expression profiling-based method. In this section, we will discuss about the prediction of regulatory regions of protein-coding genes in eukaryotes.

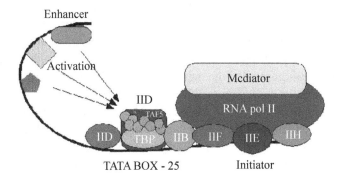

Fig. 17.5: A schematic diagram of an eukaryotic promoter with transcription factors (IIA, IIB, IIF, IIE, IIH, TBP) and RNA pol II bound to the promoter.

Ab initio method predicts promoters and other regulatory regions of the genes based on the characteristic sequence patterns of these regions. This is done by searching for the consensus elements like TATA box, Inr sequence, CpG islands and various TF-binding sites. This method mainly relies on the content information like hexamer frequency prediction. The weakness of this method is that it needs training sets and so it becomes species–specific. The conventional approach to detecting a promoter or regulatory site is matching of the query sequences to the consensus sequence patterns of known promoters and regulatory elements stored in a database by using a position-specific scoring matrix (PSSM). The shorter length of the consensus sequences results in the difficulty in differentiating true promoters from random sequence matches and generates high rates of false positives. This limitation is overcome by developing a new generation of algorithms that take into account the higher order correlation of multiple subtle features by using discriminant functions, neural networks, or Hidden Markov Models (HMMs).

Phylogenetic foot-printing based method is relied on the comparative analysis of the consensus sequences of regulatory regions between two phylogenetically closely related organisms. The identification of conserved non-coding DNA elements that serve crucial functions is known as phylogenetic foot-printing and the elements are called phylogenetic footprints. The advantage of this method is that unlike *ab initio* method it does not depend on training set and therefore, has potential to discover new regulatory motifs shared among organisms. However, caution should be taken to consider only upstream regions of the transcription start site to avoid background noise and also selection of the phylogenetically closely related organisms should be taken carefully for the proper filtering out of regulatory elements.

Expression profiling-based method depends on the profile construction of the regulatory regions of the sequences that are co-expressed, which are obtained by high throughput transcription profiling analysis like microarray analysis. Co-expressed genes are believed to have similar expression profiles due to the presence of similar promoters and regulatory regions. The method performs multiple sequence alignment of the regulatory regions of the co-expressed genes to build a profile. To avoid the problem of very shorter length of consensus sequences, the simple multiple sequence alignment is replaced by an advanced alignment-independent profile construction method such as EM (Expectation Maximization) and Gibbs motif sampling. This helps to find the subtle sequence motifs. One of the shortcomings of this method is that co-expressed genes are determined by clustering of experimentally expression data. This clustering approach is error prone and sometimes includes functionally unrelated genes. Moreover, the assumption that all co-expressed genes have similar regulatory

Computational Methods in Plant Genome Sequence Analysis

regions is not always correct. Therefore, caution should always be taken while using this method. A list of currently used programs for prediction of regulatory regions of genes is given in Table 17.6.

Table 17.6: List of programs for prediction of regulatory regions

Programs	Description	Website
Eponine	Predicts transcription start sites using PSSM matrix	http://servlet.sanger.ac.uk:8080/eponine/
Cluster-Buster	Finds clusters of regulatory binding sites using HMMs.	http://zlab.bu.edu/cluster-buster/cbust.html
McPromoter	Predicts promoters by using neural network	http://genes.mit.edu/McPromoter.html
TSSW	Distinguishes promoter from non-promoter sequences based on a combination of content and signal information	www.softberry.com/berry.phtml?topic=promoter
ConSite	Finds putative promoter elements by comparing two orthologous sequences.	http://mordor.cgb.ki.se/cgi-bin/CONSITE/consite
rVISTA	Finds promoter by close cross-species comparison	http://rvista.dcode.org/
PromH(W)	Predicts regulatory sites by pairwise sequence comparison	www.softberry.com/&subgroup=promoterrry.phtml?topic=promhw&group=program
MEME	Used for protein and DNA motif finding using EM based programs	http://meme.sdsc.edu/meme/website/meme-intro.html
AlignACE	Finds out common motifs by using the Gibbs sampling algorithm	http://atlas.med.harvard.edu/cgi-bin/alignace.pl
INCLUSive	Used for microarray data collection and sequence motif detection	www.esat.kuleuven.ac.be/<"dna/BioI/Software.html
PhyloCon	Identify regulatory motifs by phylogenetic footprinting and gene expression profiling analysis	*http://ural.wustl.edu/~twang/PhyloCon/*
Melina	Runs four individual motif-finding algorithms eg, MEME,GIBBS sampling, CONSENSUS and Coresearch, simultaneously.	http://melina.hgc.jp/
NSITE	Search for TF-binding sites or other consensus regulatory sequences	http://www.bioinformatics.vg/biolinks/bioinformatics/verbose/Promoter%2520Scan.shtml
PromoterScan	Predict promoter region based on scoring homologies with putative eukaryotic Pol II promoter sequences	http://bimas.dcrt.nih.gov/molbio/proscan/
NNPP	Promoter prediction by neural network for prokaryotes or eukaryotes	http://www.fruitfly.org/seq_tools/promoter.htm
PLACE	Plant cis-acting regulatory elements	http://www.dna.affrc.go.jp/htdocs/PLACE/
Polyadq	Locate polyadenylation sites	http://rulai.cshl.edu

Phylogenetics Basics and Tree Construction Methods

Phylogenetics is the study of evolutionary relatedness among groups of organisms using tree-like diagrams to represent pedigrees of these organisms. The tree branching patterns which represent the evolutionary divergence are known as phylogeney. In this section, we will discuss some commonly used phylogenetic tree construction methods briefly, preceded by some basic terminologies that characterize the phylogenetic tree.

Basic Terminologies in Phylogenetics

A phylogenetic tree is composed of taxa, nodes, branches and a root node (Fig. 17.6). Taxa represent the present-day species/sequences that occupy the tips of phylogenetic tree. The lines in the phylogenetic tree are known as branches. A node is a point where two adjacent branches join. The bifurcating point at the bottom of the tree is called the root node, which represents the common ancestor of all taxa of the tree. Depending on the common ancestor, a group of taxa may be monophyletic (clade) or paraphyletic. In monophyletic group, a group of taxa descended from a single common ancestor, whereas in paraphyletic group they share more than one closest common ancestor. The branch path depicting an ancestor-descendant relationship on a tree is called a lineage. A phylogenetic tree may be rooted or unrooted. A rooted phylogenetic tree provides information for common ancestor and holds a root at the bottom of the tree, while the unrooted tree provides no information of common ancestor and is without root. There are different ways to draw the phylogenetic trees, namely cladogram and phylogram (Fig.17.7). In a cladogram, the branch lengths are unscaled, meaning that they are not proportional to the number of evolutionary changes. In this case, only the relative ordering of the taxa are shown. On the other hand, in a phylogram, the branch lengths are scaled providing the amount of the evolutionary divergence. The advantage of phylogram is that it not only shows evolutionary relationships but also provide information about the relative divergence time of the branches.

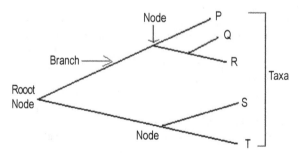

Fig. 17.6: A phylogenetic tree showing taxa, nodes, branches and root node

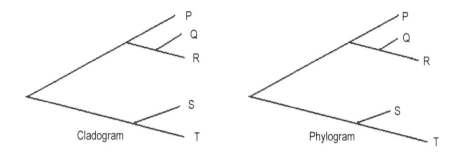

Fig. 17.7: Phylogenetic trees in the form of Cladogram and Phylogram. The branch lengths are unscaled in the cladograms and scaled in the phylograms.

Different Steps in Phylogenetic Analysis

The first step in phylogenetics is to decide whether to choose DNA sequences or protein sequences. It is recommended that for reconstructing evolutionary relationships between very closely related organisms, nucleotide sequences that evolve more rapidly than proteins can be used. In the contrary, for studying the evolution of more widely divergent groups of organisms it is advisable to choose either slowly evolving nucleotide sequences, namely ribosomal RNA or protein sequences. In most cases, protein sequences are preferable as they are more conserved and allow more sensitive alignment compared to DNA. Alignment is the second step in phylogenetic analysis. Correct alignment is necessary to build correct tree. Most multiple sequence alignment programs such as ClustalW, T-COFFEE, MAFFT-2, MAFFT-I etc may introduce errors in the alignment. The programs, such as Rascal (ftp://ftp-igbmc.u-strasbg.fr/pub/RASCAL), NorMD etc have been developed to improve alignment quality by correcting alignment errors and removing potentially unrelated or highly divergent sequences. Moreover, the program like Gblocks (http://woody.embl-heidelberg.de/phylo/) can help to detect and eliminate the poorly aligned positions and divergent regions in order to generate the alignment more suitable for phylogenetic analysis. The next step is to choose an appropriate substitution model/evolutionary model that gives estimation of the true evolutionary event by taking into account multiple substitution events. For constructing DNA phylogenies, the nucleotide substitution model Jules-Cantor model can be used. This model assumes that all nucleotides are substituted with equal probability. Kimura two-parameter model is another model to correct evolutionary distances, where mutation rates for transitions and transversions are considered to be different. In case of protein sequences, the evolutionary distances from the alignment can be corrected using PAM or JTT amino acid substitution matrix. The 4th step is to determine a tree building method to construct a phylogenetic tree. Finally, the reliability and quality of the tree is evaluated by using resampling strategies (bootstrapping, jackknifing) and conventional statistical tests.

Different Tree Construction Methods

The most popular and frequently used phylogenetic tree building methods can be broadly classified into two major categories: Distance-based method and Character-based method. In the distance-based method, all the sequences under study are assumed to be homogeneous and the distance between two taxa is equivalent to the sum of all branch lengths connecting them. Using different evolutionary models evolutionary distances between sequences are evaluated and the distances can be used to construct a distance matrix, in which values in the cells represent distances between all individual pairs of taxa. The algorithms used for distance-based method are of two types: clustering-based and optimality-based. The popular distance-based tree building methods, namely an unweighted pair group method using arithmetic average (UPGMA) [45] and neighbor joining (NJ) [46] exploit the clustering-based algorithm. The optimally-based algorithm includes the Fitch-Margoliash and minimum evolution algorithms. Character-based methods are based on sequence character, rather than the pairwise distance between sequences. It calculates the number of mutational events accumulated in the sequences. Two most popular character-based methods are the maximum parsimony (MP) and maximum likelihood (ML) methods.

Distance Based Method

UPGMA method: The UPGMA, a simplest method, employs a sequential clustering algorithm to construct the phylogenetic tree. The method assumes that all taxa evolve at a constant evolutionary rate and that they are equally distant from the root, indicating that it assumes a molecular clock hypothesis. From a given distance matrix, the method begins by joining two closest taxa together and a node is placed at the midpoint between them. The newly clustered pair is treated as a single new composite taxon. A reduced matrix is created by evaluating the distances between this new composite taxon and all remaining taxa. The process continues until all taxa are placed on the tree. The last taxon added is considered the outgroup creating a rooted tree. UPGMA method is fast and can handle many sequences. However, it can lead to an erroneous tree if the rates of mutation in the branches of the tree are not uniform.

NJ method

The NJ method does not make the assumption of molecular clock hypothesis i.e., it does not assume that all the taxa are equidistant from the root. It employs an evolutionary rate correction step before tree building. The first step of the NJ method involves the calculation of r-value (evolutionary rate) and r'-value (corrected evolutionary rate) which is done by using the following formulae:

$r_i = \Sigma d_{ij}$ where i,j are two different taxa and d_{ij} is the distance between i,j taxa,

$r_i' = r_i/n-2$ where n is the total number of taxa.

Based on these r-values the corrected distances are obtained using the formula:

$d_{ij} = d_{ij} - 1/2 \times (r_i + r_j)$ where d_{ij} represents the corrected distance between i and j, d_{ij} indicates the actual evolutionary distance between i and j.

Then a new distance matrix is formed using these corrected distances. Before tree construction, all possible nodes are collapsed into a star tree by joining all taxa onto a single node. A pair of taxa that shows the shortest corrected distances is considered as the closest neighbors and joined first as a node. Then the newly created cluster reduces the matrix by one taxon and next most closely related taxon would be joined to the first node. The process continues until all internal nodes are resolved. The advantage of NJ method is that it is computationally fast and generally gives better results than UPGMA method. The disadvantages of NJ method are that it generates only one possible tree, strongly dependent on the model of evolution used. NJ method produces unrooted tree, unlike UPGMA. To overcome the limitations of regular NJ method a generalized NJ method has been developed. In this method several NJ trees with different initial taxon groupings are generated. A best tree is then chosen from pool of regular NJ trees which best fit the actual evolutionary distances.

Fitch-Margoliash (FM) & Minimum Evolution (ME)

The FM selects a best tree among all possible trees which has the lowest square deviation of actual distances and calculated tree branch lengths [47]. Minimum evolution (ME) searches a tree among all possible trees with a minimum overall branch length. FM and ME perform the best in the group of distance-based methods. But they are computationally slow and can not be used for a large dataset (>12).

Distances matrix methods are found in the programs like ClustalW, Phylo_win, Paup; Paupsearch, Distances (in GCG package) and DNADIST, PROTDIST, FITCH, KITCH, NEIGHBOR (in Phylip package) etc. In Phylip package,

DNADIST evaluates distances among DNA sequences and PROTDIST for protein sequences for the construction of phylogenetic tree. FITCH and KITSCH construct phylogenies using Fitch-Margoliash method. The latter assumes a molecular clock whereas the former does not. NEIGHBOR estimates phylogenetic tree using neighbor-joining or UPGMA method.

Maximum Parsimony

Maximum parsimony involves the identification of a tree topology that requires the minimum number of evolutionary changes to explain the observed differences among the taxa under study [48&49]. In MP method, the sites which have richest phylogenetic information, known as informative sites, are utilized in tree determination. The informative site is a site that has at least two different kinds of characters, each appearing at least in two of the sequences under study (Fig. 17.8). The other sites, known as noninformative sites, are discarded from the analysis. The method begins with the identification of informative sites. Next step is to calculate minimum number of substitutions at each informative site and sum up the number of changes in all the informative sites for each possible tree. Finally, the method chooses a tree with smallest number of changes as a maximum parsimony tree. From MP method ancestral sequences can be inferred. It is suitable for similar sequences which have small amount of variations. However, this method provides little information about the branch lengths. It may lead to an incorrect tree if the amount of evolutionary change is adequately divergent in different branches. Moreover, MP only considers informative sites and ignores other sites, thus certain phylogenetic signals may be lost. It is very sensitive to the "long-branch attraction" artifacts. For Maximum Parsimony method of phylogenetic analysis Phylip package includes separate programs for DNA and protein sequences: DNAPARS, DNAPENNY, DNACOMP, DNAMOVE for DNA sequences and PROTPARS for protein sequences.

sites / taxa	1	2	3	4	5	6	7	8
I	T	A	A	T	A	T	A	T
II	C	A	A	C	G	G	C	G
III	T	G	A	G	C	G	C	T
IV	T	C	A	T	A	G	C	A
V	T	G	A	T	C	G	C	G
VI	T	C	A	A	G	G	C	A

Fig. 17.8: Example of identification of informative sites used in MP. Sites 2, 5 and 8 within rectangular boxes are informative sites.

Maximum Likelihood

ML is an exhaustive method which considers all positions in an alignment and evaluates all possible tree topologies. Using probabilistic models the method searches a best tree among all possible tree topologies [50]. The best tree has the maximum likelihood of reproducing the observed data. ML calculates the total likelihood of ancestral sequences evolving to internal nodes and then to existing sequences by using a particular substitution model. Among all approaches ML is the most mathematically rigorous as it uses full sequence information. However, it can create an incorrect tree if the substitution model is not chosen correctly. Moreover, due to exhaustive nature of this method, it cannot work when the number of taxa increases to a modest size. To overcome these limitations different approaches like quartet puzzling, genetic algorithms (GAs), Bayesian inference have been introduced. ML is available in the programs like Phylip, Paup or Puzzle. For ML phylogenetic analysis Phylip package includes two programs: DNAML and DNAMLK, both estimate phylogenies from DNA sequences, the only difference is that the later assumes a molecular clock hypothesis.

Summary

The development of databases and computational methods/tools are the core areas in Bioinformatics. Biological databases are rich sources of biological data collected from scientific experiments, published literature, and computational analyses. These databases comprise data from different research areas including genomics, proteomics, metabolomics, microarray gene expression, phylogenetics *etc*. Different plant genome resources mentioned in this chapter would be beneficial for plant genome researchers. Different computational methods on sequence alignment, identification of protein-coding genes & regulatory regions and phylogenetic analysis discussed in this chapter are useful for genome sequence analysis of various organisms including plants. Gene identification is a major challenge in the field of computational biology. There are two different approaches to computational gene prediction: *ab-initio* or intrinsic methods that use statistical features to distinguish coding from non-coding regions and extrinsic methods that attempt to find similarities between genomic sequences and known proteins. Phylogenetic analysis provides evolutionary relatedness among groups of organisms. Every method has its own merits and demerits. Though these methods are meeting the needs of plant genome researchers there is room for improvement in prediction accuracy. Researchers are trying to modify the existing algorithms and develop new algorithms for analysis of experimental data for better prediction of not only the evolutionary relationship of different organisms but also the location, function and structure of macromolecules of biological interest.

References

1. Lander, E.S., Linton, M., Birren, B., Nusbaum, C., Zody, C., Baldwin, J., Devon, K., Dewar, K., Doyel, M., Fitzhugh, W. *et al.* (2001) Initial sequencing and analysis of the human genome, Nature, 409:860–921.
2. Venter, J.C., Adams, M., Myers, E., Li, P., Mural, R., Sutton, G., Smith, H., Yandell, M., Evans, C., Holt, R.A. *et al.* (2001) The Sequence of the Human Genome, Science, 291:1304–1351.
3. The Arabidopsis Genome Initiative (2000) Analysis of the genome sequence of the flowering plant *Arabidopsis thaliana*, Nature, 408:796-815.
4. Tuskan, G.A., Difazio, S., Jansson, S., Bohlmann, J., Grigoriev, I., Hellsten, U., Putnam, N., Ralph, S., Rombauts, S., Salamov, A., *et al.* (2006) The genome of black cottonwood. *Populus trichocarpa* (Torr. & amp: Gray), Science, 313:1596-1604.
5. Yu, J., Hu, S., Wang, J., Wong, G.K., Li, S., Liu, B., Deng, Y., Dai, L., Zhou, Y., Zhang, X. , *et al.* (2002) A draft sequence of the rice genome (Oryza sativa L.ssp.indica), Science, 296:79-92.
6. Hughes, A.E. (2001) Sequence databases and the Internet, Methods Mol. Biol., 167:215-223.
7. Rhee, S.Y. and Crosby, B. (2005) Biological Databases for Plant Research, Plant Physiol., 138:1–3.
8. Matthews, D.E., Lazo, G.R. and Anderson, O.D. (2009) Plant and crop databases, Methods Mol. Biol., 513:243-262.
9. Needleman, S.B. and Wunsch, C.D. (1970) A general method applicable to the search for similarities in the amino acid sequence of two proteins, J. Mol. Biol., 48:443-453.
10. Smith, T.F. and Waterman, M.S. (1981) Identification of Common Molecular Subsequences, J. Mol. Biol., 147:195–197.
11. Gibbs, A.J. and McIntyre, G.A. (1970) The diagram, a method for comparing sequences. Its use with amino acid and nucleotide sequences, Eur. J. Biochem., 16:1–11.
12. Doolittle, R.F. (1986) Of URFs and ORFs: A primer on how to analyze derived amino acid sequences. University Science Books, Mill Valley, California.
13. States, D.J. and Boguski, M.S. (1991) Similarity and homology. In Sequence analysis primer (Ed. M. Gribskov and J. Devereux), pp. 92–124. Stockton Press, New York.
14. Pearson, W.R. and Miller, W. (1992) Dynamic programming algorithms for biological sequence comparison, Methods Enzymol., 210:575-601.
15. Lipman, D.J. and Pearson, W.R. (1985) Rapid and Sensitive Protein Similarity Searches, Science, 227:1435-1441
16. Pearson, W.R. and Lipman, D.J. (1988) Improved tools for biological sequence comparison, Proc. Natl. Acad. Sci USA, 85:2444-2448.
17. Altschul, S.F., Gish, W., Miller, W., Myers, E.W. and Lipman, D.J. (1990) Basic local alignment search tool, J. Mol. Bio., 215:403-410.
18. Dayhoff, M.O. (1978) Survey of new data and computer methods of analysis. In: Atlas of protein sequence and structure, vol. 5, suppl. 3. National Biomedical Research Foundation, Georgetown University, Washington, D.C.
19. Dayhoff, M.O., Barker, W.C. and Hunt, L.T. (1983) Establishing homologies in protein sequences, Methods Enzymol., 91: 524–545.
20. Henikoff, S. and Henikoff, J.G. (1991) Automated assembly of protein blocks for database searching, Nucleic Acids Res., 19:6565–6572.
21. Henikoff, S. and Henikoff, J.G. (1992) Amino acid substitution matrices from protein blocks, Proc. Natl. Acad. Sci., 89:10915–10919.
22. Henikoff, S. and Henikoff, J.G. (1993) Performance evaluation of amino acid substitution matrices, Proteins Struct. Funct. Genet., 17:49–61.

23. Jukes, T.H. and Cantor, C.R. (1969) Evolution of Protein Molecules. New York: Academic Press. pp. 21–132.
24. Li, W. and Graur, D. (1991) Fundamentals of molecular evolution, Sinauer Associates, Sunderland, Massachusetts.
25. Kimura, M (1980) A simple method for estimating evolutionary rates of base substitutions through comparative studies of nucleotide sequences, J. Mol. Evol. 16:111–120.
26. Feng, D.F. and Doolittle, R.F. (1987) Progressive sequence alignment as a prerequisite to correct phylogenetic trees, J. Mol. Evol., 25:351–360.
27. Taylor, W.R. (1987) Multiple sequence alignment by a pairwise algorithm, Comput. Appl. Biosci. 3:81–87.
28. Lipman, D.J., Altschul, S.F. and Kececioglu, J.D. (1989) A tool for multiple sequence alignment, Proc. Natl. Acad. Sci. USA, 86:4412–4415.
29. Gotoh, O. (1990) Consistency of optimal sequence alignments, Bull. Math. Biol., 52: 509-525.
30. Van Walle, I., Lasters, I. and Wyns, L. (2004) Align-m–a new algorithm for multiple alignment of highly divergent sequences, Bioinformatics, 20:1428–1435.
31. Thompson, J.D., Higgins, D.G. and Gibson, T.J. (1994) CLUSTAL W: Improving the sensitivity of progressive multiple sequence alignment through sequence weighting, position-specific gap penaltiesand weight matrix choice, Nucleic Acids Res., 22: 4673-4680.
32. Thompson, J.D., Gibson, T.J., Plewniak, F., Jeanmougin, F. and Higgins, D.G. (1997) The CLUSTAL X windows interface: Flexible strategies for multiple sequence alignment aided by quality analysis tools, Nucleic Acids Res., 25:4876–4882.
33. Heringa, J. (1999) Two strategies for sequence comparison: Profile-preprocessed and secondary structure induced multiple alignment, Comput. Chem., 23: 341–364.
34. Gotoh, O. (1996) Significant improvement in accuracy of multiple protein sequence alignments by iterative refinement as assessed by reference to structural alignments, J. Mol. Biol., 264:823–838.
35. Morgenstern, B., Dress, A. and Werner, T. (1996) Multiple DNA and protein sequence alignment based on segment-to-segment comparison, Proc. Natl. Acad. Sci. USA, 93: 12098–12103.
36. Do, J.H. and Choi, D-K. (2006) Computational approaches to gene prediction, J. Microbiol., 44: 137-144.
37. Mathe, C., Sagot M-F., Schiex T. and Rouze P (2002) Current methods of gene prediction, their strengths and weaknesses, Nucleic Acids Res., 30:4103-4117.
38. Sleator, R.D. (2010) An overview of the current status of eukaryote gene prediction strategies, Gene, 461:1-4.
39. Staden, R. and McLachlan, A.D. (1982) Codon preference and its use in identifying protein coding regions in long DNA sequences, Nucleic Acids Res., 10:141-156.
40. Gelfand, M.S., Mironov, A.A. and Pevzner, P.A. (1996) Gene recognition via spliced sequence alignment, Proc. Natl. Acad. Sci. USA, 93:9061-9066.
41. Dong, S. and Searls, D.B. (1994) Gene structure prediction by linguistic methods, Genomics, 23:540-551.
42. Hannenhalli, S. and Levy, S. (2001) Promoter prediction in the human genome, Bioinformatics, 17(Suppl):S90-96.
43. Hehl, R. and Wingender, E. (2001) Database-assisted promoter analysis, Trends Plant Sci., 6:251–255.
44. Ohler, U. and Niemann, H. (2001) Identification and analysis of eukaryotic promoters: Recent computational approaches, Trends Genet., 17:56–60.

45. Sneath, P.H.A. and Sokal, R.R. (1973) in Numerical taxonomy, pp.230-234, W. H. Freeman and Company, San Francisco, California, USA.
46. Saitou, N. and Nei, M. (1987) The neighbor-joining method: a new method for reconstructing phylogenetic trees, Mol. Biol. Evol., 4:406-425.
47. Fitch, W.M. and Margoliash, E. (1967) Construction of phylogenetic trees, Science, 155: 279-284.
48. Fitch, W.M. (1971) Toward defining the course of evolution: minimum change for a specific tree topology, Syst. Zool., 20:406-416.
49. Hartigan, J.A. (1973) Minimum evolution fits to a given tree, Biometrics, 29:53-65.
50. Felsenstein, J. (2004) Inferring Phylogenies Sinauer Associates: Sunderland, Massachusetts

Index

A

A.thaliana 192
AAT 376
AATAAA hexamer 374
Ab initio Approaches 108
Ab initio Gene-finders 377
Ab initio structure prediction 234
ABC transporters 135
ABRC 95
Abstract Syntax Notation 1 (ASN.1) 37
Ac/Ds tag line 95
ACC deaminase 300
ACD/ChemSketch 137
AceDB 361
Acidobacteria 301, 307
Acronychia pedunculata 139
Acronycine 139
Active X 252
ADDA 30
Aadenine 340
Advanced Visual Systems 252
Aegle marmelos 139
AFLPs 60, 273, 351
Agarase 297
AgBase 80, 110
Aging-related research 347
Agricultural biotechnology 9
Agricultural GDP 2
Agriculturally important traits 156
AGRIS 95
Agro-biodiversity 51
Agro-technique 92
Agrobacterium tumefaciens 304
AGRON-OMICS 350
Alcohol 297
Alcohol oxidoreductase 297

Algorithms 342
AlignACE 383
Alignment 100, 218
Allelic data 61
Allicin 139
Allium sativum 139
ALN 376
ALN/ClustalW2 Format 35
Alpha (a) diversity 320
Alpha numeric string 13
AltGenExpress microarray data 200
Alu PCR sequences 12
Alzheimer's 347
AMBER 226
AMDIS 119
AMOS 98
Amylase 297
Amyloplasts 104
Analysis of Molecular Marker Data 61
Andrographis paniculata 139
Andrographolide 139
Angiosperm mitochondria 264
Angiosperm mitogenome 263
Angiosperms 263
Aniseed 272
Annotated 74
Anti-tumor Components 133
Antibiotics 295
Antidiabetic 93
Antisense Expression 191
APIRS 146
Application of Phylogenetics 84
Applied Biosystems SOLiD 97
Arabidopsis 107
Arabidopsis brassica 80
Arabidopsis genome 150
Arabidopsis Genomic RNAi 114

Arabidopsis MPSS 95
Arabidopsis thaliana 20, 68, 75, 172, 194, 195, 197, 200
Arachne 98
AraCyc 20, 95
Archaeoglobus fulgidus 359
Ariadne Genomics 125
ARLEQUIN 61
Arlequin software package 102
Aromatic amino acid biosynthesis 117
ArrayExpress 115
Arthrobacter 300
Artificial intelligence (AI) 169, 170
Artificial Neural Network (ANN) 170, 172
Aspergillus oryzae 376
Assembly 100
ASVM 205
ATGC 362
AtGDB 362
Atlas of Protein Sequences 322
Atomic Reconstruction of Metabolism (ARM) 134
ATP 260
ATTED-II 116
AUGUSTUS 380
Automated Arabidopsis Plant Root Cell Segmentation 194
Avena sativa [oat] 83
Average magnitude difference function (AMDF) 171
AVS 253
Azoarcus 300
Azoarcus grass endophytes 301
Azospirillum 300

B

Bacillus 300
Bacillus coli 326
Backcross Mapping Population 161
Backcrosses 160
Bacteria 26
Bacterial Artificial Chromosome (BAC) 298
Bacterial chromosomes 17
Bacterial Genome Databases 17
Bacteriodetes 307
Bacterium coli 326
Baicalin 139
Balancing 53
Barley 167

Barley SNP 363
BASE 115
Base caller 100
Basic Perl Programming 67
Basic Terminologies 384
Basics of population genetics 52
Basil 272
Bay leaf 272
Bayesian Networks 170
Berberine bridge enzyme (BBE) 113
Best linear unbiased estimates (BLUE) 103
Best linear unbiased predictions (BLUP) 103
Beta (b) diversity 320
Beyond 344
BIAdb 93
BIANA 125
Bibliographic data 15
Bbibliographic information 92
Bibliographic references 31
Binary data 61
Binary indicator sequences 174
Binomial nomenclature 333
Bio-molecular Relations in Information Transmission 87
Bioactive compounds 297
Biocatalysts 297
Bioconductor 115
BioCyc databases 131
Biodiversity 6
Biodiversity and Biological Collection Web Server 331
Biodiversity Data 327
Biodiversity Database Integration 330
Biodiversity extinction rate 332
Biodiversity Informatics 85, 276, 319, 323, 329
Biodiversity Information Network – Agenda 21 325
Biodiversity Research Center 325
Biofilm associations 303
Biofilms on Plant roots 303
Bioinformatics 348, 389
Bioinformatics and Systems Biology (IBIS) 22
Bioinformatics Database Search Engines 40
Bioinformatics sequence deposition 99
Biological Data 10
Biological Data Types 328
Biological databases 389
Biological heritage 319
Biomed Central 14
Bioremediation 209, 295

Index 395

Biosphere 296
Biostatistics 348
Biosynthetic Pathway of Camptothecin 134
Biotin production 297
Biotransformation 345
BioWeka 203
Bird's Eye 272
Biskit 232
BisoGenet 125
Black pepper (*Piper nigrum* L.) 272, 274, 275
BLAST 20, 21, 24, 41, 42, 74, 144, 220, 342, 369
BLAST E-values 195
BLAST searches 307
BLAST-hits 309
BLASTN 370
BLASTP 370, 379
BLASTX 370, 375, 379
BLOCKS 13, 106
Blocks Amino Acid Substitution Matrices (BLOSUM) 42, 62, 74, 220, 370, 371
Bobscript 253
BOLD-ECS (External Connectivity) 101
BOLD-IDS (Identification Engine) 101
BOLD-MAS (Management And Analysis) 101
Boolean operators 13
Boot strapping 85
Borrelia burgdorferi 359
Bottom up (clone or contig maps) approach 352
Boulder 321
Brachypodium 76
Brachypodium distachyon 75, 167
Brassica 110
Brassicaceae 80
Brazil's Base de Dados Tropical (BDT) 325
BRENDA 40
BrGP 363
Brucella abortus 326
Brucella suis 326
Bulk Segregant Analysis 164
Burkholderia 300

C

C-to-U RNA Editing Sites 198
C. roseus 104, 134
C. elegans 145
CaArray 115
CABS 232

Caenorhabditis elegans 19, 172, 359, 376
Calvin cycle 117
Cambridge Crystallographic Data Centre 37
Cambridge Structural Database (CSD) 37
Camptotheca 139
Camptotheca acuninata 139
Camptothecin 139
Candidate genes 274
Canscora decussate 139
CAP3 98
Caper 272
Capsicum 103, 272, 273
Cardamom 274
Cardamom (small) 272
CardCC 277
Carica papaya 75
CASP4 245
Cassava 167
Catalogue of Life Project 326
Catalytic Site Atlas (CSA) 39
Catharanthus roseus 112
Cation transporter 297
cDNA 375, 379
cDNA sequences 69, 144
Celera Genomics Corporation 360
Celery 272
Cell programming 129
CellDesigner 121
CellML 124
Cellular integrity 345
Cellulase 297
Cellware 121
CEM (Conserved Exon Method) 376
Centimorgan (cM) 351
Centromeres 15
CephBase 326
Cereal Genomic Databases 81
Character-based method 386
CHARMM 226
Chemical Reaction 123
Cheminformatics 5, 6, 275
Chemometrics 199
Chilli 272
Chime 253
Chimera 253
Chimeric REpresspr gene Silencing Technology (CRES) 114
Chitinase 277, 297
Chloroplast Genome 71
ChloroplastDB 104, 110

CHMM (class HMM) 378
Choice of Macromolecular Sequences 85
ChromDB 362
Chromoplasts 104
Chromosomal or cytogenetic maps 101
Chromosome (BAC) 298
Chromosome Map of *A.thaliana* 78
CIB/DDBJ 24
Cinnamate–monolignol pathway 117
Cinnamomum 272
Cinnamon 272
CIS-acting regulatory elements 71
CITES Plants 93
Cladogram 384
Class Architecture Topology Homology (CATH) 38
Classification 190
Classification of Bioinformatics Databases 14
Clearing House Mechanism (CHM) 325
Clinical effects of mutations 360
Clinical pathologies 346
Clustal 372
Clustal alignment 372
ClustalW 222, 385, 387
Cluster 51, 115
Cluster-Buster 383
Clustering algorithms 195
Clusters of Orthologous Groups of proteins (COGs) 308
CMKb 93
CMR (comprehensive microbial Resource) 17
Cn3D 253
Co-expressed Genes 195
Co-expression analysis 116
Co-response analysis (FANCY) 119
Cocoa 167
Coding Potential Prediction (CPP) 108
Coding regions 15
Codon Bias Data 88
CODONUSAGE 377
COG 106
Coiled-coils structure prediction 236
Combined Methods 379
Commercial software 233
Committee on Data for Science and Technology (CODATA) 86
Commonly Observed Features 70
Comparative analysis 21
Comparative Anchor Tagged Sequence (CATS) 108

Comparative genetics 5
Comparative Genomics 166, 272, 274, 340
Comparative, loop 218
Comparative modelling 241
Comparative protein modeling 212
Complementary DNA 10
Composite Databases 12, 167
Compositional diversity pattern 321
Comprehensive Yeast Genome Database (CYGD) 23
Computational analysis 332
Computational biology 348
Computational gene prediction 389
Computational Methods 359
Computational software programs 271
Computational techniques 9
Computer-Aided Drug Discovery and Development 136
Computer-based clinical information systems 326
CONABIO 325
Conformational Search Approaches 224
Conservation biology 9
Conserved non-coding sequence (CNS) 377
ConSite 383
Consortium 274
Consortium for the Barcode of Life (CBOL) 334
Constraints 329
Construction of loops 222
Content sensor 373
Continuous DNA sequence 74
Contour plots 177
Control regions 15
Convention on Biological Diversity (CBD) 320, 322, 325
CONVERT 61
COPASI 128
Coriander 272
Corrections for Sampling Error 58
Correlation structures 171
Cosmid/BAC/YAC end sequences 12
Cost and Benefits 327
Cotton 167
Cox2 264
CPCRI 8
CpG islands 382
CPHModel 232
Crossing over 157
Crustacea 260

Index

397

CSA 39
Cucumis sativus 75
Cucurbitaceae 110
CuGenDB 110
Culturable bacteria 296
Cumin 272
Cupressus pyramidalis 139
Curator-evaluated computational analysis 26
Cycas mitochondria 264
Cyclone 125
Cygwin 202
Cysteines 28
Cystic fibrosis 11
Cytochrome 265
Cytochrome c oxidase I (COI) 333
Cytogenetic mapping 159
Cytoscape 125
Cytosine 174, 340

D

DALI 44, 361
Data Repository 121
Data Collection 121
Data Exchange 124
Data integration 348
Database 363
Database (GMOD) 20
DATF 95
Datura 103
DbClustal 372
dbEST 273, 375
dbEST databases 376
DBGET / LinkDB 41
dbSNP database 116
Dchip 115
Dchip software 114
DDBJ 23, 361
de novo modeling 212
De Novo Protein Modeling 243
Deaminase 300
Degree of linkage 158
DELTA (Description Language for Taxonomy) 16
Dendropoma 265
Detoxification 321
DFT 171
DHGP (German Human Genome Project) 23
DIALIGN2 373
DiaMedBase 93

Dideoxynucleotides 299
Digital signal processing 169, 170
Digitalization 92
Dino 253
Diploid 76
Directional 53
Discriminant Analysis 377, 379
Disease resistance genes 148
Distance based methods 238, 386
Distributional information 325
DIVERSITAS 331
Diversity Array Technology (DArT) 114
DNA Barcoding 101, 332
DNA barcoding protocol 334
DNA Geometric Flexibility 198
DNA in prokaryotes 10
DNA libraries 166
DNA microarray analysis 116
DNA of eukaryotes 10
DNA polymerase 10
DNA sequences 30, 88, 101, 160, 194, 364
DNA strider 366
DNA structure 67
DNA taxonomy 332
DNA transcript 340
DNACOMP 388
DNADIST 387, 388
DNAML 389
DNAMLK 389
DNAMOVE 388
DNAMOVE for DNA sequences 388
DNAPARS 388
DNAPENNY 388
DNAse 297
DNP 93
DOCK 137
DOIs (Digital Object Identifiers) 331
Doryteuthis pealeii 326
Dot matrix method 364
double-stranded RNA (dsRNA) 113
Doubled Haploids (DHs) 161
DP, IMM 380
DP, NN 380
Drawing System (LDS1) 249
Drosophila dipterans 19, 20
Drosophila melanogaster 19, 172, 359
Drosophila researchers 20
Drug design 210
Drug target identification 138

DrugBank 137
Dual Protein Targeting 201
Dynamic Approach 342
Dynamic Programming 108, 170, 241, 366, 377, 378

E

E-Cell 128
E. californica 113
E.coli 134
EasyModeller 232
EBI 21, 83
EBI genome databases 17
EBI Genomes 15
Ecological diversity 320
ECOPARSE 378
Ecosystem diversity 6
Ecosystems 319
Effective number of alleles 55
Effective Population Size 56
Effective wavelengths (EWs) 200
EGAN 125
EGassembler 98
EIIP values 172
Electron Transport System 120
Electron-Ion Interaction Potential (EIIP) 169, 174
Electrophoresis analysis 105
EM (Expectation Maximization) 382
EMBL (Europen Molecular Biology Laboratory) 23, 24, 71, 273, 361
EMBL Nucleotide Sequence Database 24
EMBOSS 97
Emboss software package 366, 369
Emodin 139
eMOTIF 43
Encyclopedia 93
Energy function 241
Energy minimization (EM) 226
Energy production 295
Enological Parameters 193
Ensembl 15, 21
Ensembl Plants 110
Enterobacter 300
Entrez 40
Entrez Genome 146, 148
Entrez Structure 37
Entries in GenBank file 31
Environmental Resources Information Network (ERIN) 324

Enzyme catalytic sites 28
Enzyme Commission (EC) 131
Enzyme Databases 123
Enzyme reverse transcriptase 10
EP:NG 115
Epimedium species 102
Eponine 383
ERRAT Program 230
Erwinia 302
Erwinia carotovora 304
Escherichia coli 326, 359
EST 144, 351, 379
EST (Expressed Sequence Tags) maps 352
EST analysis 348
EST collections for comparison purposes. 108
EST processing 102
ESTs 11, 95
ESyPred3D 232
Etioplasts 104
Eu.Gene Analyzer 126
Eucalyptus globulus 69
EuGène 109
Euglena 326
Eukaryotes 74
Eukaryotic Gene Structure 373
Eukaryotic Promoter Database (EPD) 112
Eukaryotic Sequences 374
eukaryotic systems 72
Evolutionary history 267
Evolutionary Insight 262
Evolutionary relationship detection 206
EW-LS-SVM models 200
Exon trapped genomic sequences 12
Exon-Intron Database (EID) 172
Exons 170
ExPASy 97
ExPASy Proteomics Server 97
Expected Heterozygosity 56
ExpressDB 115
Expressed Sequence Tags (ESTs) 3, 6, 10, 11, 69, 102, 107
Expression correlation 195
Extrinsic Gene-finders 375

F

F_2 derived F_3 population 160, 161
F_2 population 160
Fabaceae 80
Family 92

Index 399

FASTA 24, 41, 42, 74, 369, 370
FASTA Format 34
FASTA suite of programs 369
FASTX 370
FASTY 370
FBA 132
Feature Extraction 190
Feature Selection 189, 190
Fennel 272
Fenugreek 272
Fermentative approaches 345
Feulgen densitometry 75
FFAS 241, 242
FGENES (Find Genes) 379
FGENESH 109, 379, 380
FGenesh 378
Filters 190
Firmicutes 307
FishBase 326
FITCH 387, 388
Fitch-Margoliash (FM) 387
Flavonoid biosynthesis genes 117
Flow Cytometry 75
Fluorescent *In Situ* Hybridization (FISH) 101, 352
Flux Balance Analysis 131
FlyBase 19, 361
Food chain 321
Forensics 9
Forward and Reverse Genetics 113
Four data types 15
FOX line 95
FrameD 75
Francis 360
Free literature 13
FRLHT 146
FRODO 249
From Metabolite to Metabolite (FMM) 134
Full-length cDNA clones 95
Function-based analysis 299
Functional annotation 309
Functional Genomics 107, 275, 340
Functome 348
Fusarium 302

G

GABI (Genome Analysis in Plants) 23
GabiPD 110
GacA 303

Gamma (g) diversity 320
GAP4 98
Garlic 272
Gaussian distribution 175
Gbrowse 20
GC-MS 4
GC-Skew 198
GCG software package 366
GCG/MSF Format 35
GCOS 115
GDA 63
GDR 110
GEM System 131
Gen-omics 341
GenAlEx 63
GenBank 23, 24, 71, 330
GenBank EST's 69
GenBank Format 31
Gene dictionary 82
Gene discovery 2, 107
Gene Expression 149, 360
Gene finding and Annotation 143
Gene flow 53
Gene frequencies 54
Gene functional annotation 360
Gene identification methods 360, 373
Gene identifier (GI) numbers 12
Gene localization 360
Gene Ontology (GO) Consortium 80
Gene prediction 21
Gene Predictor Programs 374
Gene predictors 309
Gene regulation 267
Gene Traffic 115
Gene-Expression Data 195
Gene3D 30
Genehacker 75
GeneID 378, 380
GeneId3 109
GeneMark 75, 109, 377, 378
GENEPOP 63
General Databases 82
Generalized hidden markov model 378
Generation Challenge Programme 103
Generic Model Organism Database (GMOD) 20
GENES database 22
GeneSeqer 376, 380
GeneSeqer@ 108
GeneSilico 232

400   Agriculture Bioinformatics

GeneSplicer 109
GeneSpring 115
Genetic Algorithm for Rule-set Prediction
(GARP) 325
Genetic algorithms (GAs) 389
Genetic and functional studies 206
Genetic Distance 59, 159
Genetic Diversity 51, 320
Genetic diversity analysis 52
Genetic drift 53
Genetic fingerprinting 155
Genetic Linkage Mapping 101, 158
Genetic Maps 159
Genetic Markers 273
Genetic resource stock information 82
Geneview program 378
GeneWise 376
GeneX 115
GeNGe 126
Genie 378
GenLang 380
GenMAPP 126
Geno3D 232
Geno3D homology 29
Genome Annotation 68
Genome Biology 15
Genome Browsers 21
Genome Databases 14, 15, 122
Genome evolution 259
Genome Gateway 110
Genome Mapping 101, 155, 210, 351
Genome Mapping Methods 156
Genome Sequence Analysis 95, 170, 359
Genome Sizes 75
Genome-based Modeling (GEM) System
129, 130
Genomes OnLine Database (GOLD) 17
Genomic DNA 10, 375
Genomic DNA Comparison 376
Genomic inference methods 144
Genomic Survey Sequences (GSSs) 10, 11
Genomics 10, 344, 389
Genotype–phenotype 326
GenScan 75, 109, 379
GenThreader 241
GenView2 377
GEO 115
Geographical distribution 92
Geometric Information 328
GEPASI 128

GHMM 380
Gibbs motif sampling 382
Ginger 272, 273, 274
Gingko leaves 321
Gist 203
GLIDE 137
GLIMMER 378
Glimmer 309, 378
GlimmerHMM 109, 378
GlimmerM 378
Global Databases 331
Global Musa Genome 274
Global Ocean Sampling Expedition (GOS) 27
Gluconacetobacter 300
Glycine 80
Glycine max 75, 112, 192
Glycine max [soybean] 83
Glycomics 341
Golgi-resident Proteins 197
GolgiP 197
GPDT 204
GPI 110
Grail 380
GrainGenes 110, 147
Gramene 81, 110, 145, 147, 363
Gramineae 80, 110
Grammar rule 380
Grape 167
Graphics systems 248
Graphs 328
Grasp 253
Greater galanga 272
GreenPhylDB 146
GRIN 110, 145
GROMOS 226
Plant Growth Promoting Bacteria (PGPB) 300
GRPL 378
GSA program (Gene Structure Assembly) 379
Guanine 174, 340
Gymnosperm Cycas 264

H

Haemophilus influenzae 17, 359
Hardy-Weinberg Principle 54
Helicobacter pylori 359
Helicos HeliScope 97
Hemolysis 297
Herb Research Foundation 321
Herbal drugs 91

Index

Herbaspirillum 300
Heterozygous 52
Hpred 232
Hidden Markov Model (HMM) 13, 74, 108, 170, 378, 382
High-dimensional Biology (HDB) 339, 354
High-dimensional data 328
High-throughput Sequencing (HTS) 7
Higher determination coefficients for validation 193
History of Biodiversity Informatics 324
Homologs 361
Homology Model 214
Homology modeling 211, 212
Homozygosity 161
Homozygous 52
Hardeum 80
Hordeum vulgare [barley] 83
Horizontal gene transfer (HGT) 263
Horizontal transfer of genes 85
Horse radish 272
Horticulture 1
HPV sequence database 16
HTML 96
Human cancer cells 346
Human Genome Project 211
Huperzia 96
Huperziaceae family 96
Hydrophobic packing 239
Hydrophobic protein 265
Hypothetical Proteins 355

I

ICTVdb 16
ICTVdB (Virus Database) 331
ICuGI 363
Identification of duplicates 155
Identification of regulatory regions 360
IGGP 363
IIM, GHMM 380
IISR 8, 276
Illumina GAIIx 97
IMM, WWAM 380
IMPPDS 93
In silico 276
In silico gene prediction methods 374
In silico mapping 102
In Situ Hybridization 159
In vitro 301

INBio 325
Inbreeding and Relatedness 57
INCLUSive 383
Indian agriculture 1
Indian Bioinformatics Scenario 7
Indirubin 297
INFO (Interruption Finder and Organizer) 375
Information Standards and Protocols for Interchanges 325
Infrared Spectroscopy 199
Ingenuity Pathway Analysis (IPA) 133
Inheritance in Man (OMIM) 14
INIBAP 167
Inr sequence 382
Integrated database 95
Integrated Microbial Genomes (IMG/M) system 310
Integrated Taxonomic Information Systems (ITIS) 86, 326
Interactomics 4, 341, 347
Intergenic regions 170
International Committee on Taxonomy of Viruses 16
International Crop Information System 81
International Human Genome Sequencing Consortium 360
International Nucleotide Sequence Databases 27
International Protein Index (IPI) 26
International Rice Information System (IRIS) 81
International Union of Biological Sciences (IUBS) 86
Internet Resources 9
Interpolated Markov models (IMMs) 378
Interpopulation diversity 59
InterPro 30, 106
Intrapopulation genic diversity 59
Intron-exon 170
Intron-exon boundaries 143
Introns 170
INVDOCK 138
Inverse PCR 307
Invertebrates 266
Ionomics 135
IPGC 363
Isotope-coded affinity tags (ICAT) 104
ISSRs 273
ITSs 273
IWGSC 363

J

JavaScript 96, 252
JCell 126
JGI 82
JNets 126
JOIN MAP 165
Joinmap 160
JTT 385
Jukes-Cantor model 371
Juniper berry 272
"JUNK" DNA 17
Juzbox 278

K

κ tuple (word) method 369
KEGG 22, 82, 131, 144
KEGG Atlas 126
KEGG Bioinformatics for Plant Genomics Research 87
KEGG EDRUG 87
KEGG ENZYME Database 39, 40
KEGG PLANT Resource 87, 111
Kerala State Council for Science Technology and Environment 334
Kernel Based Machine Learning Algorithm 197
Kilobases (Kb) 351
Kimura Model 371
Kimura two-parameter model 385
KINEMAGE 249
KINSIM 128
KITSCH 387, 388
Klebsiella 300
Knock-out Line Analysis (AGRIKOLA) 114
Knowledge assembly protocols 348
KOG 82
Kyoto Encyclopedia of Genes and Genomes (KEGG) 22

L

LAMP (Linux-Apache-MySQL-PHP) 96
Large miRNA mediated regulatory interactions 201
Large scale EST 67
LC-MS 4
LEARNSC 205
Least-squares Support Vector Machines 193
Legume genera 264

LegumeTFDB 112
LIBRA I 232
LIBSVM (Library for Support Vector Machines) 203
LIGAND 87
LIGAND database 22
LINE [Long Interspersed Nuclear Elements] 70
Line of no discrimination 179
Linear discriminant analysis (LDA) 379
Linear Pathway 120
Linguistic Methods 377, 379
Linkage mapping 3
Linkage1 160
Linkages 323
Linux 202
Lipase/esterase 297
Lipidomics 341
Liquid Chromatography (LC) 118
Literature Databases 13, 123
Liverwort Marchantia polymorpha 264
Local extinctions 92
Lod score 158
Loligo pealeii 326
LOMETS 232
Loop Modeling Method 223
Loss of genetic diversity 92
Lotus 80
Lotus japonica 112, 167
Low-resolutions models 226
Lower root-mean-square error of validation (RMSEP) 193
LS-SVMlab 205
LSVM (Lagrangian Support Vector Machine) 205
Lucy 98
Lycopersicon 80
Lycopersicum [tomato] 83
Lytic enzyme 302

M

M. truncatula RNAi Database 114
Machine Learning 195
Machine learning algorithms 188
Machine Learning Approach 196
Macroarrays 348
Macromolecular 3D structures 360
Macromolecular modelling software 239
Macromolecular sequences 85, 330

Index

Macronutrients 275
MAFFT-2 385
MAFFT-I 385
MAGE 253
Maize 167, 262
Maize Genetics site 65
Maize Hybrid Performance 201
Maize mitogenome 266
MaizeGDB 111, 362
Maiztedb 362
Malus domestica 75
Mammals 26
Manually-annotated records 26
MAPA 93
MAPK3 198
MAPMAKER 165
MapMaker/QTL 160, 164
Mapmanager 160
MAPPA 93
Mapping Polygenes (Quantitative Trait Loci) 162
Mapping Populations 162
Mapping Strategies 165
MapQTL 164
MapView 78
MapViewer 20
Marine gastropods 265
Marker assisted breeding 156
Marker assisted selection (MAS) 3
Markov Classification Model 200
Markov model (MM) 377
MAS 7
Mass spectrometry 103, 105, 118, 341, 348
Match-Box 373
MATDB 361
Mathematical Genetics and Bioinformatics Site 65
MATLAB 134
MATLAB SVM Toolbox 204
Matthews Correlation Coefficient (MCC) 178, 181
Maximum likelihood (ML) 389, 386
Maximum parsimony (MP) 388, 386
MBGD 18
MCell 128
McPromoter 383
Medicago 80
Medicago truncatula 75, 112, 167, 192
Medicinal Plant Research 91, 93, 94
Medicinal properties 92

Medicine 9
MEDLINE 13, 71, 330
MEGA 62, 97
Meiosis 157
Melina 383
MEME 383
Mendelian genetics 52
Metabolic Data Classification 118
Metabolic flux analysis (MFA) 349
Metabolic map 95
Metabolic Pathway 120, 200
Metabolic pathway analysis 2
Metabolic pathway data 360
Metabolic Pathway Prediction 132
Metabolite 297
Metabolite fingerprinting 119
Metabolite profiling 116, 119
Metabolites 345
Metabolome 4
Metabolomics 4, 10, 117, 124, 341, 344, 345, 389
Metabolomics society 119
Metabonomics 345
Metacrop 150
MetaFIND 119
Metagenomic data 308
Metagenomic libraries 295
Metagenomics 295
MetAlign 119
Metalloprotease 297
MetaModel 128
Metaphase chromosome bands 15
Metaproteomics 308
MetaRoute 126
Metatranscriptomics 308
Methanococcus jannaschii 359
Method of Selecting a Model from Library 240
Methodology 172
Methods of Threading Modelling 238
MetNetAligner 127
MG-RAST (MetaGenome Rapid Annotation using Subsystem Technology 310
MGED 115
mGENE 75
MICE 253
Micro - RNA Target Prediction 191
Microarray Analysis 10, 95, 114, 389
Microarray gene expression 10, 389
Microbial Genome Database (MBGD) 18
microbial genomes 296

MICROSAT 102
MICROSAT program 102
Microsatellites 70
Migration 53, 54
Mineral phosphate 302
Miniature inverted-repeat transposable elements 80
Minimum Evolution (ME) 387
Mint 272
MIPS 22
MIPS PlantDB 145
MIPS-GSF 23
MIRNA genes 191
miRNA target prediction 206
MITE: [Miniature Inverted-repeat Transposable Elements 70
Mitochondria 104, 201
Mitochondrial and Chloroplast DNA 10
Mitochondrial DNA (mt DNA) 85, 260, 333
Mitochondrial DNA recombination 72
Mitochondrial Gene Rearrangements 259
Mitochondrial Genome 72
Mitochondrial polypeptides 73
Mitochondrial RPS13 polypeptide 264
Mitochondriomics 346
Mitochondrion 265
Mitogen Activated Protein Kinase3 (MAPK3) 198
Mitogenomes 261, 263, 265
Mitogenomics 259
MKDOM2 29
MMDB 37
ModBldReview.htm 253
Model builder 253
Model evaluation 218, 227, 237
Model Plant species in Genomics Research 76
Model species 76
MODELLER 232
Modelling 348
Model's energy calculation 227
MOE 137
Molecular Biology 9, 156
Molecular cytogenetic markers 166, 274
Molecular functions - Enzymatic Catalysis 39
Molecular maps 159
Molecular Markers 51, 95, 158
Molecular Modelling Database (MMDB) 37, 41
Molecular Phylogenetics 84
Molinspiration 137

MOLSCRIPT 29, 250, 253
Monomeric proteins 211
Monophyletic group 384
Morphological description 325
MOTIF 41
Motif analysis 20
Motif-based Search Engines 43
MP method 388
mRNA protein profiling 67
MSFACTS 119
mSplicer 75
MSS (Mass survey system) 24
MSU rice 145
mt DNA sequence 262
Mucuna pruriens 139
Multiple Sequence Alignments (MSA) 221
Multivariate Analysis 60
MUMmer 377
Munich Information Center for Protein Sequence (MI) 147
Mustard 272, 273
Mutant images 82
Mutant lines 95
Mycoplasma genitalium 359
MyMBGD 18
Myriapoda 260
MySQL 21, 96
mySVM 202
MZEF (Michel Zhang's Exon Finder) 379
MZmine 2 119

N

N and C metabolism 349
N-acyl homoserine lactone (N-AHSL) 304
N-AHSL degradation enzymes (N-AHSLases) 304
N-AHSL-inactivating enzymes 304
Nardostachys 139
NASC proteome database 95
National Agricultural Bioinformatics Grid (NABG) 8
National Center for Biotechnology Information (NCB) 77, 82, 326
National Forum on Biological Diversity 320
National Research Council (NRC) 320
Natural selection 156
Natural variation 95
Nature of Biological Data 328
NCBI 40

Index 405

NCBI GenBank 11
NCBI Map Viewer 22
Near Infrared Spectroscopy 193
Near-isogenic Lines (NILs) 160
Needleman Wunsch algorithm 367
NEIGHBOR 387
Neighbor joining (NJ) 386
NetBuilder 121
NetGene2 109
NetPlantGene 109
NetStart 379
Neural Networks (NN) 108, 377, 366, 378
Neurodegenerative diseases 96
Neurospora crassa genome (MNCDB) 23
New Themes in Plant Bioinformatics 85
Next Generation Sequencing 97
NJ method 387
NLM Catalog 13
NMR 4, 345
NMR analysis 246
NMR constraints 237
NMR spectroscopy 211
NNPP 383
Non coding DNA 10, 68
Non-coding DNA sequences 112
Non-coding RNA 95
Non-Linear Pathway 120
Non-ribosomal protein synthesis 303
Noncoding regions 15
Nordborg lab 95
Normality indices 227
Novel enzymes 211, 295
NRDB 12
NSITE 383
Nuclear Magnetic Resonance (NMR) 118
Nucleosomes 10
Nucleotide Sequence Features 198
Nucleotides 340
Nucleus 104
Nutmeg 272
Nutraceutical properties 5
Nutritional genomics 6
Nutritional genomics biotechnology 210

O

Observed Heterozygosity 56
Oilpalm 167
Omics 339
'Omics' technologies 67

OMIM/OMIA 18
Oncoproteomics 353
Online Mendelian Inheritance in Animals (OMIA) 14
Online Mendelian Inheritance in Man (OMIM) 116
Onychophora 260
Oomycetes 7
Open reading frames (ORF) 307
Operating characteristic curves (ROC) 178
Ophiorrhiza pumila 134
Oregano 272
ORF [Open Reading Frames] 74
ORF Finder 75
Organic Chemistry Portal 137
Organism Specific Genome Databases 18
Oroxylum indicum 139
Orphan crops 5, 272
Orphan Protein-coding Genes 191
Orthologs 361
Oryza 80
Oryza databases 147
Oryza sativa 70, 192
Oryza sativa ssp. *indica* 75
Oryza sativa ssp. *japonica* 75
Oryzabase 82, 361, 363
Osprey 127
OWL database 376
Oxidation of polyois 297
Oxidative phosphorylation 262
Oxygenase 297

P

P. colubrinum 275
Paenibacillus 300
PAIR 196
Pair-wise Pearson correlation coefficients (PCCs) 117
Paired and weighted spectral rotation (PWSR) 171
Paired spectral content 171
Pairwise Sequence Alignment 364
PAL database 277
PAM 42, 385
PANTHER 30, 82, 106
Papaver somniferum 113
Papaya 167
Paprika 272
Paraphyletic group 384

406   Agriculture Bioinformatics

Pareto-optimal Clusters 195
Parsley 272
Partial least-squares (PLS) 193
PASSCOM 277
Pasteuria penetrans 301
Pathway Databases 22, 122
Pathway Studies 118
PATIKA 127
Pattern Matching 20
Pattern Recognition 118
Patterns 329
PAUP 62, 97
Paupsearch 387
PCR 158, 271
PCR primers 371
PDB (Protein Data Bank) 361
PDB-BLAST 220
PDBe 37
PDBj 37
PDBREPORT database 225
PDBs 39
Peach 167
Peanut (*Arachis hypogea*) 108
Pepper long 272
Peranema 326
Percent Accepted Mutation (PAM) 370
Perl program MISA 102
Peroxisomes 104
Pfam 30, 361
PGPB-mediated antifungals 300
Pharmaceutical industry 9
Phenome-genome 343
Phenome-genome networks 343
Phenomes 4
Phenotypes 210
Phlegmariurus 96
PhoshAt 95
PhosPhAt dataset of phosphoserines 196
Phosphoylation site Database 107
Phrap 309
PHT 127
PHYLIP 97
PHYLIP package 102
PhyloCon 383
Phylogenetic Analysis 96, 385, 389
Phylogenetic Markers 260
Phylogenetic relationships 96, 259
Phylogenetic studies 267
Phylogenetics 10, 384, 389
Phylogenomics 342

Phylogentic tree 306
Phylogeny 261
Phylogeny Programs 65
Physcomitrella patens 75, 192
Physical distance 159
Physical map 101
Physical Mapping 351
Physiomics 341
Phytfinder 277
Phytic acid 301
Phyto-mellitus 94
Phytophthora 274, 302
Phytophthora e-lab 278
Phytophthora infestans 274
Phytoplasmas 7
PhytoWeb 277
Phytozome 82, 111
Piperbase 277
PipMaker 377
PIR Format 34
PIRSF 30
PKS proteins 131
PLACE 383
Plant related Database 215
Plant Biology 143, 278
Plant genome assembly 98
Plant Genome Databases 360
Plant genome sequencing 150
Plant Genomics 67
Plant improvement 156
Plant Info 94
Plant Mitogenomes 262
Plant Omics 348
Plant Organelles 201
Plant Oriented Resources at NCBI 83
Plant Pathogens and Disease Resistance Genes 148
Plant Protein Phosphorylation Database 104
Plant Specific Databases 145
Plant-Omics 350
PlantCARE 71, 112, 362
PlantGDB 69, 108, 111, 147, 362
PlantMarkers 111
PlantMiRNAPred 192
PlantProm DB 112, 362
Plants 26, 93
Plants Database 94
PlantSat 111
PlantsDB 362

Index

PlantsP 362
PlantTFDB 148
PLASBID 277
Plasma membrane 104
Plasmid DNA 10
PlasmoDB 19
Plasmodium species 19
Plastids 201
PLEXdb 146, 362
PLoS 14
PMRD 111
Polyadq 383
Polyketide biosynthesis 303
Polyketide synthase 297
Polyketide Synthases (PKSs) 131
Polymorphic information content (PIC) 102
Polymorphism 55
Pomegranate 272
POPGENE 63
Poplar 167
Poppy seed 272
Population genetics 52
Populus trichocarpa 75, 192
Position weight matrices (PWMs) 378
Position-specific scoring matrix (PSSM) 382
Potato 167
Potato Genome Sequencing Consortium (PGSC) 7
POWERMARKER 62
PPAP 146
PPDB 95
PPMdb 362
PRABI 29
Practical Extraction and Reporting Language 87
Prediction-based methods (PBM) 238
Predictive Metabolic Engineering 124
Preliminary map 159
PRELUDE 94
Pressure and Threats 327
PRGdb 148
Primary accession numbers 12
Primary Databases 12
Primary Structure Databases 36
PRIMe 119
Principal components (PCs) 193
Principle Component Analysis (PCA) 193
PRINTS 13, 30, 106, 361
Probabilistic model 169
Probabilistic profile 74

ProbeSet 41
PROCHECK 227, 228
Procruster 375
ProDom 29, 30
Program GeneParser 378
Project Pursuit Regression 193
Prokaryotes 74, 305
PromH (W) 383
Promoter Region 70, 74
Promoter sequences 70
PromoterScan 383
PROSITE 13, 28, 29, 30, 43, 106, 361
PROSPECTOR 242
Prostethic group attachment sites 28
PROTDIST 387, 388
Prote-omics 341
Protein and Protein-related Databases 122
Protein Data Bank (PDB) 36, 38, 39, 212, 252
Protein databases 105
Protein Explorer 253
Protein ID numbers 12
Protein identification 105
Protein Information Resource (PIR) 25
Protein microarrays 341
Protein Modeling 212
Protein Phosphorylation Site Prediction 196
Protein Protein Interactions 149
Protein Sequence Information 198, 388
Protein sequences Databases 25
Protein Structure Prediction 209, 239
Protein Structure Visualization 247
Protein Threading 213, 237
Protein threading modelling 241
Protein-coding genes 73
Proteobacteria 307
ProteoLens 127
Proteome 124
Proteome profile 95
Proteomics 3, 10, 103, 275, 344, 389
Proteomics Standards Initiative Molecular Interact 124
PROTPARS for protein sequences 388
Proviz 127
PRRN 372
Pseudo – Test Cross Approach 165
Pseudo Plant pre-miRNAs 192
Pseudogenes poly (A) features 21
Pseudomonas 300, 304
PSI-BLAST 29, 39, 220
PSVM 205

PubChem 137
PubMed Central 13, 330
Purdue Ionomics Information Management System 135
PyMOL 254
Pyrosequencing technology 299
PySCeS 128

Q

QGene 164
QTL Cartographer 164
QTL Fine Mapping 163
QTL Localization 163
QTL softwares 159
Quadratic discriminant analysis (QDA) 379
Qualitative Modeling 129
Quanta 254
Quantitative modeling 129
Quantitative structure-activity/property relations 136
Quantitative Trait Loci (QTL) 78, 102, 103, 145
Query Sequence 239, 240
QuickPDB 252
Quorum sensing (QS) 304

R

Radial Basis Function (RBF) 169, 171, 175, 193
Radiated Rice 199
Radiation hybrid (RH) maps 101
Radio-labelled isotopes 349
RAFL clones 95
Rahnuma 127
Ramsar Convention 322
Random "single pass read" genome survey sequences 12
RankProd 115
RAP-DB 111, 147
RAPD 60, 67, 351
RAPDs 273
Rapid 277
Rapid DNA amplification techniques 271
Rapid sequencing methods 271
RARGE 95
RARTF 95
RasMol 41, 250, 253
RasTop 254
RBF kernels 175
RCSB 37

RCSB PDB 252
rDNA 80
Readthru 278
Receiver Operating Characteristic (ROC) 169
Recombinant DNA (rDNA) 10, 11
Recombinant Inbred Lines (RILs) 160, 161
Recombinant Vaccines 11
Reductionist Approach 342
RefSeq 82
Region Merging 194
Regular Expressions in Plant Genome 70
Regulatory elements 70
Regulatory Elements in Plant Genome 69
Regulatory genes 69
Regulatory interaction prediction 206
Regulatory Pathway 120
Regulatory regions 73, 383
Regulatory Sequence Analysis Tools 112
Relational database management system (RDBMS) 96
Repeats 15
Repetitive DNA 166
Repetitive DNA sequences 81
Repetitive elements 70
Research Collaboratory for Structural Bioinformati 252
Resequencing Microarrays 200
Resources for Plant Genomics 79
Retro-transposon repeat sequences 273
Retrotransposons 71
RGAP 363
RGP 363
Rheum palmatum 139
Rhizoctonia solani 303
Rhizosphere 312
Rhizosphere Microbial Diversity 300
Ribosomal Genome 71, 72
Rice 76, 167
Rice as a model plant 79
Rice genome 103
Rice proteome database 104
Rice rhizosphere metagenome 306
RIS - Rice Information System (BGI-RIS) 82
RMBNToolbox 127
RMSD 244
RNA editing 263
RNA encoding genes 68
RNA Interference 113
RNA polymerase 112, 381
RNA viruses 10

Index 409

ROBETTA 232
Roche 454 FLX 97
Root colonization 303
Rosaceae 110
Rosemary 272
ROSETTA 376
rRNA 85
RSLpred 192
rVISTA 383

S

S. cerevisiae 145
Saccharomyces cerevisiae 18, 262, 359
Saffron 272
SAGE 272
SAKURA 24
Salutaridinol 7-Oacetyltransferase (SalAT) 114
Satellite DNA 70
SBML 129
Scalar and vector fields 328
SCAMP 128
ScanProsite 43
Schizophrenia 347
Science 94
SCOP 106
Scutellaria baicalensis 139
Secondary Databases 13
Secondary metabolites 303
Secondary Structure Databases 38
Secondary structure prediction 235
Seed storage 92
Segregation Distortion of Markers 162
Selvita 233
Senile dementia 321
Sensitivity (Sn) 178, 179
Sensitivity analysis 124
SeqClean 98, 102
Sequence Alignment 361
Sequence comparison 99
Sequence Databases 23
Sequence deposition 97, 99
Sequence File Formats 30
Sequence Information 97
Sequence maps 101
Sequence motif analysis 67
Sequence motifs Databases 28
Sequence retrieval 99
Sequence Retrieval Server (SRS) 41
Sequence Similarity database [SSDB] 87

Sequence Tagged Sites 101
Sequence-based 105
Sequence-based (Sequence Similarity) Search Engine 42
Sequence-based approach 299
Sequence-tagged sites 15
Sequenced crop 95
Sequenced genome 77
Sequences 328
SeqViewer 20
Serial Analysis of Gene Expression (SAGE) 114
Serratae 96
Serratia 300
SGD 18
SGN 363
SGP-1 (Syntenic Gene Prediction) 376
Shannon index 57
Shepherd's purse 167
Short-Wave Near Infrared Spectroscopy 199
Sickle cell anemia 11
Side-chain modelling 218
Siderophore formation 302
Sign-O-Bacteria 277
Signal sensor 373
Signal Transduction Pathway 120
Signature Database 30
Silicon Graphics 249
SIM4 376, 380
Similarity Based Methods 73
Simple Object Access Protocol 40
Simple Sequence Repeats (SSRs) 3, 102
Simulation 124
SINE [Short Interspersed Nuclear Elements] 70
Single nucleotide polymorphisms (SNPs) 3, 200
Single-genome Comparative Genomics 199
siRNAs 191
SMART 30
SmartLab 204
SMD 115
Smith-Waterman algorithm 367
Snazer 128
SNP 351
SNPs 273
Software programs 61
Software tools 277
SOL Genomics Network (SGN) 103
Solanaceae 80, 275

Solanaceae Genomics Network 275
Solanum 80
Solanum genome initiative 167
Solanum tuberosum 71
Solaris 202
Soluble Solids Content 199
SORFIND 377
Sorghum 76, 80, 167
Sorghum bicolor 75, 192
Sources and Contacts 327
SoyBase 363
Soybean 167
Soybean proteome database 104
Spatial information 329
SpecAlign 119
SPECIES 2000 331
Species 2000 86
Species Analyst (TSA) 325
Species diversity 320
Specific DNA markers 156
Specificity (Sp) metrics 178
Specificity values 179
Spice bibliography 277
Spice Bioinformatics 271
Spice genes 277
Spice Prop 277
Spicepat 277
SpicEST 277
Spider 204
Spidey 108, 376
Splice signal identification 206
Splice Sites 169
Spliceosome 174
Splicing-site Recognition 194
SSR detection 102
SSRs 273, 351
Star anise 272
Stationary-phase 303
Statistical genetics and Bioinformatics Site 65
Stereochemistry 228
Stoichiometric network analysis 124
Structural bioinformatics 212
Structural Classification of Proteins (SCOP) 38
Structural Genomics 97
Structural model 240
Structural rRNA genes 71
Structure 64, 166
Structure and Derived Databases 36
Structure of Plant Genome 68
Structure of tRNAs 267

Structure-based (Structure Similarity) Search
   Engi 43
STS 351
STS-content maps 352
Sub viral database 16
SUBA 95
Subcellular localization 95
Substitution matrix 366. 370
Supperfamily 30
Supervised Learning 195
Support Vector Machine (SVM) 169, 170, 171,
   188, 194, 195, 202
Support vectors 171
Suppressive Soil 302
Sustainable management 327
SVM based method 170
SVM Classification 194
SVM classifier 175
SVM kernel engineering 171
SVM parameters 172
SVM Toolbox 206
SVMlight 202
SVMstruct 202
SVMTorch 203
Sweet flag 272
Systems Biology 4, 87, 149, 343
Systems Biology Markup Language (SBML)
   124

T

T-Coffee 372, 385
T-DNA line 95
TAIR 20, 82, 95, 145, 149, 362
Tamarind 272
Targeting induced local lesions in genomes 113
Tarragon 272
TASSEL 166
TATA binding protein 71
TATA-less Promoters 198
Taxonomic Database Working Group 329
Taxonomic Information System 331
Taxonomic Problems 325
Taxonomical hierarchy 306
Taxonomy 9
TBLASTN 370
TBLASTX 370
TCMGeneDIT 94
Telomeres 15
Telomeric DNA 70

Index

411

Template Search 217
Terpenoid Indole Alkaloids (TIAs) 134
Tetrahedron representation 171
Text-based Search Engines 40
TF-binding sites 382
TFASTX 370
TFASTY 370
TFGPA 64
The Arabidopsis Information Resource (TAIR) 147
Threading/Fold recognition 233
Three-dimensional (3D) structure 210
Thyme 272
Thymine 174, 340
TIA metabolism 134
TIGR Plant Repeat Databases 80
TIGRFAMs 30
TILING Array 113
TILLING 95
TinySVM 204
TOF-MS 4
Tomato 167, 274
Tools and Softwares 125
Top down (restriction mapping) approach 352
Total diversity 59
Tranmembrane helix and signal peptide prediction 236
Transcript-omics 341
Transcription factor (TF) 112, 195, 381
Transcription factor data 95
Transcription of genes 266
Transcription start sites (TSSs) 112
Transcriptome 4, 124
Transcriptomics 4, 114, 344
TRANSFAC 71
Transgenics 156
Transport Systems Study 135
Transposable elements 10, 80
Transposon-tagged sequences 12, 70, 71
Tree Construction Methods 384, 386
Tree of life 84
Tree topology 388
TREEVIEW 114
TrEMBL (Database of Translated Nucleic Acid Sequences 25, 361
Tri hybrid crosses 157
Triticum 80
Triticum aestivum [wheat] 83
tRNA Arg 264
tRNALeu 264

Tryptophan 73
TSS clusters 112
TSSW 383
Turbomycin 297
Turmeric 272, 273, 274
Two-dimensional polyacrylamide gel electrophoresis 103
Type III polyketide synthases 197

U

UCSC Genome Bioinformatics website 23
Unified Medical Language System (UMLS) 326
UniProt (Universal Protein Resource) 25, 82
UniProt Archive (UniParc) 26
UniProt Knowledgebase (UniProtKB) 25, 26, 27
UniProt Metagenomic and Environmental Sequences 26, 27
UniProt Reference Clusters (UniRef) 26, 27
UniProtKB 29
UniRef100 database 27
UniRef50 27
UniRef90 27
United Nations Earth Summit, 1992 320
Universal Biological Indexer and Organizer 326
Universal Virus Database 16
University of California, Santa Cruz (UCSC) Genome 23
University of Kansas Natural History Museum 325
UPGMA 60, 387
UPGMA method 386, 388
Useful Internet Resources 64
Utility of Homology Modeling 231
Utilization of bioinformatics 2
UV-B light 149

V

Vacuoles 104
Validation of Model 226
Vanilla 272
Variant detection 100
Variation data 95
VAST 43
VCell 128
VDW exclusion test 225

Vector NTI software package 366
VEGA Genome Browser 21
VEIL 378
Vernacular names 92
Verrucomicrobia 301
Version numbers 12
Vertebrate mt DNA 262
Vertebrates 266
VGDB 16
Violacein 297
Viral Bioinformatics Resource Centre (VBRC) 16
Viral Genome Database 16
VirGen 16
VisANT 127
Visualization Software 250
VitaPad 127
Vitis vinifera 75
VMD 254
Vxlnsight 20

W

WABA 377
Web-based plant proteome-related databases 103
Web-Based Visualizing Software 252
Webliography 94
WebMol 253
Weka 203
Wet-lab 344
WGCNA 127
Wheat 167

Wheat Genetic and Genomic Resources Center 82
Whole Genome Tiling Microarrays 191
Windows 202
Withania somnifera 133, 139
World Heritage Convention 322
WormBase 19
Wrappers 190

X

X-ray 246
X-ray crystallography 211
X-ray diffraction 219
Xanthomonas 302
XML-like language 124
XtalView crystallography package 250, 254

Y

Youden's Index 178, 181
Youden's Index (J) 181

Z

Z. scores 241
Z. zerumbet 275
Z. curve 309
Zea 80
Zea mays [corn] 75, 83, 192
ZINC 137
Zingiber officinale 96
Zingiberaceae 274